GENETICS AND
THE ORIGIN OF SPECIES

Columbia Classics in Evolution Series

Niles Eldredge and Stephen Jay Gould, editors

COLUMBIA CLASSICS IN EVOLUTION SERIES

Genetics and the Origin of Species   THEODOSIUS DOBZHANSKY

Systematics and the Origin of Species   ERNST MAYR

GENETICS AND
THE ORIGIN OF SPECIES

BY

THEODOSIUS DOBZHANSKY

With an Introduction by Stephen Jay Gould

COLUMBIA UNIVERSITY PRESS

NEW YORK

Library of Congress Cataloging-in-Publication Data

Dobzhansky, Theodosius Grigorievich, 1900–1975.
   Genetics and the origin of species.

   (Columbia classics in evolution series)
   Reprint. Originally published: New York:
Columbia University Press, 1937. (Columbia
biological series; no. 11)
   Bibliography: p.
   Includes index.
   1. Evolution.   2. Genetics.   I. Title.
II. Series.   III. Series: Columbia biological
series; no. 11.
QH366.2.D59   1982   575   82-4278
ISBN 978-0-231-05475-1 (pbk.)   AACR2

Columbia University Press
New York   Chichester, West Sussex
Printed in the United States of America

CONTENTS

ILLUSTRATIONS

A NOTE ON THE SERIES

EVOLUTION, the proposition that all organisms are related by descent, is the central organizing principle of biology. The facts of physiology and molecular biology, while fascinating in their own right, present a pattern only explicable, ultimately, in the context of biological evolution. Like all vital subjects, the history of evolutionary thought has been marked by periods of intense debate and other times of general satisfaction. As we enter a time of debate after three decades or so of relative quiet, history assumes a new importance and the documents that inspired previous agreement demand scrutiny. The only sound guide to where we are going is a firm knowledge of where we have been.

The current renaissance of interest in evolutionary theory stems from something more profound than the inevitable swing of the pendulum of biological fashion. The Modern Synthesis—a reconciliation and fusion of data and concepts from such disparate fields as genetics, systematics, and paleontology—was born in the 1930s, matured in the 40s, and was polished to its ultimate form in the 50s. By the time of the centennial celebration of Darwin's (1859) *Origin of Species*, evolutionary biologists could confidently assert that their science had achieved an integrated, complete theory of evolution. Widely hailed as a great scientific achievement, the synthetic theory stands today as an all-embracing statement of how the evolutionary process actually works.

Final formulation for a body of thought would have the unhappy consequence of robbing a subject of any ongoing scientific interest beyond mere caretaking. But nothing, to date, in human intellectual history has survived as the final word. The modern synthesis is now being scrutinized at a level of intensity unmatched since its birth. Once again, evolutionary biology is throbbing; it has become the object of intense scientific inquiry.

It is more than interest in intellectual history *per se* that prompts the initiation of this series of reprints. If a paradigm is to be discussed critically, it must be characterized accurately, fairly, and completely. What is the Modern Synthesis, anyway? What did its proponents really say? Did they all agree on every point? Hardly! Was there more agreement initially, or later on in the theory's development? Though some analyses of these questions have already appeared, it is important that scientists currently engaged in the critical examination, and perhaps reformulation, of evolutionary theory—as well as the formidable ranks of the *defensores fidei*—be thoroughly conversant with the issues, as well as the roots, of established theory.

Hence this series. We shall provide introductions, critical evaluations of the text that attempt to place each book in its appropriate historical context. We shall also comment on the changes in thought that each author experienced as the original work was revised or supplanted by later efforts. Our critical biases are apparent, and we do not try to hide them, if only because they express our best guess about probable truth. This series exists to reprint and make available once again the great documents that set modern evolutionary theory. These books are not historical curiosities: they are the fundamental building blocks of the major evolutionary world view still with us. We can progress beyond only by fully comprehending what has come before.

NILES ELDREDGE
The American Museum of
   Natural History
New York, New York

STEPHEN JAY GOULD
Museum of Comparative Zoology
Harvard University
Cambridge, Massachusetts

FOREWORD

IT IS fitting that the Columbia Biological Series should resume publication with a discussion of the same problem to which the first volume in this series was devoted. Forty years ago evolution was a problem in history, and in *From' the Greeks to Darwin* H. F. Osborn traced the origins and development of the idea of gradual change as the means by which the world and its inhabitants had assumed their present forms. It was not only the ideas and theories which were studied as history; the facts of evolution itself were shown to be determined by changes in the past history of the earth. The essence of Darwin's theory was just this, and the efforts of biologists were concentrated on describing as completely as possible this record of the past.

But description and reconstruction fail to satisfy for long even the most ardent historians of nature and it was Darwin's second great service to have focused the attention of biologists upon the forces that caused the changes and particularly upon the problem of what agencies brought about and maintained the diversification of animals and plants into distinct species. His own theory of natural selection, by appealing to two occurrences of which the details were quite unknown—the origin of new variations in animals and plants and the perpetuation of these by heredity—served as a great stimulus and provided much of the motivation for biological research in the period which followed 1859. There were at that time, however, no reliable methods by which these two problems could be studied. The history of genetics since the rediscovery of Mendel's principles in 1900 is a history of the development of just such methods, and it is possible now to take stock of what these methods have done to improve our understanding of what has been, in spite of all research, so great a mystery—the origin of species.

This is what the present volume attempts to do. When one considers that this is 1937, a hundred years after the germ of the theory of natural selection first stirred in Darwin's mind, and nearly forty years after Mendel's theory and methods came to recognition, one might be tempted to suppose that such a reassessment of evolution was long overdue. But a glance at the sources of Professor Dobzhansky's material shows that this is not the case. The works cited in his bibliography are largely products of the twenties and especially the thirties of the present century. This expresses very well the recency of much of our knowledge of the actual factors involved in the differentiation of species.

The reasons for this are not far to seek. Variation and heredity had first to be studied for their own sakes and genetics grew up in answer to the interest in these problems and to the need for rigorous methods for testing by experiment all ideas we might hold about them. The requirements of this search drove genetics into the laboratory, along an apparently narrow alley hedged in by culture bottles of Drosophila and other insects, by the breeding cages of captive rodents, and by maize and snapdragons and other plants. Biologists not native to this alley thought sometimes that those who trod along it could not or would not look over the hedge; they admitted that the alley was paved with honest intentions but at its end they thought they could see a red light and a sign "The Gene: Dead End."

That condition, if it ever existed to any marked degree, is again changing, and Professor Dobzhansky's book signalizes very clearly something which can only be called the Back-to-Nature movement. The methods learned in the laboratory are good enough now to be put to the test in the open and applied in that ultimate laboratory of biology, free nature itself. Throughout this book we are reminded that the problems of evolution are given not by academic discussion and speculation, but by the existence of the great variety of living animals and plants. The facts and relationships found in nature have to be examined from many points of view and by the aid of many different methods. Evolution, in the author's words, is a change in the genetic composition of populations, and populations follow laws which may be derived by mathematical reasoning as extrapolations of the known behavior of the fundamental units of reproduction—genes and chromosomes. It is a kind of tour de force that in this book the recent work in this field is fitted into its important place in a way which does not offend the sensibilities of those who are repelled by mathematical formulas. Cytology too has become an essential weapon of those who would attack the species problem, and full advantage has been taken of the newer methods of studying chromosomes in Chapter four. Here too one may read a cytological detective story from the author's own experience, which, while it may shock the older makers of phylogenies, will convince the modern man that the riddle of speciation is by no means hopeless. Still other methods are brought to bear on the questions of sterility in species hybrids and of the mechanisms which are effective in keeping species separate, a field which has been recently illuminated by Professor Dobzhansky's original contributions.

In all this, the author appears not only as geneticist and as student of natural history, but as one who received his training in both fields in Russia. English-speaking biologists have special cause to be grateful for this last fact, for it has enabled Professor Dobzhansky to make available to us many important contributions from workers in the Soviet Union, where researches in this field have been actively prosecuted.

There was need for such a summary and synthesis of the new experimental evidence, and for reassessment of the older theories. It is of less importance that of these latter natural selection has survived the ordeal than that both the theory and the underlying reality, the species, have taken on new, and as one may guess, more fruitful meanings.

L. C. Dunn
Columbia University
August 1937

PREFACE

THE PROBLEM of evolution may be approached in two different ways. First, the sequence of the evolutionary events as they
have actually taken place in the past history of various organisms may
be traced. Second, the mechanisms that bring about evolutionary
changes may be studied. The first approach deals with historical
problems, and the second with physiological. The importance of
genetics for a critical evaluation of theories concerning the mechanics
of evolution is fairly generally recognized. The present book is devoted to a discussion of the mechanisms of species formation in terms
of the known facts and theories of genetics. Some writers have contended that evolution involves more than species formation, that
macro- and micro-evolutionary changes may be distinguished. This
may or may not be true; such a duality of the evolutionary process is
by no means established. In any case, a geneticist has no choice but
to confine himself to the micro-evolutionary phenomena that lie
within reach of his method, and to see how much of evolution in
general can be adequately understood on this basis.

Considerations of space have forced us to refrain from a detailed
discussion of some of the objections that have been advanced against
the genetic treatment of evolutionary problems. Thus, Lamarckian
doctrines find but a brief mention. The treatment had to be made
assertive rather than polemic, dogmatic rather than apologetic.

This book is based on a series of lectures delivered at Columbia
University, New York City, in October, 1936. Each lecture was followed by a discussion in which representatives of various biological
disciplines took part. To these colleagues, as well as to Drs. Edgar
Anderson, A. F. Blakeslee, M. Demerec, L. C. Dunn, T. H. Morgan, A. H. Sturtevant, and Sewall Wright, the writer is much in-
debted for many valuable suggestions and criticisms. The help of Mrs.
N. P. Sivertzev-Dobzhansky in the preparation of the manuscript is
gratefully acknowledged.

THEODOSIUS DOBZHANSKY
Pasadena, California
July 1937

INTRODUCTION

GENETICISTS AND NATURALISTS

IN THE SECOND edition of the *Descent of Man*, Darwin stated, as he did so often in the face of constant confusion, that he had written the *Origin of Species* with two separate goals in mind: first, to establish the fact of evolution itself; second, to propose the theory of natural selection as a mechanism for evolutionary change.

I had two distinct objects in view; firstly to show that species had not been separately created, and secondly, that natural selection had been the chief agent of change. . . . Some of those who admit the principle of evolution, but reject natural selection, seem to forget, when criticising my book, that I had the above two objects in view; hence if I have erred in giving to natural selection great power, which I am very far from admitting, or in having exaggerated its power, which is in itself probable, I have at least, as I hope, done good service in aiding to overthrow the dogma of separate creations.

Darwin presented this clarification again and again for a simple reason: he felt that he had triumphed in his first goal; virtually all educated people now accepted the fact of evolution. But he knew that his second goal stood far from realization, with no prospect for success in his lifetime. Darwin feared that people might confuse fact with mechanism, and cite the unresolved debate about natural selection as a denigration of his greatest achievement in establishing the fact of evolution. (In surveying the literature of so-called "scientific creationism" today, we understand that Darwin's fear was judicious and not paranoid; for these latter-day antediluvians are making propaganda by willfully distorting our healthy contemporary debates about evolutionary mechanisms into phony claims that even scientists have begun to question the fact of evolution itself.)

The absence of consensus about mechanisms remained the primary frustration of evolutionary theory from Darwin's day well into our own century. Since many scientists had assumed that ignorance of heredity underlay much of the frustration, it is ironic (and still not widely appreciated by biologists, though often stated by historians) that the rediscovery of Mendel's laws initially resolved nothing, but only heightened confusion. Battles of the late nineteenth century had pitted two primary contenders against each other: 1) Darwinian natural selection, with its insistence upon random variation as raw material and selection as a creative force; and 2) a host of otherwise disparate alternatives, including neo-Lamarckism and various styles of orthogenesis and vitalism, that proposed a creative role for variation itself and relegated natural selection to an executioner's task as eliminator of the unfit.

Since Mendelism made its initial impact upon evolutionary theory via de Vries' notion of the saltatory origin of species through macromutation, it managed only to increase the chaos by adding a third alternative, yet another anti-Darwinian theory that made sudden, fortuitous mutation the creative force in evolution and again demoted natural selection to a headsman.

William Bateson, a saltationist of long standing (Bateson 1894), lamented the disreputable state of evolutionary theory in a famous address in 1922: "Less and less was heard about evolution in genetical circles and now the topic is dropped. When students of other sciences ask us what is now currently believed about the origin of species, we have no clear answer to give. Faith has given place to agnosticism."

We all know the official outline of the subsequent, heroic tale (always be wary of the official and the heroic; be almost certain that their combination indicates oversimplification): Laboratory geneticists proved that small, continuous variability also had a mutational basis, and that micromutations could serve admirably as just the kind of raw material that Darwin had envisioned. Population geneticists then showed that tiny selection pressures could be effective at all evolutionary levels, thus obviating the need either for variation inherently directed toward adaptation (Lamarckism, some forms of vitalism, and orthogenesis), or for saltatory origin of new forms. This emerging, central core of evolutionary genetics won the hearts of more conservative natural historians. Discipline after discipline fell under its sway like a row of dominoes before an irresistible initiating force. Dobzhansky (1937—the volume reprinted here) set the outlines, and Mayr (1942 for systematics), Simpson (1944 for paleontology), Rensch (1947 for morphology), and Stebbins (1950 for botany) followed. The modern synthesis had emerged, with population genetics at its center and orthodox Darwinism throughout its body. Darwin smiled benignly from Westminster Abbey. He had waited one hundred years since the *Beagle* docked, but he had been vindicated.

Parts of this tale are true, and insufficiently appreciated by many professional evolutionists. Since the orthodox modern synthesists out-Darwined Darwin in their reliance upon natural selection, we often tend to trace an unbroken line from Darwin to our own times and to forget how recently the chaos described by Bateson was dispelled. Ernst Mayr (personal communication), for example, reports that he and Rensch received their training from Lamarckians and accepted the reality and importance of soft inheritance well into the 1930s. We cannot appreciate the excitement generated by the modern synthesis, and the importance of what it achieved, until we recognize that it did not cap a long and complacent Darwinian tradition, but rather replaced a set of cogent alternatives and imposed fruitful order upon a chaotic field.

But I believe that one important part of the conventional story is misleading, if not downright false. Our view of science has long been distorted by the reductionist bias that establishes an ordering of prestige by valuing quantitative and experimental disciplines that probe the basic, constituent parts of objects, and by downgrading holistic and historical fields. In this context, genetics is superior to natural history—and the modern synthesis must represent an imposition of genetic truths upon a static and musty (if not downright benighted) group of taxonomists. Thus, the modern synthesis is often portrayed as a unilinear transfer of truth, an irresistable genetic proclamation of Darwinism. Since Dobzhansky was a geneticist, and since his 1937 book was the founding document of the modern synthesis, this conventional tale gains force.

Ernst Mayr, not a disinterested observer to be sure, has challenged this view in a comprehensive analysis of the modern synthesis (Mayr and Provine 1980). He argues that the synthesis was a true interaction, not a one-way transfer. Mayr contrasts "experimental geneticists" with "population-naturalists" and writes:

They differed in their studies of causation (proximate and ultimate), in the level of the evolutionary hierarchy with which they were concerned, and in the dimensions they studied. They truly represented two very different research traditions. . . . What happened during the evolutionary synthesis was a fusion of these widely different traditions. Such an event occurs only rarely in the history of science. . . . What happened in 1937–1947 was precisely such a synthesis between research traditions, that had previously been unable to communicate.

Among the insights and approaches that naturalists supplied to geneticists in forging the modern synthesis, Mayr cites population thinking, the polytypic species, the biological species concept with its emphasis on reproductive isolation, and the role of behavior and change of function in the origin of evolutionary novelties. He writes: "It was the naturalists who solved the great species problem, a problem which the geneticists either side-stepped altogether or answered in a typological manner."

I recognize, of course, that Mayr's view is partly a partisan defense of his own profession and personal life's work. But I believe that his reconstruction is more just and accurate than the usual unilinear view. I also believe that the best defense of Mayr's position lies in considering the life and work of the man whose greatest book, here reprinted, served as the pivot and founding document of the modern synthesis. Why could Dobzhansky accomplish what no one else had done in a comprehensive way? Why could he fuse the traditions of genetics and natural history to establish a framework for the modern synthesis? Such extensive questions have no simple solution, but an outline can be sketched. Dobzhansky brought to America a research tradition that had not flowered here, and that the hope of synthesis required. He was, in short, both a geneticist by later choice and a naturalist by primary training. He combined in himself both the traditions that stood apart in America, often in mutual ignorance, if not outright hostility.

DOBZHANSKY AND THE RUSSIAN CONNECTION

In 1974, Ernst Mayr brought together all living founders of the modern synthesis (Simpson and Rensch could not attend) to assess and evaluate their work in a meeting with historians of biology. I had the privilege of attending this gathering of all my heroes. These men could not be kept on the topic of history; they were too excited about modern developments, and wanted to argue about the latest papers in

Evolution or the *American Naturalist*. The finest minds are never stilled. When they could be drawn back into their pasts, one distinct memory emerged. The great works of the synthesis had not been a set of independent volumes, each drawing its separate inspiration from Fisher, Haldane, and Wright. Instead, all the gathered authors looked at Dobzhansky (who clearly enjoyed the accolades) and said that they had drawn primary inspiration from *Genetics and the Origin of Species* (1937). Dobzhansky had not simply been first, by good fortune, in an inevitable line; his book had been the direct instigator of all volumes that followed.

If we view the modern synthesis primarily as a Darwinian fusion of three research traditions—theoretical population genetics,* experimental genetics, and taxonomy and natural history—then why was Dobzhansky able to serve as the primary catalyst?

In the United States, the traditions of experimental genetics and natural history were so separate that union from American hands could not be expected. Dobzhansky originally left Russia to work with Thomas Hunt Morgan, America's premier experimental geneticist. In remarks at Mayr's conference, Dobzhansky recalled Morgan's attitude to natural history:

"Naturalist" was a word almost of contempt with him, the antonym of "scientist." Yet Morgan himself was an excellent naturalist, not only knowing animals and plants but esthetically enjoying them. . . . Morgan was profoundly sceptical about species as biological and evolutionary realities. The species problem simply did not interest him. . . . Biology had to be strictly reductionistic. Biological phenomena had to be explained in terms of chemistry and physics. Morgan himself knew little chemistry, but the less he knew the more he was fascinated by the powers he believed chemistry to possess. There was no surer way to impress him than to talk about biological phenomena in ostensibly chemical terms.

Dobzhansky also remembered that Morgan "liked to say that genetics can be studied without any reference to evolution." Could the synthesis have taken root in such soil?

---

* As an ironic footnote, I believe that none of the major synthesists, with the exception of Simpson, ever read (or could have read) much of Fisher, Haldane, and Wright in the mathematical original. Both Dobzhansky and Mayr cheerfully admitted this to me. The primary conclusions could be, and were, divulged by their authors in plain English, and were therefore readily available for general consumption.

But things were dramatically different in Dobzhansky's native Russia, for there experimental genetics developed in the context of natural history, and was pursued largely by naturalists who had to learn their genetics, rather than by laboratory geneticists who unlearned their natural history or despised what they had never been taught.

Dobzhansky was born on January 25, 1900, in Nemirov, 130 miles SE of Kiev, in the Ukraine. He graduated from the University of Kiev in 1921 and was, until 1924, assistant professor of zoology at the Polytechnic Institute, also in Kiev. He spent the rest of his Russian career in Leningrad, lecturing at the university and working closely with the great evolutionist, Iurii Filipchenko. Nikolai Vavilov, the most famous martyr to Lysenkoism, also worked in Leningrad and was at the height of his organizational powers.

Dobzhansky left Russia in 1927, after receiving a fellowship from the Rockefeller Foundation to work in Morgan's laboratory at Columbia. He followed Morgan to CalTech and taught there in the department of genetics from 1929 to 1940. Dobzhansky then moved to Columbia University, and spent the rest of his career there, at Rockefeller University after 1962, and finally at the University of California in Davis. He died in 1975, in the midst of numerous experiments and future plans. His work never stilled; his commitment never abated. He was the finest evolutionary geneticist of our times. A complete bibliography to 1970 can be found in Levene et al. (1970).

In Russia—and this is the ill-appreciated point serving as a focus for this essay—Dobzhansky was not trained primarily as a geneticist. He began his scientific career as a classical systematist, a specialist on the Coccinellidae, or ladybird beetles. Most of his early papers are descriptive and taxonomic. Later, his interests enlarged to include genetics and evolution, but coccinellids remained his primary focus of research. He published on physiological problems of geographic variation and polymorphism in his favorite beetles. Only two papers from his Russian years deal with *Drosophila*, including an important 1927 study of the manifold effects of individual genes.

While working with Morgan in America (1927–1933), Dobzhansky devoted himself primarily to problems in classical genetics, making little explicit appeal to evolutionary principles or interests. He studied the mechanism of sex determination in *Drosophila*, chromosome aberrations, and position effects. He investigated the role of translocations in preventing crossovers, and used translocations to present one of the first cytological proofs that genes are arranged linearly on chromosomes. In the mid-1930s, Dobzhansky's older interest in evolutionary problems began to revive. He published his first paper on *Drosophila pseudoobscura* in 1933, a study of reproductive isolation between what he then called "races" A and B, now known to be sibling species. In 1936, in a celebrated paper, Dobzhansky showed how phylogenies of inversions could be constructed.

Thus, when Columbia University invited Dobzhansky to give the Jesup lecture series in October 1936 (published in 1937 as the book here reprinted), he felt ready to attempt a synthesis of the foci of his own personal life's work: evolutionary systematics and experimental genetics. Dobzhansky stood in a unique position to inaugurate such a synthesis because he had emerged from the one research tradition—the Russian—that had developed experimental genetics in the context of natural history, and he had brought these concerns to America, where he had the opportunity to apply them in the world's finest laboratory and research group in experimental genetics.

The specter of Lysenko haunts the history of Russian genetics so dramatically that many American evolutionists do not recognize how sophisticated Russian genetics had been before the debacle. First of all, Darwinism had a long tradition of favorable reception in Russia. To be sure, the reasons were not all scientific. Darwinism had become a rallying point for Russian left-wing intellectuals during the nineteenth century, and its prestige had endured in Russia, despite low popularity in the rest of Europe. Russian genetics therefore developed in a Darwinian context. More specifically, Moscow's Kol'tsov Institute became the center for Russian experimental and population genetics. Kol'tsov, though neither a naturalist nor geneticist himself, recognized the value of a link and insisted that all his geneticists be well trained in natural history (Adams 1980).

Dobzhansky, to be sure, was not situated in Moscow, but in Leningrad, where his chief supervisor Filipchenko was a more Western-style non-Darwinian. Filipchenko espoused a general view that stood as the chief impediment to any synthesis of evolutionary theory under the principles of classical genetics: he believed that micro- and macroevolution were distinct in principle, and could be traced to different underlying genetic mechanisms.* Dobzhansky recalled: "Filipchenko distinctly believed that the characteristics of orders, classes, and phyla are non-Mendelian, presumably located somehow or other in the cytoplasm. The characteristics of varieties and species are in the Mendelian genes on the chromosomes."

But Dobzhansky was also in close contact with his Darwinian colleagues in Moscow (at Mayr's conference, he reminisced nostalgically about his frequent overnight train journeys to Moscow). Although Adams (1980) reports that Dobzhansky's letters at the time betray ambiguity, if not tentative support for Filipchenko's view, Dobzhansky (perhaps with the inaccuracy of long retrospect) remembers himself as at least a closet Darwinian during his Russian years. He said: "I remember Conklin's lecture in Woods Hole in the summer of 1928, where among other things he presented this idea [of the unbridgeable genetic gulf between micro- and macroevolution] to my utter horror, because I had been fighting that with my friend Filipchenko for years."

During the 1920s, the Kol'tsov Institute may well have been the world's largest center for *Drosophila* research. Sergei Chetverikov presided over the major group of twelve, which included, among others, Timofeef-Ressovsky, who soon left for Germany and worked there until the end of the war. (Timofeef-Ressovsky helped to translate, literally and figuratively, the work of Chetverikov's school into language more familiar to Western scientists. He was arrested after the war and spent many years in Russian prison camps. He is the unnamed geneticist whom Solzhenitzyn met at the gulag and who figures prominently in his books). Serebrovsky had his own *Drosophila* group at the Institute; it included Dubinin, who did much

---

* Contemporary critics, myself included, of the synthesists' premise that gradual change of gene frequencies within populations serves, by extrapolation, as an adequate model for all evolutionary events do not accept this council of despair from an earlier era. We believe that a common set of genetic principles produces different patterns of change at various levels of the evolutionary process, and that several bulwarks of traditional microevolution—change by gradual and sequential allelic substitution, each with small effect, and the adaptive nature of virtually all change, for example—do not always apply to macroevolution.

to reestablish Soviet genetics after Lysenko's fall from favor. The Kol'stov Institute had been blessed with a precious gift in 1922, when visitor H. J. Muller brought *D. melanogaster* cultures containing 32 of the famous mutants identified by Morgan's school. These flies brought Russia into the mainstream of *Drosophila* research, and Chetverikov's group made several important discoveries. As an interesting footnote to drosophology, the American species *D. pseudoobscura* was first recognized by two geneticists at the Kol'tsov (Frolova and Astaurov), who were unable to cross supposed American "*D. obscura*" with European strains, discovered the chromosomal distinctions in the American form, and named the new species.

But the major distinction of Chetverikov's school was its avowedly evolutionary focus and its emphasis upon the genetics of *natural* populations. Kol'tsov insisted (as Dobzhansky remembered) that his experimentalists know classical subjects in natural history "like multiplication tables." Chetverikov himself began his career as a butterfly taxonomist and learned his genetics later. In 1926, Chetverikov wrote a seminal paper (reprinted in English in 1961) attempting to integrate biometry, genetics, and natural history in a Darwinian framework. The Kol'tsov drosophologists performed the first systematic studies of genetic variation in wild populations of *D. melanogaster*. When they found several of Morgan's mutations in natural demes, the common idea that mutations were laboratory aberrations without evolutionary significance faded away. They formulated the concept of the gene pool (Adams 1979) and developed the theory of genetic drift independently of Sewall Wright. They made the first determination of the frequency of lethals in natural populations.

Dobzhansky followed this work avidly, visited Moscow frequently, and was friendly with all leading workers at the Kol'tsov. If Lysenko and World War II had not dismantled this center, the history of twentieth-century evolutionary biology might have been very different. But Dobzhansky preserved the legacy and brought to America both the evolutionary focus of Russian genetics, and his personal experience as a taxonomist and naturalist turned geneticist. No one in the United States had this winning combination. Dobzhansky was preadapted to initiate the synthesis.

GENETICS AND THE ORIGIN OF SPECIES

Like most great books, Dobzhansky's volume is not merely a review and categorization of existing data. It is a long argument for a general attitude toward nature and a specific approach that might unite the disparate elements of evolutionary theory. The argument, of a striking but subtle originality in its time, contains two major parts:

1. The central problem of evolution is the origin of discontinuity between species. It sounds a commonplace today, but only because Dobzhansky and the synthesis made it so. In 1937, it was a surprising statement, which, if followed, promised to reorient the focus of evolutionary study. Morgan and virtually all experimentalists would have argued that the origin and nature of variation, or the way in which variation spreads through populations, are the key issues in evolutionary theory. Morgan disavowed the species problem as, at best, the hang-up of dull taxonomists and, at worst, a bogus issue, because species are not real in the flow of nature and we name them only because our poor minds cannot handle continuity. Dobzhansky didn't deny that both the origin of variation and its spread through populations were central questions in biology. But he argued that evolution works on a series of hierarchical levels, and that the primary focus of natural history lay not at these lower levels, but at a higher one—the origin of species itself (Darwin's title, after all). Diversity is the primary fact of nature (and the first topic of chapter 1). Diversity arises by the splitting of lineages—by speciation. Speciation leads to discontinuity in nature (the second topic of chapter 1). How can the continuous process of genetic change yield discontinuity in nature? The origin of discontinuities between species is the key problem of evolutionary theory. Only a naturalist (better yet, a trained systematist) could have reset the stage for synthesis in such a fruitful way. The issue that experimentalists usually laughed at became the central focus for study:

The origin of hereditary variations is, however, only a part of the mechanism of evolution. If we possessed a complete knowledge of the physiological causes producing gene mutations and chromosomal changes, as well as a knowledge of the rates with which these changes arise, there would still remain much to be learned about evolution. These variations may be compared with building materials, but the presence of an unlimited supply of materials does not in itself give assurance that a building is going to be constructed. . . . Mutations and chromosomal changes are constantly arising at a finite rate, presumably in all organisms. But in nature we do not find a single greatly variable population of living beings which becomes more and more variable as time goes on; instead, the organic world is segregated into more than a million separate species, each of which possesses its own limited supply of variability which it does not share with the others. . . . The origin of species . . . constitutes a problem which is logically distinct from that of the origin of hereditary variation. (p. 119)

2. Principles of genetics, accessible through experimental work in the laboratory or short-term study of natural populations, suffice to explain evolution at all levels, despite the distinctness of problems at different levels. A unified evolutionary theory, based upon phenomena that can be comprehended directly (not merely inferred imperfectly by considering fossilized results over eons of time), is within our grasp.

Dobzhansky's original probe toward synthesis was more a methodological claim for knowability than a strong substantive advocacy of any particular genetic argument. It held, contrary to his own mentor Filipchenko in Russia or H. F. Osborn in America, that the methods of experimental genetics can provide enough principles to encompass evolution at all levels, but it did not play favorites among the admitted set of legitimate principles. It did not, in particular, proclaim the pervasive power of natural selection leading to adaptation as a mechanism of evolutionary change. Macroevolution, it stressed, is not a thing apart, unknowable in principle from experimental work on laboratory and natural populations, and requiring different (and perhaps unfathomable) modes of genetic change. Dobzhansky writes cautiously, emphasizing the hope for complete knowability based on microevolutionary genetics:

Experience seems to show, however, that there is no way toward an understanding of the mechanisms of macroevolutionary changes, which require time on a geological scale, other than through a full comprehension of the microevolutionary processes observable within the span of a human lifetime and often controlled by man's will. For this reason we are compelled at the present level of knowledge reluctantly to put a sign of equality between the mechanisms of macro- and microevolution, and, proceeding on this assumption, to push our investigations as far ahead as this working hypothesis will permit. (p. 12)

I do not, of course, assert that Dobzhansky stated no substantive preferences in prescribing a set of legitimate genetic principles. The focus of his argument is Darwinian, particularly when he insists that mutation supplies raw material only, and does not by itself impart direction to evolutionary change. Lamarckism, fundamentally a theory of adaptively directed variability, was still a live, if fading, issue in the 1930s, and Dobzhansky attacks it explicitly in several passages—for example, on page 171 where he rejects Lamarckian explanations for phenocopies.

The irony, almost a catch-22, of reading most great books in science lies in their capacity to render themselves obsolete by their own persuasive power. Much of Dobzhansky's book seems overstated or commonplace now; in some passages, he tilts at windmills. But then we realize that this book set the tone and style of an entire profession; the cascade of fruitful work that it engendered quickly converted Dobzhansky's original statements into conventional claims. Some inkling of the chaotic and depressed state of evolutionary theory before the synthesis can be glimpsed in a simple list of previously popular arguments that Dobzhansky regarded as sufficiently important to refute, claims that denied his hope for synthesis by suggesting that Mendelian processes observed in the laboratory do not represent the genetic style of "important" evolutionary change in nature. Among claims requiring refutation, Dobzhansky included: continuous variation in nature is non-Mendelian, and different in kind from discrete mutational variation in laboratory stocks (p. 57); Mendelian variation can only underlie differences between taxa of low rank (races to genera), while higher taxa owe their distinctions to another (and unknown) genetic process (p. 68); chromosomal changes are always destructive and can only lead to degeneration of stocks (p. 83); differences between taxa of low rank are directly induced by the environment and have no genetic or evolutionary basis (p. 146); Johannsen's experiments on pure lines showed the ineffectiveness of natural selection as a mechanism of evolutionary change (p. 150); selection is too slow in large populations to render evolution, even in geological time (p. 178); genetic principles cannot account for the origin of re-productive isolation (p. 255).

Précis may be the dullest of literary forms, but a sequential account of Dobzhansky's argument captures his strategy for synthesis better than any discursive analysis I could present.

The initial short chapter on "organic diversity" establishes the basic premise that evolutionary theory must account not only for genetic variation and change in populations, but also for the discontinuity that arises in speciation and generates diversity as a result. Evolution works on at least three levels—change in individuals (by mutations and recombination), change in populations, and the origin of repro-ductive isolation in speciation (the problem of discontinuity). Events at all three levels shall be explained by a single set of genetic princi-ples inferred from experimental studies of laboratory and natural populations.

Chapter 2, on "gene mutation," argues that Mendelian mutation produces the kind of variability that Darwin required as raw material to make natural selection a primary agent of evolutionary change. Mutations affect all parts, the fundamental as well as the superficial. They produce all effects, lethal to beneficial; they are not always harmful. They yield phenotypic results of all magnitudes, not just major (and deleterious) alterations; in fact, mutations with small ef-fects are by far the most common. They are haphazard in direction and do not produce evolutionary change by their differential occur-rence alone; x-rays may induce a higher frequency of mutation, but they do not cause mutations to occur in definite directions. Muta-tions are frequent in occurrence; a paucity of variation will not serve as an important brake upon Darwinian processes. In short, mutations are copious, usually small in effect, and random in direction.

In the third chapter, Dobzhansky demonstrates that these muta-tions are not only phenomena of the laboratory; they also exist in wild populations, where they form the basis of continuous variability and serve as the raw material that natural selection requires. This had been the major finding of Chetverikov's school in Russia. Not only do mutations, including some of Morgan's most famous, exist as variants within populations, but they also serve, in fixed or prevalent form, as the major genetic distinctions between races and species.

Thus we find complete continuity between the study of mutant individuals in the laboratory and the foundation of genetic differences between species in nature. The origin of species is not intrinsically mysterious or based upon processes that cannot be observed or manipulated in our experiments. The synthesis stressed continuity as a guarantee of knowability. Dobzhansky writes:

Race formation begins with the frequency with which a certain gene or genes becomes slightly different in one part of a population from what it is in other parts. If the differentiation is allowed to proceed unimpeded, most or all of the individuals of one race may come to possess certain genes which those of the other race do not. Finally, mechanisms preventing the interbreeding of races may develop, splitting what used to be a single collective genotype into two or more separate ones. When such mechanisms have developed and the prevention of interbreeding is more or less complete, we are dealing with separate species. (p. 62)

But laboratory genetics deals with another class of variants beyond point mutations: differences in chromosomal structure, arrangement, and number. In chapter 4, Dobzhansky demonstrates that these chromosomal variants, like point mutations, also exist in nature and also underlie differences between races and species. (He stresses the evidence for translocation as the basis of racial differentiation in the jimson-weed *Datura*, and inversions for *Drosophila*.) Again, he argues for continuity between experimental genetics in the laboratory and evolution in nature: "It is undeniable that recent investigations have bridged the gap between the chromosomal aberrations observed in the laboratory and the variations of the chromosome structure found in the wild state" (p. 114). In a closing section on "position effects," Dobzhansky presents an eloquent defense of holistic approaches by arguing that chromosomes cannot simply be strings for a row of independent beads (or "a sort of sausage stuffed with a definite number of layers of genes, arranged in a definite but fortuitous linear order" (p. 115), to use his terms), for then the mere rearrangement of sequence (inversion) or alteration of place (translocation) could not yield major effects upon phenotypes.

The fifth chapter, on "variation in natural populations," stresses the pluralism of the early synthesis. Observable genetic phenomena are the source of all evolution; continuity from studies in the laboratory, to variation within natural populations, to formation of races and species is crucial to this central claim:

It is now clear that gene mutations and structural and numerical chromosomal changes are the principal sources of variation. Studies of these phenomena have been of necessity confined mainly to the laboratory and to organisms that are satisfactory as laboratory objects. Nevertheless, there can be no reasonable doubt that the same agencies have supplied the materials for the actual historical process of evolution. This is attested by the fact that the organic diversity existing in nature, the differences between individuals, races, and species, are experimentally resolvable into genic and chromosomal changes that arise in the laboratory. (p. 118)

But what forces shape and preserve this variation in nature? Dobzhansky stresses natural selection (p. 120), but he does not grant it the dominant role that later "hard" versions of the synthesis would confer upon it. He emphasizes genetic drift (which he calls "scattering of the variability") as a fundamental process in nature, not as an odd phenomenon occurring in populations too small to have an evolutionary legacy. He argues that local races can form without the influence of natural selection, and supports Crampton's (1916, 1932) interpretation of the nonadaptive and indeterminate character of substantial racial differentiation in the Pacific land snail *Partula*. He emphasizes that evolutionary dynamics depend, in large measure, upon the size of populations *because* selection is not always in control (and we therefore need information about numbers of individuals and their mobility in order to assess the effects of drift, migration, and isolation). He coins the term "microgeographic race" and argues that most are nonadaptive and genetically based, contrary to many naturalists who regard them as adaptive and nongenetic.

The sixth chapter then treats natural selection explicitly. Dobzhansky begins by clearing away some Mendelian misconceptions about the impotence of natural selection (logical errors in interpretating Johannsen's experiments on pure lines, for example). He then poses a central question: Darwin devised the theory of natural selection to explain adaptation; admitting Darwin's success in this area, may we then extrapolate and argue that selection controls the direction of all evolutionary change (p. 150). Dobzhansky answers no, criticizes the strict selectionism of Fisher (p. 151), and has some good words for a book that would later be castigated by synthesists as a remnant of older and unproductive ways of thought—Robson and Richards (1936).

The main body of the chapter discusses observational and inferential evidence for natural selection: experimental studies in the laboratory, historical changes in natural populations (including the inevitable industrial melanism), protective and warning coloration, and laws of geographic distribution. He also includes, in a surprisingly short section (pp. 176–185), the book's only explicit discussion of the theoretical population genetics of Fisher, Haldane, and Wright, including their central defense of small selection pressures as sufficient to render evolution in geological time.

A long concluding section (pp. 185–191) supports Wright's "island model" of selection among semi-isolated demes occupying different peaks of an adaptive landscape. He pleads for more study of "the physiology of populations" since Wright's model proclaims three factors as important in different ways, but does not grant predominance to any: genetic drift, migration, and natural selection:

Since evolution as a biogenic process obviously involves an interaction of all of the above agents, the problem of the relative importance of the different agents unavoidably presents itself. For years this problem has been the subject of discussion. The results of this discussion so far are notoriously inconclusive; the "theories of evolution" arrived at by different investigators seem to depend upon the personal predilections of the theorist. (p. 186)

Dobzhansky does, however, suggest that Wright's model may explain the conviction of naturalists that the morphological differentia of races and species so often seem to be nonadaptive.

The seventh chapter, on polyploidy, continues the theme of pluralism by stressing the importance of this saltatory mechanism of speciation in plants:

The process of species formation is apparently a slow and gradual one, consuming time on at least a quasi-geological scale. It is highly remarkable, therefore, that alongside this slow method of species formation there should exist in nature a quite distinct mechanism causing a rapid, sudden, cataclysmic emergence of new species. (p. 192)

Dobzhansky invented the term "isolating mechanism," and his eighth chapter takes up the subject, so neglected by geneticists, that he regarded as central to evolution: the origin of discontinuities between populations in nature. Although he denies any claim for universal allopatry, if only because he regards host races of insects as cases of incipient (or completed) sympatric speciation (p. 257), his classification of isolating mechanisms does grope toward the primacy of geographic separation that Mayr (1942) would soon enshrine in his theory of allopatric (or geographic) speciation. (The terms "allopatry" and "sympatry" are Mayr's). He divides isolating mechanisms into two categories: geographic and physiological (the latter itself divisible into premating and postmating mechanisms). There is surely an important confusion here in mixing phenomena of such different significance—for geographic isolation is a facilitator of speciation (or a precondition for those who deny sympatric speciation), while physiological isolation is a sign that speciation has already occurred and a device for maintaining the distinction. Georgraphic isolation should not be classified as an isolating "mechanism" at all, but this confusion would not be resolved until Mayr devised his useful terminology of "allopatric" and "sympatric" speciation. Still, Dobzhansky seems to know that his has mixed two different things, but lacks a framework for their proper separation. He does imply that geographic and physiological isolation are different in kind, and he does identify the presence of physiological isolating mechanisms as the criterion for recognizing separate species in nature.

Dobzhansky, the naturalist and taxonomist, deplores the "appallingly insufficient attention" (p. 254) that his geneticist colleagues had devoted to the problem of isolation. It is this dearth of information that permits evolutionists like Richard Goldschmidt to argue for a fundamental distinction between causes of variation within and among populations, and thus to preclude any possible synthesis based on the traditions of laboratory and experimental genetics (p. 232).

Dobzhansky's passionate desire to stir his genetical colleagues out of their apathy toward the problem of isolation explains his ninth chapter, on hybrid sterility, the most technical and detailed in his book, and otherwise an anomaly of sorts for such a general treatise.

Hybrid sterility is the one isolating mechanism that has been analyzed genetically; an account of the evolutionary value of these studies might encourage others, and further a synthesis based on genetic principles:

A wealth of data on the occurrence of various isolating mechanisms in different subdivisions of the animal and plant kingdoms is scattered through biological literature. The genetic analysis of isolating mechanisms, with the possible exception of hybrid sterility, has however been left in abeyance. It is a fair presumption that the pessimistic attitude of some biologists (e.g., Goldschmidt, 1933), who believe that genetics has learned a good deal about the origin of variations within a species, but next to nothing about that of the species themselves, is due to the dearth of information on the genetics of isolating mechanisms. (p. 232)

In the last chapter, "species as natural units," Dobzhansky the taxonomist again urges his colleagues in genetics to study evolution at the crucial level of speciation. Species, he argues, are real units in nature (for sexually reproducing organisms at least), not categories of convenience. Nature's discontinuities are produced by speciation and maintained by the physiological isolating mechanisms that prevent amalgamation of populations. If we define species as groups that have become irrevocably separate by the evolution of physiological isolating mechanisms (pp. 312–313), then the species of the naturalist and systematist will coincide and a common foundation for a unified evolutionary theory will be set.

This précis does little justice to the richness and subtlety of Dobzhansky's presentation. It is merely a bare-bones outline of the central argument. But it does illustrate that the synthesis was launched by a geneticist steeped in the traditions of systematics and natural history. A conventional American experimental geneticist would have had neither the interest nor the training to ask questions about discontinuity in nature and to identify the genesis of isolating mechanisms as a keystone of evolution. The synthesis was a true fusion of genetics and natural history, not an imposition of one progressive field upon another hidebound profession.

THE SYNTHESIS HARDENS

Genetics and the Origin of Species went through three editions (1937, 1941, and 1951). As in the various versions of Darwin's *Origin,* the differences are not trivial or cosmetic, but represent a major change in emphasis—a change that set the research program for most of evolutionary biology throughout the 1960s and 1970s. The original works of the synthesis, Dobzhansky's and others, held fast to the claim for knowability at all levels based upon principles of genetics inferred directly from studies of laboratory and natural populations. Their substantive emphasis was strongly Darwinian, but their approach to assessing the relative importance of admissible evolutionary forces was pluralistic. In particular, they did not grant a dominant or nearly exclusive role to adaptation produced by natural selection. As the synthesis developed, however, the adaptationist program grew in influence and prestige (Gould and Lewontin 1979), and other modes of evolutionary change were denied, neglected, or redefined as locally operative but unimportant in the overall picture.

Dobzhansky's third edition (1951) clearly reflects this hardening. He still insists, of course, that not all change is adaptive. He attributes the frequency of some traits to equilibrium between opposed mutation rates (1951: 156) and doubts the adaptive nature of racial variation in blood group frequencies. He asserts the importance of genetic drift (1951: 165, 176), and does not accept as proof of panselectionism one of the centerpieces of the adaptationist program—Cain's work on frequencies of banding morphs in the British land snail *Cepaea* (1951: 170).

But inserted passages and shifting coverage have, as their common focus, Dobzhansky's increasing faith in the scope and power of natural selection, and in the adaptive nature of most evolutionary change. He deletes the two chapters that contained most material on nonadaptive or nonselected change (polyploidy and chromosomal changes, though he includes their material, in reduced form, within other chapters). He adds a new chapter on "adaptive polymorphism" (1951: 108–134). He argues that anagenesis, or "progressive" evolution, works only through the optimizing, winnowing agency of selection based on competitive deaths; species adapting by increased fecundity in unpredictably fluctuating environments do not contribute to anagenesis (1951: 283).

But the most remarkable addition occurs right at the beginning. I label it remarkable because I doubt that Dobzhansky really believed what he literally said; I feel sure that he would have retracted or modified it had anyone pointed out that he had allowed a fascination for adaptationism to displace the oldest of evolutionary truths.

In this addition to the introductory material on discontinuity in nature, Dobzhansky poses the key question of why morphological space is so "clumped"—why a cluster of so many cats, another of dogs, a third of bears, and so much unoccupied morphological space between? He begins by transferring Wright's model of the "adaptive landscape" to a hierarchical level where Wright did not intend it to apply (Eldredge and Cracraft 1980: 251 ff). In so doing, Dobzhansky subtly shifts the meaning of the model from an explanation for nonoptimality (with important aspects of nonadaptation) to an adaptationist argument based on best solutions. Wright devised the model to explain differentiation among demes *within* a species. He proposed it to justify a fundamentally nonadaptationist claim: if a "best solution" exists for a ·species' phenotype (the highest peak in the landscape), why do not all demes evolve it? When the model is "upgraded" to encompass differences *between* species within a clade, then it becomes a framework for strict adaptationism. Each peak is now the optimal form for a single species (not the nonoptimal form for some demes within a species), and related peaks represent a set of best solutions for different adaptations of separate evolutionary entities within the clade.

Dobzhansky then attempts to solve the problem of clumping with a peculiar adaptationist argument based upon the organization of ecological space into preexisting optimal "places" where good design may find a successful home. If evolution has produced a cluster of cats, this is because an "adaptive range," studded with adjacent peaks, exists in the economy of nature, waiting, if you will, for creatures to discover and exploit it.

The enormous diversity of organisms may be envisaged as correlated with the immense variety of environments and of ecological niches which exist on earth. But the variety of ecological niches is not only immense, it is also discontinuous. . . .

The adaptive peaks and valleys are not interspersed at random. "Adjacent" adaptive peaks are arranged in groups, which may be likened to mountain ranges in which the separate pinnacles are divided by relatively shallow notches. Thus, the ecological niche occupied by the species "lion" is relatively much closer to those occupied by tiger, puma, and leopard than to those occupied by wolf, coyote, and jackal. The feline adaptive peaks form a group different from the group of the canine "peaks." But the feline, canine, ursine, musteline, and certain other groups of peaks form together the adaptive "range" of carnivores, which is separated by deep adaptive valleys from the "ranges" of rodents, bats, ungulates, primates, and others. In turn, these "ranges" are again members of the adaptive system of mammals, which are ecologically and biologically segregated, as a group, from the adaptive systems of birds, reptiles, etc. The hierarchic nature of the biological classification reflects the objectively ascertainable discontinuity of adaptive niches, in other words the discontinuity of ways and means by which organisms that inhabit the world derive their livelihood from the environment. (1951, pp. 9–10)

Thus, Dobzhansky renders the hierarchical structure of taxonomy as a fitting of clades into ecological spaces. Discontinuity is not so much a function of history as a reflection of adaptive topography. But this cannot be: surely, the cluster of cats exists primarily as a result of homology and historical constraint. All felines are alike because they arose from a common ancestor shared with no other clade. That ancestor was well adapted, and all its descendants may be. But the cluster and the gap reflect history, not the current organization of ecological topography. All feline species have inherited the unique cat *Bauplan*, and cannot deviate far from it as they adapt, each in its own particular (yet superficial) way. Genealogy, not current adaptation, is the primary source of clumped distribution in morphological space. To ignore history and treat clumping as an optimal mapping of form upon adaptive topography is about as skewed as the claim of a colleague who once seriously proposed to me that tetrapods have four legs because four is an optimal solution to problems of movement and support in a world ruled by gravitational forces. Did he forget who tetrapods evolved from, I ask incredulously.

In increasing his allegiance to selection and adaptation, Dobzhansky reflected a trend pervading the entire synthesis (or had he, continuing an innovative role, created the trend in the first place). Comparison of original and later works by other major authors reveal the same pattern of change (see my 1980 analysis of G. G. Simpson, 1944 vs. 1952). In 1953, for example, Simpson recast "quantum evolution," the distinct and largely nonadaptive process that he had advocated in 1944 for the origin of higher taxa, as merely a rapid form of adaptive phyletic evolution. The absolutely "inadaptive phase" of 1944 becomes the "relatively inadaptive" phase of 1953— relative, that is, with respect to ancestors who are living elsewhere and descendants who do not live yet. The "relatively inadaptive" phase is still adapted in its own environment. Indeed, Simpson (1953) asserts that quantum evolution is more strictly adaptive than any other mode because selection must work with unusual severity upon such major alterations of form.

I do not fully understand why this hardening occurred, but I regard it as an important topic for historical research since its result so dominated the research program of evolutionary biology for many years. In part, the "ecological genetics" of E. B. Ford and his pan-selectionist school in England must have had a major effect. Their commitment to adaptationist explanations of everything and their discovery of strong selection coefficients in nature buoyed strict Darwinian faith. Dobzhansky must also have gained confidence from the claims of macroevolutionary colleagues like Simpson, Rensch, and Schmalhausen, who held that selection and adaptation might triumph in the very realm that had once denied it most strongly. Dobzhansky's 1951 argument for continuity between micro- and macroevolution (p. 17) is much stronger than the 1937 version. Finally, Dobzhansky's own empirical work increased his belief in the power of selection. In 1937, he tended to attribute inversion frequencies in natural populations of *Drosophila* to genetic drift, but he then discovered that these frequencies fluctuate in a regular and repeatable way from season to season, and decided (with evident justice) that they must be adaptive.

EPILOGUE

The homeotic mutants of *Drosophila* generated quite a literature among geneticists during the 1920s and 1930s. I recently had occasion to read through the original articles of discovery and found them disquieting in their failure to draw from this fascinating topic all the rich implications so clearly inherent in it. They analyzed the genetics elegantly and made some superficial comments about the morphological peculiarity involved. But they did not pursue the bizarre morphology to extract any evolutionary messages. Then I read Bridges and Dobzhansky's (1933) description of the discovery of *proboscipedia*, the mutation that puts part of a foot in a fly's mouth. The formal analysis was as elegant as ever, but it was now accompanied by a rich discussion of broader meanings. The authors, for example, noted that the teratology of leg for mouth parts was accompanied by a set of subtler modifications to other mouth parts, all tending to reproduce in flies with *proboscipedia* an approach to the structure of biting insects from which flies had evolved. Then I realized the reason for the difference—Dobzhansky, not Bridges. Only a trained taxonomist and entomologist could have seen what Dobzhansky noted, and all his genetical colleagues had missed. There is no substitute for a thorough knowledge of natural history.

At Mayr's conference, Dobzhansky began his personal recollections by stating: "I was, if you please, an entomologist specializing in taxonomy of Coccinellidae, lady beetles. Seeing a lady beetle still produces in me a flow of a love hormone. The first love is not easily forgotten." I will forego my usual penchant for tedious analysis, and simply let this comment stand as the mark of a great man and scientist.

STEPHEN JAY GOULD
Museum of Comparative Zoology
Harvard University
Cambridge, Massachusetts

WORKS NOT CITED IN THE EARLIER EDITIONS

Adams, M. B. 1979. From "gene fund" to "gene pool": On the evolution of evolutionary language. *Studies Hist. Biol.* 3:241–285.

Adams, M. B. 1980. Sergei Chetverikov, the Kol'tsov Institute, and the evolutionary synthesis. In E. Mayr and W. Provine, eds., *The Evolutionary Synthesis*, pp. 242–278. Cambridge: Harvard University Press.

Bateson, G. 1894. *Materials for the Study of Variation*. London: MacMillan.

Bridges, C. B. and Th. Dobzhansky. 1933. The mutant *Proboscipedia* in *Drosophila melanogaster*: A case of hereditary homoösis. W. *Roux' Archiv fur Entwicklungsmechanik* 127:575–590.

Cain, A. J. and P. M. Shepard. 1950. Selection in the polymorphic land snail *Cepaea nemoralis*. *Heredity* 4:275–294.

Chetverikov, B. 1926. Eng. tr. 1961. On certain aspects of the evolutionary process from the standpoint of modern genetics. *Proc. Am. Phil. Soc.* 195:167–195.

Darwin, C. 1874. *The Descent of Man*. 2d. ed. London: John Murray.

Dobzhansky, Th. 1941. *Genetics and the Origin of Species*. 2d. ed. New York: Columbia University Press.

Dobzhansky, Th. 1951. *Genetics and the Origin of Species*. 3d ed. New York: Columbia University Press.

Eldredge, N. and J. Cracraft. 1980. *Phylogenetic Patterns and the Evolutionary Process*. New York: Columbia University Press.

Gould, S. J. 1980. G. G. Simpson, paleontology, and the modern synthesis. In E. Mayr and W. Provine, eds., *The Evolutionary Synthesis*, pp. 153–172. Cambridge: Harvard University Press.

Gould, S. J. and R. C. Lewontin. 1979. The spandrels of San Marco and the Panglossian paradigm: A critique of the adaptationist programme. *Proc. Roy. Soc. London B* 205:581–598.

Levene, H., L. Ehrman, and R. Richmond. 1970. Theodosius Dobzhansky up to now. In M. K. Hecht and W. C. Steere, eds., *Essays in Genetics and Evolution in Honor of Theodosius Dobzhansky*, pp. 1–41. New York: Appleton-Century-Crofts.

Mayr, E. 1942. *Systematics and the Origin of Species*. New York: Columbia University Press. Reprint edition, 1981.

Mayr, E. and W. B. Provine, eds. 1980. *The Evolutionary Synthesis*. Cambridge: Harvard University Press.

Rensch, B. 1947. *Neuere Probleme der Abstammungslehre.* Stuttgart: Enke. Updated and translated into English as *Evolution Above the Species Level.* New York: Columbia University Press, 1960.

Simpson, G. G. 1944. *Tempo and Mode in Evolution.* New York: Columbia University Press.

Simpson, G. G. 1953. *The Major Features of Evolution.* New York: Columbia University Press.

Stebbins, G. L. 1950. *Variation and Evolution in Plants.* New York: Columbia University Press.

GENETICS AND
THE ORIGIN OF SPECIES

I: ORGANIC DIVERSITY

FOR CENTURIES man has been interested in the diversity of
living beings. The multitude of the distinct "kinds" or species
of organisms is seemingly endless, and within a species no uniformity
prevails. In the case of man himself it is generally taken for granted
that every individual is unique, different from every other one who
now lives or has lived. The same is probably true for individuals of
species other than man, although our methods of observation are
frequently inadequate to show this. Attempts to understand the
causes and significance of organic diversity have been made ever
since antiquity; the problem seems to possess an irresistible aesthetic
appeal, and biology owes its existence in part to this appeal.

The true extent of organic diversity can only be surmised at present. In 1758 Linnaeus knew 4,236 species of animals. The recent
estimates of described species (Pratt 1935) are as follows:

Arthropoda	640,000	Total of Column 1	794,500
Mollusca	70,000	Annelida	6,500
Chordata	60,000	Plathelminthes	6,000
Protozoa	15,000	Echinodermata	4,800
Coelenterata	9,500	All others	10,965
		Total	822,765

The number of described species of flowering plants is around
133,000, and of lower plants 100,000, a total of 233,000. That these
totals fall short of the actually existing number of species is clear
enough. For although in some groups—such as mammals and birds—
most species are known already, in other groups—notably among
insects, which make up more than a half of the species of animals—
many new species are described every year, and large additions may
be expected in the future. A million and a half species of animals and
plants combined is a conservative estimate. Of course, this does not

take into account the intraspecific variation, which is commensurate only with the number of living individuals.

DISCONTINUITY

Organic diversity is an observational fact more or less familiar to everyone. It is perceived by us as something apart from ourselves, a phenomenon given in experience but independent of the working of our minds. A more intimate acquaintance with the living world discloses another fact almost as striking as the diversity itself. This is the discontinuity of the organic variation.

If we assemble as many individuals living at a given time as we can, we notice at once that the observed variation does not form a single probability distribution or any other kind of continuous distribution. Instead, a multitude of separate, discrete, distributions are found. In other words, the living world is not a single array of individuals in which any two variants are connected by unbroken series of intergrades, but an array of more or less distinctly separate arrays, intermediates between which are absent or at least rare. Each array is a cluster of individuals, usually possessing some common characteristics and gravitating to a definite modal point in their variations. Small clusters are grouped together into larger secondary ones, these into still larger ones, and so on in an hierarchical order.

The discontinuity of organic variation has been exploited to devise a scientific classification of organisms. Evidently the hierarchical nature of the observed discontinuity lends itself admirably to this purpose. For the sake of convenience the discrete clusters are designated races, species, genera, families, and so forth. The classification thus arrived at is to some extent an artificial one, because it remains for the investigator to choose, within limits, which cluster is to be designated a genus, family, or order. But the same classification is natural so far as it reflects the objectively ascertainable discontinuity of variation, and the dividing lines between species, genera, and other categories are made to correspond to the gaps between the discrete clusters of living forms. Therefore the biological classification is simultaneously a man-made system of pigeonholes devised for the pragmatic purpose of recording observations in a convenient manner and an acknowledgment of the fact of organic discontinuity. A single example will suffice to illustrate this point.

Any two cats are individually distinguishable, and the same probably holds for any two lions. And yet no living individual has ever been seen about which there could be a doubt as to whether it belongs to the species-cluster of cats (*Felis domestica*) or to the species-cluster of lions (*Felis leo*). The two clusters are discrete because of the absence of intermediates, and therefore one may safely affirm that any cat is different from any lion, and that cats as a group are distinct from lions as a group. Any difficulty which may arise in defining the species *Felis domestica* and *Felis leo*, respectively, is due not to the artificiality of these species themselves, but to the fact that in common as well as in scientific parlance the words "cat" and "lion" frequently refer neither to individual animals, nor to all the existing individuals of these species, but to certain modal points toward which these species gravitate. The modal points are statistical abstractions having no existence apart from the mind of the observer. The species *Felis domestica* and *Felis leo* are evidently independent of any abstract modal points which we may contrive to make. No matter how great may be the difficulties encountered in finding the modal "cats" and "lions," the discreteness of species as naturally existing units is not thereby impaired.

What has been said above with respect to the species *Felis domestica* and *Felis leo* holds for innumerable pairs of species, genera, and other groups. Discrete groups are encountered among animals as well as plants, in those that are structurally simple as well as in those that are very complex. Formation of discrete groups is so nearly universal that it must be regarded as a fundamental characteristic of organic diversity. An adequate solution of the problem of organic diversity must consequently include, first, a description of the extent, nature, and origin of the differences between living beings, and, second, an analysis of the nature and the origin of the discrete groups into which the living world is differentiated.

MORPHOLOGICAL AND PHYSIOLOGICAL METHODS

A scientific study of organic diversity may proceed in two methodologically distinct ways. First, one may describe the diversity by recording as accurately as possible the multitudinous structures and functions of the beings now living and of those preserved as fossils; the descriptions are then catalogued and the regularities revealed in the process are formulated and generalized. Second, an analysis of causes underlying the diversity and determining its properties may be made. These two methods are known as the generalizing and the exact induction, respectively (Hartmann).

At the beginning of its existence as a science, biology was forced to take cognizance of the seemingly boundless variety of living things, for no exact study of life phenomena was possible until the apparent chaos of the distinct kinds of organisms had been reduced to a rational system. Systematics and morphology, two predominantly descriptive and observational disciplines, took precedence among biological sciences during the eighteenth and nineteenth centuries. More recently physiology has come to the foreground, accompanied by the introduction of quantitative methods and by a shift from the observationalism of the past to a predominance of experimentation. The great significance of this change is readily apparent and has been stressed in the writings of many authors, some of whom went to the length of ascribing to the quantitative and experimental methods almost magical virtues. Another characteristic of modern biology, which has been emphasized perhaps less than its due, is the prevalence of interest in the common properties of living things instead of in the peculiarities of separate species. This attitude is important, for it centers the attention of investigators on the fundamental unity of all organisms.

The problem of organic diversity falls within the provinces of morphological as well as of physiological biology, but it is treated differently by the two. Morphology is predominantly an order-creating and historical discipline. It is concerned first of all with the recording of the fact of organic diversity as it appears to our senses, and with the description of this diversity in terms of ideal prototypes (the modal points of the species, genera, families, etc.). Next, morphology traces the development of individuals (ontogeny) and of groups (phylogeny), striving to secure an understanding of the present status of the living world through a knowledge of its past. Genetics, being a branch of physiology concerned in part with the problem of organic diversity, is a nomothetic (law-creating) science. To a geneticist organic diversity is one of the most general and fundamental properties of living matter; diversity is here considered, so to speak, as an aspect of unity through a study of the mechanisms which may be responsible for the production and maintenance of variation, an analysis of the conflicting forces tending to increase or to level off the differences between organisms. The aim of the present book is to review the genetic information bearing on the problem of organic diversity; it is not concerned with the morphological aspect of the problem.

EVOLUTION

Since Darwin any discussion of organic diversity unavoidably involves a consideration of the theory of evolution, which represents the greatest generalization advanced in this field. Here again morphological and physiological biology are interested in different aspects of the matter.

The theory of evolution asserts that the beings now living have descended from different beings which have lived in the past; that the discontinuous variation observed at our time-level, the gaps now existing between clusters of forms, have arisen gradually, so that if we could assemble all the individuals which have ever inhabited the earth, a fairly continuous array of forms would emerge; that all these changes have taken place due to causes which now continue to be in operation and which therefore can be studied experimentally. The evolution theory was arrived at through generalization and inference from a body of predominantly morphological data and may be regarded as one of the most important achievements of morphological biology. However, the evolutionists of the morphological school have concentrated their efforts on proving the correctness of the first and second of the three assertions listed above, leaving the third rather in abeyance. They are interested primarily in demonstrating that evolution has actually taken place, in "bridging the gap," in filling up the discontinuities between the existing groups of organisms, in understanding the relationships between the branches of phylogenetic trees, and not so much in elucidating the nature of the discontinuities themselves or of the mechanisms through which they originated. As a matter of fact, Darwin was one of the very few nineteenth-century evolutionists whose major interests lay in studies of the mechanisms of evolution, in the causal rather than the historical problem. It was exactly this causal aspect of evolution which toward the close of the last century began to attract more and more attention, and which has now been taken up by genetics. In this sense genetics rather than evolutionary morphology is heir to the Darwinian tradition.

How neglected were the studies of the mechanisms of evolution is apparent from the fact that in 1922 Bateson was able to write: "In dim outline evolution is evident enough. But that particular and essential bit of the theory of evolution which is concerned with the origin and nature of species remains utterly mysterious." As recently as 1934 Goldschmidt expressed similar opinions. To most biologists these views seemed unduly pessimistic. Be that as it may, the fact remains that among the present generation no informed person entertains any doubt of the validity of the evolution theory in the sense that evolution has occurred, and yet nobody is audacious enough to believe himself in possession of the knowledge of the actual mechanisms of evolution. Evolution as an historical process is established as thoroughly as science can establish a fact witnessed by no human eye. The mass of evidence bearing on this subject does not concern us in this book; we take it for granted. But the understanding of causes which may have brought about this evolution, and which can bring about its continuation in the future, is still in its infancy. Much work has been done already to secure such an understanding and undoubtedly more remains to be done. In the following pages an attempt will be made to evaluate the present status of knowledge in this field.

THE BEARING OF GENETICS ON EVOLUTION

It should be reiterated that genetics as a discipline is not synonymous with the evolution theory, nor is the evolution theory synonymous with any subdivision of genetics. Nevertheless, it remains true that genetics has so profound a bearing on the problem of the mechanisms of evolution that any evolution theory which disregards the established genetic principles is faulty at its source.

Every individual resembles its parents in some respects but differs from them in others. Taken as a group, a population, every succeeding generation of a species resembles but is never a replica of the preceding generation. Evolution is a process resulting in the development of dissimilarities between the ancestral and the descendant populations. The mechanisms that determine the similarities and dissimilarities between parents and offspring constitute the subject matter of genetics. Genetics is physiology of inheritance and variation. This is the reason why the quest for an understanding of the mechanisms of evolutionary continuity or change has devolved upon genetics. But inheritance and variation may be studied on their own account, as general physiological functions, without reference to their bearing on the problem of organic diversity in any of the ramifications of the latter.

The signal successes of genetics to date have been in studies on the mechanisms of the transmission of hereditary characteristics from parents to offspring, that is, on the architectonics of the germ plasm of the sex cells. The germ plasm has been shown to be essentially discontinuous, composed of discrete particles known as genes. Quantitative and qualitative characters, fluctuating variability as well as discontinuous differences between individuals, normal and pathological, intraspecific and interspecific, "superficial" and "fundamental" characters of organisms are determined by the genes carried in the sex cells, or, to be more exact, in the chromosomes of these cells. Chromosomes as carriers of genes have been studied in detail, with the result that the physical basis of hereditary transmission has been revealed. The transmission of hereditary characters has been brought under human control, in the sense that in organisms which have been well studied genetically the characteristics of the offspring are frequently predictable, with a rather high degree of accuracy, from a knowledge of the characteristics of the parents. In *Drosophila melanogaster,* and to a lesser extent in certain other forms, hereditary types possessing a desired set of characteristics may, within limits, be synthesized at will, and the schemes of such "syntheses" worked out in theory are almost always realized in minute detail in the actual experiments.

The elegance and precision of methods devised by genetics to control the results of experiments involving crosses of individuals differing in many hereditary characteristics have led to claims that the problem of heredity has been solved. Although a large amount of work still remains to be done in this field, it is indeed fair to say that the genetics of the transmission of hereditary characters is, by and large, understood now. But the problem of heredity is much wider. Knowing the rules governing the distribution of the hereditary characteristics of an organism among the sex cells, one is in a position to predict what constellations of genes are likely to be present in the zygotes coming from the union of such sex cells. Between the genes of a fertilized egg and the characters of the adult organism arising from it there lies, however, the whole of individual development during which the genes exert their determining action. The mechanisms of gene action in development constitute the central problem of the second major subdivision of genetics, which has been variously labeled as the genetics of the realization of the hereditary characters, phenogenetics, or developmental genetics.

The problem of gene action is far from having been solved. It is known in some instances that the formation of adult characteristics, such as size or coloration of the body or its parts, is preceded by the appearance in the developing organisms of chemical substances of the hormone type which are operative in the processes that produce the adult characters. The great interest of such information is self-evident, although a physiologically minded biologist always would have taken it for granted that chemical precursors of the visible morphological traits must be active in development. The disappointing feature of this work is that to date it has failed to give a clear insight into the gene action proper. A gene is a particle located in a chromosome in the cell nucleus. Its effects on development must of necessity start with intracellular processes, which are subsequently translated into more or less long chains of reactions, culminating in the appearance of visible adult characters. Regarding those intracellular processes, the first links in the reaction chains, nothing is known at present. Are all genes continuously active, or does each gene exert its determining function at a certain period of development and remain in a quiescent state at other periods? Is gene action merely a by-product of the process of self-reproduction of the genes in the course of cell division? Are genes specialized, in the sense of being concerned each with a single or a few reactions taking place in the body, or is their action of a more general sort? No answers are available to these and many related questions, and, what is still more important for further work, no reliable methods have been devised for investigations in this field.

The genetics of the transmission, and the genetics of the realization of hereditary materials are concerned with individuals as units. The former establishes the rules governing the formation of the gene constellation in individual zygotes, and the latter deals with the mechanisms of gene action in ontogeny. The third subdivision of genetics has as its province the processes taking place in groups of individuals—in populations—and therefore is called the genetics of populations. A population may be said to possess a definite genetic constitution, which is evidently a function of the constitutions of the individuals composing the group, just as the chemical composition of a rock is a function of that of the minerals entering into its make-up. The rules governing the genetic structure of a population are, nevertheless, distinct from those governing the genetics of individuals, just as rules of sociology are distinct from physiological ones, in spite of being merely integrated forms of the latter. Imagine, for example, that some factors have arisen in the environment which discriminate against too tall or too short individuals of a species. From the standpoint of an individual, some growth genes would have acquired lethal properties, and the effects of these genes might be described adequately by stating the precise nature of the physiological reactions leading to death. From the viewpoint of population genetics, death of this category of individuals is merely the beginning of a complex chain of consequences; the relative frequencies of individuals homozygous and heterozygous for certain growth genes and for genes located in the same chromosomes would be altered; some genetic factors which previously were being eliminated because of their harmfulness might become neutral or even favorable; after some generations the genetic constitution of the whole species may be changed.

Since evolution is a change in the genetic composition of populations, the mechanisms of evolution constitute problems of population genetics. Of course changes observed in populations may be of very different orders of magnitude, from those induced in a herd of domestic animals by the introduction of a new sire to phylogenetic changes leading to the origin of new classes of organisms. The former are obviously trifling in scale compared with the latter, and it may not be convenient to have all of them subsumed under the name "evolution." Experience seems to show, however, that there is no way toward an understanding of the mechanisms of macro- evolutionary changes, which require time on a geological scale, other than through a full comprehension of the micro-evolutionary processes observable within the span of a human lifetime and often controlled by man's will. For this reason we are compelled at the present level of knowledge reluctantly to put a sign of equality between the mechanisms of macro- and micro-evolution, and, proceeding on this assumption, to push our investigations as far ahead as this working hypothesis will permit.

EVOLUTIONARY STATICS AND EVOLUTIONARY DYNAMICS

Since the middle of the last century the organic diversity observed in nature has been considered a result of the evolutionary process. The principal tenet of the evolutionary doctrine is that the living world as seen by us at present has not always been as it is now; what we study at our time level is a cross section of the phylogenetic lines, the beginnings of which are lost in the dim past. But evolution is a process of change or movement. Description of any movement may be logically and conveniently divided in two parts; statics, which treats of the forces producing a motion and the equilibrium of these forces, and dynamics, which deals with the motion itself and the action of forces producing it. Following this scheme, we shall discuss, first, the forces which may come under consideration as possible factors bringing about changes in the genetic composition of populations (evolutionary statics), and second, the interactions of these forces in race and species formation and disintegration (evolutionary dynamics).

In bare outline, the mechanisms of evolution as seen by a geneticist appear as follows. Gene changes, mutations, are the most obvious source of evolutionary changes and of diversity in general. Next come the changes of a grosser mechanical kind involving rearrangements of the genic materials within the chromosomes. It seems probable at present that such rearrangements may at least occasionally entail changes in the functioning of the genes themselves (position effects), since the effects of a gene on development are determined not only by the structure of that gene itself but also by that of its neighbors. Finally, reduplications and losses of whole chromosome sets (polyploidy) are important as evolutionary forces, especially among some plants.

Mutations and chromosomal changes arise in every sufficiently studied organism with a certain finite frequency, and thus constantly and unremittingly supply the raw materials for evolution. But evolution involves something more than origin of mutations. Mutations and chromosomal changes are only the first stage, or level, of the evolutionary process, governed entirely by the laws of the physiology of individuals. Once produced, mutations are injected in the genetic composition of the population, where their further fate is determined by the dynamic regularities of the physiology of populations. A mutation may be lost or increased in frequency in generations immediately following its origin, and this (in the case of recessive mutations) without regard to the beneficial or deleterious effects of the mutation. The influences of selection, migration, and geographical isolation then mold the genetic structure of populations into new shapes, in conformity with the secular environment and the ecology, especially the breeding habits, of the species. This is the second level of the evolutionary process, on which the impact of the environment produces historical changes in the living population.

Finally, the third level is a realm of fixation of the diversity already attained on the preceding two levels. Races and species as discrete arrays of individuals may exist only so long as the genetic structures of their populations are preserved distinct by some mechanisms which prevent their interbreeding. Unlimited interbreeding of two or more initially different populations unavoidably results in an exchange of genes between them and a consequent fusion of the once distinct groups into a single greatly variable array. A number of mechanisms encountered in nature (ecological isolation, sexual isolation, hybrid sterility, and others) guard against such a fusion of the discrete arrays and the consequent decay of discontinuous variability. The origin and functioning of the isolating mechanisms constitute one of the most important problems of the genetics of populations.

In the following chapters an attempt will be made to summarize the available evidence bearing on these three levels of the evolutionary process. Variation-producing agents (gene mutation, chromosomal changes) are known from laboratory experiments. But it is not a priori certain that these agents observed under laboratory conditions have been effective in nature and are responsible for the organic diversity empirically observable out-of-doors. We assume this to be the case because no other equally satisfactory working hypothesis has been proposed. Nevertheless, the validity of the working hypothesis must be rigorously tested by examining whether the differences between forms encountered in nature can be resolved into the elements whose origin is known in experiments. Thus evolutionary statistics will be covered in Chapters II-IV. Next the dynamic stage, the evolutionary process in the strict sense, comes under consideration in Chapters V-X. Here one must necessarily proceed by inference, for, with very few important exceptions, the experimentalist is still not in a position to reproduce in the laboratory the historical processes that have taken place in nature. Nonetheless, a rapidly growing body of both observational and purely experimental evidence gives at least a promise that an adequate analysis of evolutionary dynamics will be possible in a not too distant future.

II: GENE MUTATION

HEREDITARY AND ENVIRONMENTAL VARIATION

DIFFERENCES among individuals of a species are observed in any natural or artificial population—in man, in domestic animals and plants, and in wild species. The cause of individual variation is twofold. The genetic constitution of an individual, its genotype, determines its reaction to the environment; the appearance or phenotype is the resultant of the interaction between the genotype and the environment. Much discussion has centered around the question of the relative importance of genotype and environment in determining the end result of the development of an organism. Most of this discussion has been due to faulty thinking: in its general form the question is meaningless. The appearance (phenotype) of a plant, such as *Limnophila heterophylla,* grown under water is so different from that of the same plant grown on land that it seems inconceivable that the two belong to the same species. On the other hand, no system of feeding and exercise will suffice to make a draft horse able to race successfully against a thoroughbred. The number of phenotypes that can develop on the basis of the same genotype is infinite, because the environment is infinitely variable; but the variation in the genotype seems to be likewise unlimited.

Since no two individuals develop in absolutely identical environments, phenotypic variation is always present. On the other hand, differences between individuals in a population may also be due to differences between their genotypes. In nature the two types of variation are almost always blended. The genotypic or hereditary variation manifests itself in a correlation between the characteristics of parents and offspring. The parent-offspring correlation can be eliminated by a continued inbreeding for a sufficient number of generations. In this way populations can be obtained in which all the individuals are genotypically alike, and variation is due solely to environmental agents. The environment being carefully controlled, the variation can be reduced to a minimum, and brothers and sisters may be no more similar to each other and to their parents than to more distant relatives. In inbred strains of flies, guinea pigs, or corn plant all individuals may be extremely similar in most respects.

The genotype of an individual can be transmitted without change to its offspring and is potentially capable of being reincarnated in any number of individuals. In asexually reproducing plants, large colonies may consist of genotypically identical individuals (Anderson 1936, in Iris). The genotype possesses tremendous self-regulatory powers, and can withstand unchanged the impact of most environmental agencies. Heredity is essentially a conservative force. Evolution is possible only because heredity is counteracted by another force opposite in effect, namely, mutation. In inbred lines or other genotypically homogeneous populations, hereditary variation sooner or later arises *de novo,* and the parent-offspring correlation reappears. This indicates that the population is no longer genotypically homogeneous, that two or more hereditary forms are found where only one was present before, and that the genotype has been changed by mutation somewhere in the process of the transmission from parents to offspring.

One of the best examples of mutation in a genetically pure population has been reported by Blakeslee, Morrison, and Avery (1927) in the Jimson weed, *Datura stramonium.* Haploid individuals of this plant have in their cells twelve chromosomes, instead of the twenty-four present in the normal diploid. A certain proportion of the egg cells and pollen produced by the haploid contain all the twelve chromosomes (non-reduction). As a result of selfing, some diploid offspring may be produced. Diploids derived from a haploid, as well as the progeny of these diploids, must be strictly identical as to genotype. The presence of residual heterozygosity, not eliminated by inbreeding, is here excluded. Nevertheless, among 173 diploid plants tested, four proved to be heterozygous for mutant genes showing Mendelian segregation. Chromosomal aberrations have likewise been observed.

DEFINITION OF MUTATION

The term mutation is used in two different senses. In the wide sense, any change in the genotype which is not due to recombination of Mendelian factors is called mutation. The term subsumes, then, such widely different phenomena as alterations of the structure of individual genes, reduplications and losses of chromosomes or of their parts (duplication and deficiency), multiplication of whole sets of chromosomes (polyploidy), and rearrangements of the genic materials within the chromosomes (translocations, inversions). In the narrow sense, mutation is a change in the structure of a gene and currently such changes are supposed to be chemical rather than mechanical in nature. Other classes of genetic changes enumerated above are mainly mechanical, and are sometimes described as chromosomal aberrations.

Restricting the term mutation solely to gene changes is undoubtedly logical, but it leads to serious difficulties in practice. Experimentally a gene mutation is identified as a variation in the morphological or physiological characters of the organism which is inherited in Mendelian fashion in crosses between the mutant and the ancestral form. Heritable variations may be due, however, to mechanical changes in chromosomes as well as to changes in individual genes. A majority of the "mutations" obtained by De Vries in his classical investigations on Oenothera have been subsequently shown to be chromosomal aberrations of one type or another. Duplications or losses of single genes or of groups of them may also be inherited as simple Mendelian factors. Only in Drosophila, where recent studies on the giant salivary gland chromosomes have furnished a method for detection of minute changes in the gross structure of the chromosomes, can gene mutation be distinguished from mechanical alterations in the chromosomes. But even in Drosophila one can not be positive that very small changes are not overlooked even with a most careful cytological examination. Stadler (1932) has correctly emphasized that what in practice is described as gene mutation is merely the residue left after the elimination of all classes of hereditary changes for which a mechanical basis is proven. Studies on position effects seem, in fact, to bridge the gap between gene mutations and chromosomal aberrations (cf. Chapter IV).

The roles played in evolution by gene mutations and by chromosomal aberrations are probably different. For this reason we prefer to discuss the two phenomena separately. Unless otherwise specified, the term "mutation" will be used to mean a gene mutation, a Mendelian change which is not known to represent a chromosomal aberration.

TYPES OF CHANGES PRODUCED BY MUTATION

Since genes are supposed to be concerned as one of the variables in all developmental processes, gene mutations may be expected to affect all parts and physiological characteristics of the organism. The available experimental evidence seems, so far as it goes, to confirm this inference.

In the vinegar fly *Drosophila melanogaster,* in which more mutations have been observed than in any other organism, the variety of characters affected is enormous. Compilation of a complete catalogue of such characters is impossible at present, since a majority of mutations have never been scrutinized from this point of view. Descriptions of mutations are made with emphasis on easily visible characteristics that may be useful for classifying flies in crossing experiments, while other, and especially physiological traits, are generally disregarded. The following list gives an idea of the diversity of mutational changes.

Some mutations affect the coloration of all external parts of the body, of the eyes, ocelli, testicular envelope, and Malpighian vessels; others, the length, diameter, and shape of the bristles; definite bristles or sets of bristles may be absent or reduplicated. Still other mutations influence the size of the eyes, antennae, legs and their parts, the form and arrangement of the ommatidia, chitinization and arrangement of body sclerites. A very interesting class of mutations causes transformation of some organs into others, revealing the homology between the two. Here belong the transformation of the balancers into a second pair of wings, of the antennae into legs, and of the mouth parts of the fly in the direction of those of lower insects. Sex organs and secondary sexual characters are affected: changes have been observed in the sex-combs, external male genitalia, number and shape of the spermathecae, number of egg-strings and egg chambers in the ovaries, shape and appearance of the eggs. Developmental stages may be changed: size, shape, and weight of the larvae, pupae, and adults. Changes in the internal organs are observed mostly in connection with external changes; thus, the brain is changed in connection with the eye size and the presence or absence of the eyes. Purely physiological characteristics are frequently affected: reaction to light and gravity, sex-determining factors, longevity, number of eggs deposited, length of development, manner of growth, number of larval instars. Mutations are known that produce tumorlike growths in various organs and at various developmental stages. The large class of lethal mutations has not received even a fraction of the attention it deserves. It is known, however, that lethals may cause death of the organism at any stage of development, including the early embryonic ones. The causes of death are obscure in almost all instances; the data of Poulson (1937), according to which gene deficiencies cause death through radical modifications of the normal course of cleavage, gastrulation, and organ formation, mark a first step toward clearing up our ignorance on this point.

A number of misconceptions exist regarding the kind of mutations observed in Drosophila. We are forced to give a brief consideration to some of these misconceptions, since they have figured prominently in certain discussions of the role of the mutation factor in species formation and evolution. It has been contended, for instance, that mutations involve only "superficial" characteristics, leaving the "fundamental" ones unaffected. Those making such assertions have wisely refrained from revealing their criteria for the discrimination between superficial and fundamental traits.

The presence of one pair of wings and one pair of balancers, as opposed to two pairs of wings, is one of the most striking distinguishing marks of the order of flies (Diptera). One may ask then, is the appearance of a four-winged Drosophila a fundamental or a superficial change? Is a mutation diverting the embryonic development to a wrong course and thus causing death, fundamental or superficial? Those who wish to see a mutant fly devoid of the alimentary canal, or with the location of the heart and the nerve chain exchanged, overlook the fact that a mutation with such a change could not survive and hence could not be detected. More important still, there is no one-to-one relationship between individual genes and separate organs. Finally, those who would call "fundamental" only a change that would transform an individual of *Drosophila melanogaster* into

a representative of another species, genus, or family, fail to understand the nature of the specific differences which are due always to
coöperation of many genes and hence can not arise through the mutation of a single gene.

MUTATION AND VIABILITY

Another type of criticism advanced against the mutation theory
asserts that the mutations observed in Drosophila and in other organisms produce deteriorations of viability, pathological changes
and monstrosities, and therefore can not serve as evolutionary building blocks. This assertion has been made so many times that it has
gained credit by sheer force of repetition. It is a fact that most
mutations dealt with in the laboratory decrease the viability of the
fly under the environmental conditions in which *Drosophila melanogaster* is usually kept. But mutations in Drosophila as well as in
other organisms form a spectrum, ranging from lethals to changes
which are neutral or even favorable to viability. It is the middle part
of this spectrum, comprising mutations producing striking phenotypic effects and from a slight to an appreciable depression of the
viability, that is mainly dealt with in genetic experiments in which
mutant genes are used as "chromosome markers." The great and
insufficiently studied class of lethals is not a homogeneous group;
some of them cause death in early developmental stages, others in
relatively late ones, and still others permit the adult fly to hatch if the
culture conditions are especially favorable. The last group of lethals
are sometimes classed as semi-lethals, and semi-lethals intergrade
with mutations that produce medium, slight, or no decrease of viability.

Timofeeff-Ressovsky (1934c, d, 1935b) has demonstrated that
the effects produced by a given mutation on viability are a function
both of the environmental conditions and of the rest of the genetic
structure of an individual (Table 1). The mutation "eversae" in
Drosophila funebris is at 15°-16° and at 28°-30° inferior, and
at 24°-25° superior to the wild type in viability. The viability of
the mutations venae abnormes and miniature is only slightly inferior
to that of the wild-type at 15°-16°, and much inferior at 28°-30°.
On the contrary, the viability of bobbed is low at 15°-16° and ap-

proaches normal at 28°-30°. Overpopulation of the cultures decreases the relative viability in the mutations eversae, venae abnormes, and miniature, but has an opposite effect on bobbed. Combinations of venae abnormes and lozenge, each of which decreases the viability, produce a summation of the deleterious effects; combination of miniature and bobbed gives a compound which is more viable than either mutation by itself (Table 1).

TABLE 1

VIABILITY OF SOME MUTATIONS AND THEIR COMBINATIONS IN *Drosophila funebris*, EXPRESSED IN PERCENTAGE OF THE VIABILITY OF WILD TYPE

(after Timofeeff-Ressovsky)

MUTATION	TEMPERATURE			COMBINATION	TEMPERATURE
	15-16°	24-25°	28-30°		24-25°
eversae	98.3	104.0	98.5	eversae singed	103.1
singed		79.0		eversae abnormes	83.7
abnormes	96.2	88.9	80.7	eversae bobbed	85.5
miniature	91.3	69.0	63.7	singed abnormes	76.6
bobbed	75.3	85.1	93.7	singed miniature	67.1
lozenge		73.8		abnormes miniature	82.7
				abnormes lozenge	59.3
				abnormes bobbed	78.7
				miniature bobbed	96.6
				lozenge bobbed	69.2

Analogous data have been reported by Kühn (1932) for the moth *Ephestia kühniella*. Mutant genes known in this animal modify the external characteristics, such as eye color, color of the wing scales, or wing pattern; some of them have also manifold effects (see below) reducing the viability of the carrier. A combination of two of these genes, which taken separately produce unfavorable viability effects, is equal in viability to the wild type.

Such data are suggestive when compared with those on the relative viability of various geographical races of *Drosophila funebris* (Timofeeff-Ressovsky 1935a) and *Drosophila pseudoobscura* (Dobzhansky 1935d). Strains of the same species coming from different localities may reverse their relative viability at different temperatures and culture conditions. The same is true for species: Timofeeff-Ressovsky (1933d) finds that *Drosophila melanogaster* is superior to *Drosophila funebris* at high, but inferior at low temperatures.

Mutations improving viability can be observed especially in cultures which already contain a mutant gene depressing the viability relative to the wild type. The gene stubbloid in *Drosophila melanogaster* decreases the length of the bristles, and causes crumpled wings, bent legs, and decreased viability. In inbred lines of stubbloid kept in the laboratory for a long time the crumpling of the wings and the bending of legs become sometimes less pronounced as the time goes on, and the general viability increases. If, however, such an "improved" line is outcrossed to an unrelated wild type strain, the stubbloid flies appearing in the F_2 generation show again the unfavorable wing, leg, and viability characteristics (unpublished observations of the writer). Similar results were obtained in other mutants by Marshall and Muller (1917) and by other observers. The best interpretation of such facts is that in culture, mutations take place to genes which suppress the effects of stubbloid, and that the suppressors are favored by selection and the strain becomes homozygous for them. An alternative interpretation is, of course, that such suppressors are always present in heterozygous condition in any strain, and are acted upon by natural selection in stubbloid strains.

The fact that mutations which improve viability are rare in normal strains under normal conditions is not surprising. Mutation is a random change of the gene structure. A random change is vastly more likely to constitute a deterioration than an improvement, since the gene structure found in "normal" representatives of a species is the product of a long historical process of natural selection. Furthermore, mutations arising in the laboratory have arisen probably innumerable times in nature. Any mutation which constitutes an improvement over the "normal" condition has had a chance to become established in the species, and the wild type with which the laboratory experiments are started may be expected to be homozygous for such mutations. The situation changes if the organism is placed in an environment different from that to which the species is adapted. Banta and Wood (1927) observed a mutation in the crustacean *Daphnia longispina,* which normally has the temperature optimum at 20° and which is unable to survive more than a few

days at 27°. The mutant has an optimum between 25° and 30°
and does not survive at 20°. Classification of mutations into favorable and harmful ones is meaningless if the nature of the environment is not stated.

A mutation inducing changes in the structure of the mouth parts
of Drosophila which prevents the access of food into the alimentary
canal (the mutant proboscipedia) is a monstrosity, and one may be
unable to imagine an environment in which such a mutation may be
favorable; yet whole families exist in different orders of insects
whose members have mouth parts unfit for feeding. Reduction or
disappearance of the eyes is a deterioration, but many insects have
no eyes. The mutation rotated abdomen has the male genitalia not
in the plane of the body symmetry, so that males homozygous for
the mutant gene can not copulate and are therefore sterile. Nevertheless, some families of flies (*Syrphidae* and others) turn out to
have twisted genitalia as one of the distinguishing characteristics.
A perusal of books on the natural history of insects will show
anyone how spurious is the value of anthropomorphic judgments
on what constitutes a malformation. Some forms existing in nature,
and to all appearances fully successful in the struggle for existence,
have characters that seem monstrous indeed. The fly family *Diop-
sidae,* and some representatives of *Ortalidae,* have the sides of the
head extended into a pair of styliform processes, on the tip of which
are found the eyes and the antennae. Non-transparent processes of
the thorax covering the eyes are observed in some beetles. Certain
species of the homopterous family *Membracidae* have outgrowths on
the thorax attaining monstrous shapes and projecting far above and
to the sides of the body. The related family *Fulgoridae* is given to
formation of almost equally uncanny outgrowths on the head. The
number of examples of this kind could be increased indefinitely. It
stands to reason that if mutations resembling these strange products
of evolution appeared in Drosophila they would be classed as pathological phenomena.

EXTENT OF CHANGES PRODUCED BY MUTATIONS

In Drosophila as elsewhere, mutations form a spectrum with respect to the extent of the effects produced, ranging from drastic

changes that are lethal to the early developmental stages to changes that are so minute that their detection presents a serious technical problem. This spectrum is not only similar to the spectrum of the mutational effects on viability (see above), but there is a loose correlation between the positions of a given mutation in the two spectra: namely, mutations producing great departures from the normal phenotype usually have an adverse effect on the viability, and vice versa.

To determine the relative frequency of the different types of mutations is not an easy problem. Evidently, the greater the phenotypical change, the easier is the detection of a mutation. Since striking mutations are more valuable as chromosome markers than the weak ones, the former are picked out and studied. The published descriptions of mutations give a grossly distorted picture of the characteristics of the mutation process.

Timofeeff-Ressovsky (1934b, and especially 1934d and 1935b) has done pioneering work in attempting to clarify the situation. He treated with X-rays wild type males of *Drosophila melanogaster* from a thoroughly inbred strain, and crossed them to ClB females which previously had been repeatedly outcrossed to the same inbred strain. The ClB females have in one of their X chromosomes a lethal gene, a marker producing a visible effect (Bar), and an inversion which eliminates crossing over between the ClB chromosome and the other X chromosome present in the same female. In the F$_1$ generation, females heterozygous for the ClB chromosome were selected and outcrossed to untreated wild type males. Such females have one ClB X chromosome (untreated), and one paternal X chromosome treated with X-rays. Their sons receiving the ClB chromosome die on account of the lethal contained in it; the sons receiving the other X chromosome survive, provided no lethal mutation has been induced in this chromosome by the treatment. The sex-ratio observed is, therefore, 2 females: 1 male. If a lethal mutation is induced in the treated chromosome, the sons receiving it die on account of the lethal, and the offspring are females only. This is the standard method, originally proposed by Muller, and now generally used for the detection of the sex-linked lethal mutations.

If a mutation which is not lethal but which decreases the viability

arises in the X chromosome, the resulting sex-ratio falls between
2 ♀ : 1 ♂ and 2 ♀ : 0 ♂, depending upon the degree of the deleterious
effect produced by the mutation. For technical reasons it was preferable to take into account only the daughters which do *not* carry the X
chromosome (ClB); they form about half of all females, and can
be recognized by the absence of the marking gene (Bar). The
frequencies of such females and males turned out to be 1 ♀ :0.95 ♂
if no mutation has been induced in the treated chromosome, and
1 ♀ :0 ♂ if a lethal mutation has been induced. The results of the
experiment can be seen in Table 2, showing the sex-ratio produced by
individual females in form of variation series. The control series
shows the ratios obtained in the progeny of males which have not
been treated with X-rays.

TABLE 2

THE ♀ : ♂ RATIOS OBTAINED BY TIMOFEEFF-RESSOVSKY IN HIS EXPERI-
MENTS ON MUTATIONS AFFECTING VIABILITY

SEX RATIO	1:1.15	1:1.05	1:.95	1:.85	1:.75	1:.65	1:.55	1:.45	1:.35	1:.25	1:.15	1:.05	1:0
Control (in %)	2.1	14.1	77.1	5.5	.7	—	—	.5	—	—	—	—	—
Treated (in %)	.7	10.1	44.9	8.8	7.2	5.3	4.2	1.8	1.1	.7	1.4	.9	13.6

In the treated series, 13 per cent (56 out of 432) of the cultures
produced no males, indicating that mutations having lethal effects
appeared. In 3 per cent of the cultures, semi-lethal mutations were
observed (sex-ratios between 1 : 0.30 and 1 : 0). The most remarkable fact is, however, that a large number of cultures gave
sex-ratios which can be accounted for only on the supposition that
mutations producing slight decreases of the viability have arisen.
The exact number of such cultures in the treated series is not easy
to determine, since the ratios observed in them overlap those in the
cultures of the control, but they make up no less than 20 per cent
of the total. Timofeeff-Ressovsky has repeated these experiments,
varying the technique and imposing checks on the validity of his conclusions. The conclusions withstood the repeated tests.

Mutations producing lethal effects have been considered the most
frequent class of mutation coming from X-ray treatments as well

as from the spontaneous process. The results of Timofeeff-Ressovsky show that this view is incorrect, and that the most frequent mutations, at least following X-ray treatments, produce weak, barely perceptible effects. The supposition that relative frequencies of mutations are on the whole inversely proportional to the magnitude of their effects is at present the one most likely to be true.

It must be noted that before Timofeeff-Ressovsky's work the same viewpoint had been advanced by Baur (1924, 1925) on the basis of observations on the snap-dragon (*Antirrhinum majus*) and related species, and by a number of others (Goldschmidt, Wright, Fisher, Muller, East) on more theoretical grounds. Baur finds that about 10 per cent of the offspring of *Antirrhinum majus* not treated with any mutation-producing agents are small mutations (Klein-mutationen). They differ from the original form in very slight changes in such characteristics as leaf and flower color, size of anthers and of seed, hairiness, etc. Studies on mutations of this kind are evidently most difficult, hence the figures indicating their frequencies are necessarily subject to correction. Nevertheless, it may be taken as established that small mutations do occur, and are relatively frequent.

The high frequency of mutations producing small changes in the phenotype raises a strong presumption in favor of supposing that such mutations play a greater role in evolutionary processes than mutations with grosser effects. Fisher (1930) has given an interesting mathematical argument in favor of this view. These considerations agree very well with the results of the genetic analyses of the interracial and interspecific differences (Chapter III), showing these differences to be caused in a majority of cases by coöperation of numerous genes, each of which taken separately has only slight effects on the phenotype.

MANIFOLD EFFECTS OF GENES

Mutant genes are named according to the most prominent characteristics they produce. In Drosophila, mutations of the gene "white" turn the eye color from red to white, "vestigial" makes vestigial wings, "stubbloid" causes a shortening of the bristles, etc. This system of naming is convenient, but the names are not to be taken as complete accounts of the differences between the mutants and the ancestral form, much less as indicative of the total range of the effects of the particular gene on development. Despite the inadequate descriptions of mutations that are customarily given in the genetic literature (attention being centered almost exclusively on a single, or a few easily observable, changes), it is well known that many mutants, in Drosophila as well as in other forms, differ from the ancestral types in complexes of diversified characters. The mutation white changes not only the eye color, but also that of the testicular membrane, the shape of the spermatheca, length of life, and the general viability. Vestigial reduces wing size, modifies the balancers, makes certain bristles erect instead of horizontal, changes the wing muscles, the shape of the spermatheca, the speed of growth, fecundity, and length of life. Under favorable external conditions vestigial decreases the number of ovarioles in the ovaries while it has the opposite effect under unfavorable conditions. Stubbloid modifies the bristles, wings, legs, antennae, and viability (Dobzhansky 1927, 1930a, Saveliev 1928, Alpatov 1932).

Genes that produce changes in more than one character are said to have manifold effects. The frequency of such genes is not well known at present. The published descriptions of mutants may give the impression that manifold effects are on the whole an exception rather than the rule, but these descriptions are admittedly incomplete. Dobzhansky (1927) has studied the shape of the spermatheca in a number of mutants of *Drosophila melanogaster* characterized by such differences as eye and body color, wing size, etc.; these mutants were not suspected of differing from each other in the internal anatomy of the fly in general or in the shape of the spermatheca in particular. Nevertheless, ten out of the twelve mutants studied showed differences in the shape of the spermatheca. Facts such as these suggest that most, and possibly all, genes have manifold effects.

It is certainly true that a majority of mutations produce striking effects on one or a few characters, and that their manifold effects, if any, involve changes which to our eyes seem trivial. Thus the "main" characteristic of the mutant vestigial in *Drosophila melanogaster* is the decrease of the wing size, and the other characters mentioned above appear unimportant. But to conclude on this basis that the gene vestigial is a "wing gene" rather than a "bristle gene" would be as naïve as to suppose that a change of the hydrogen ion concentration is always a "color gene" because it produces a striking effect on the color of certain indicators. In general, there is no conclusive evidence to show that genes have a circumscribed province including only one class of characters or physiological reactions; our understanding of the dynamics of individual development is still in its early infancy. Even if the descriptions of mutants were scrupulously complete, we would not gain an adequate knowledge of the roles played by each gene in development. It should be kept in mind that the mutant as well as the ancestral form possesses allelomorphs of the gene producing the mutation. Let the mutant gene be denoted a and the ancestral gene A. The normal condition of the phenotype found in the ancestor is determined by the effects of A (of course, in coöperation with all other genes composing the genotype); and the phenotype of the mutant, by the effects of a. The phenotypic differences between the mutant and the ancestral form are, then, indicative of the effects of the change $A \rightarrow a$, but not of the sum total of the effects of either A or a.

Today the only avenue of approach to the problem of total gene effects is through studies on physical losses (deficiencies) of genes. A start in this direction has been made by Poulson (1937), who found that in *Drosophila melanogaster* individuals homozygous for certain deficiencies die on account of great disturbances in the fundamental processes of the embryonic development. This agrees with the general experience in non-polyploid organisms, which shows that homozygous deficiencies even for very short sections of chromosomes are, as a rule, inviable. According to Demerec (1934) most deficiencies in Drosophila act as cell-lethals; that is, the absence of genes is fatal not only to the whole organism but also to a patch of deficient tissue surrounded by tissues in which all genes are present. This rule has exceptions; some deficiencies are not cell-lethals, and a few cases are known in which the whole organism survives the loss of some genes (Stadler 1933, Demerec 1934, Muller 1935, Demerec and Hoover 1936). These exceptions may indicate either that some genes are less important in the development than others, or that some genes are normally present at more than one locus in the chromosomes.

Manifold effects are especially interesting in connection with the problem of the so-called "neutral characters." Many writers have pointed out that differences between races, species, and genera involve mostly characters whose value in the struggle for existence is not evident. Indeed, a most fanciful thinking is necessary to see an adaptive significance in all the characteristics which are emphasized in the descriptions of new forms customarily given by systematists. Yet the prevalence of manifold effects makes caution necessary in reaching the conclusion that a given character or property of an organism is absolutely devoid of adaptive value. A seemingly neutral character may represent only a part of the total of effects of the genic difference causing it. Physiological properties concealed from the eye of an observer, but correlated with a "neutral" trait used by systematists for their own ends, may be important in the life of the organism.

PRODUCTION OF MUTATIONS BY EXTERNAL AGENTS

Mutations have been observed under controlled conditions in many animal and plant species. Sometimes they arise in strains not known to have been exposed to treatment by mutation-producing agents; such mutations are designated spontaneous ones. Mutants are on the whole rare, and moreover they arise mostly as single individuals among masses of unchanged representatives of a strain. This haphazard appearance in single specimens indicates either that the causes producing gene changes act on isolated individuals and not on the entire body of the strain or the species, or else that conditions favorable for the action of these causes are for unknown reasons created only in exceptional individuals. Mutations arise at all stages of the developmental cycle: in gametogenesis before and after the reduction division, in gametes, and in somatic tissues. An important fact is that whenever a mutation takes place in a diploid cell, only one chromosome of a pair is affected. Since a diploid cell has two chromosomes of each kind, the cause of the mutation must be so highly localized that only one of a pair of presumably exactly similar genes falls within the field of its action. Finally, although the distances between the adjacent genes in a chromosome are ultra-microscopically small, mutations in a large majority of cases affect only one gene in the gene string.

The name "spontaneous" when applied to any natural process obviously constitutes only a thinly veiled admission of the ignorance of the real causes of the phenomenon in question. This applies fully to the causes of spontaneous mutations. For years the attention of workers in the field of mutations was centered on attempts to find an external agent which would modify their frequency. Once available, the knowledge of such an agent might be expected to be a powerful tool for a causal analysis. In 1927 the work of Muller brought the first conclusive evidence that such a factor had been found, namely, X-rays. Through the work of Muller and many others it is now clear that the frequency of mutations in Drosophila and in other forms is increased up to 200-fold by short-wave radiation, from ultraviolet to the shortest known gamma rays. A relatively slight effect is produced also by temperature, an increase of 10° about doubling the mutation rate (Muller 1928).

Since the discovery of the production of mutations by X-rays a rather voluminous literature has accumulated on this subject. Excellent reviews have been published by Timofeeff-Ressovsky (1934a, 1937), Schultz (1936), and Timofeeff-Ressovsky, Zimmer, and Delbrück (1935). A presentation of the results of the work in this field would be outside the scope of our book. It will suffice to say that as far as known the short-wave radiation merely increases the frequency of mutations as compared with the spontaneous rate, but the kind of mutations observed remains unchanged. The mutation process has not been brought under human control, since with and without X-rays one is unable either to obtain mutations in specified genes, or to make the genes mutate in specified directions. Mutations remain haphazard. The supposition that spontaneous mutations are due to the short-wave radiation omnipresent in nature has been discredited, since the amount of such radiation seems to be too small to account for the frequency of spontaneous mutations. But even this conclusion can not be regarded as established, because it is based on the assumption of the direct proportionality between the amount of radiation and the mutation rate produced by it, an assumption which may or may not be fully valid. The mechanism of the action of radiation in producing mutations and the nature of the mutational changes in the genes remain unknown. This is, of course, neither surprising nor discouraging, for an understanding of the nature of mutations presupposes a knowledge of the nature of the genes as such, which remains as yet one of the distant goals of genetics.

Mutation-producing agents other than short-wave radiations are in all probability present in nature. This is a field which is being extensively explored at present and where discoveries are likely at any time. But for the moment one is forced to admit that no securely established conclusions have emerged. Goldschmidt (1929b) and Jollos (1931, 1934, 1935) have described induction of mutations in Drosophila by sublethal high temperatures, and Jollos claimed to have found in such temperature treatments a means for directing the mutation process. His results have not been confirmed by other workers and it seems wise to refrain from hasty decisions on this subject. Sacharow (1935, 1936) and Samjatina and Popowa (1934) have published preliminary data suggesting that iodine treatment may produce mutations in *Drosophila melanogaster*. Lobashov and Smirnov (1934) suspect the same to be the case for ammonia, and Lobashov (1935) for asphyxia. Stubbe (1930-32, 1935) seems to have obtained mutations in Antirrhinum by a variety of both physical and chemical treatments.

It must be kept in mind that the production of mutations by external agents has nothing to do with the theories of the so-called direct adaptation and the inheritance of acquired characters. Mutation-producing agents cause merely an increase of the spontaneous mutation rate and not a genetic transformation of masses of individuals. As to the direct adaptation, experimental data give no support for believing that such a thing exists. This question has been discussed almost *ad nauseam* in the old biological literature, and any text book of genetics may give the reader a review of the present status of this problem, so that we may refrain from the discussion of it altogether.

FREQUENCY OF MUTATIONS

A living species is constantly under a pressure of the mutation process tending to produce a change in the characteristics of the organism. The magnitude of the mutation pressure is evidently a problem of prime importance for any theory of evolution. This problem is as yet unsolved, although at present the amount of available data bearing on it is far greater than was the case only a few years ago. The difficulty of obtaining accurate quantitative data on the mutation pressure is apparent. Either one tries to determine the total frequency of mutations for all the genes the organism possesses, or a particular gene is selected and its mutability measured. In the former case mutations producing minute changes present an insuperable obstacle, for no known experimental procedure permits the detection of all such mutations, and they are suspected to be the most frequent class. If a single gene is selected, the mutation frequency is usually so low that accumulation of accurate data is technically difficult, slight mutations may be overlooked, and there is no assurance that all mutations of the gene in question produce changes in the same character (due to manifold effects).

The spontaneous and the induced mutation rates may be distinguished. The highest spontaneous over-all mutation rate recorded anywhere is that in Antirrhinum, estimated by Baur to be around 10 per cent (see above). The major part of the mutations are, of course, the very small changes whose significance in evolution has been emphasized by their discoverer. No comparable estimate is put on record for any other organism. Whether or not Antirrhinum is an exceptionally mutable form is an open question.

A great amount of data has been collected by various authors on the frequency of lethal mutations in the X chromosome of *Drosophila melanogaster*. The detection of such mutations is easy and accurate: a female carrying in one of its X chromosomes a newly arisen lethal produces a sex-ratio of 2 : 1 in its offspring. If such a female is heterozygous for the ClB chromosome, the detection of the lethal appearing in the normal X chromosome is simpler still, for such a female produces only daughters and no sons (see above). Among 26,145 untreated females thus tested, 48 proved to contain a sex-linked lethal, which indicates a mutation rate of 0.18 per cent

(Schultz 1936). That is to say, one X chromosome out of every 544 contains a newly arisen lethal in each generation. Muller (1928b) obtained in his experiment at low temperature 12 lethals among 6,286 chromosomes, and at a higher temperature 31 lethals among 6,462 chromosomes, or one mutation per 524 and 208 chromosomes, respectively, per generation. Muller recalculates the same data also in a different way, making an estimate of the relation between the lethal mutations in the X chromosome and time. At low temperatures the development of the flies takes longer than at high tempera-

TABLE 3

SPONTANEOUS LETHALS IN THE X CHROMOSOME OF DIFFERENT STRAINS OF *Drosophila melanogaster* (*Data of Demerec*)

STRAIN	CHROMOSOMES TESTED	NUMBER OF LETHALS	FREQUENCY OF LETHALS (IN %)
Florida	2,108	23	1.14
Wooster, Ohio	1,266	8	0.63
Formosa, Japan	2,054	8	0.39
Oregon-R	3,049	2	
Swedish-B	1,627	3	
California-C	708	2	
Huntsville, Texas	938	—	
Urbana, Illinois	1,016	1	
Canton, Ohio	922	—	0.103
Amherst, Massachusetts	572	1	
Woodbury, New Jersey	1,159	1	
Tuscaloosa, Alabama	545	1	
Seto, Japan	1,236	—	
Kyoto, Japan	875	1	
Lausanne, Switzerland	955	2	

tures; taking this factor into account the mutation rates can be expressed as 1 per 349 chromosome-months and 1 per 139 chromosome-months at the low and the high temperatures used in Muller's experiments. An interesting speculation is possible on the basis of these data. The development of Drosophila takes nine days, but in other living forms it takes years or even decades. If the mutability per unit time were the same in these latter forms as it is in Drosophila, their chromosomes would be full of lethals within a few generations; it follows that mutability rates are probably not equal in all organisms.

The above data are the best available estimates of the order of magnitude of the mutation rates for a certain class of changes (lethals) in Drosophila. Beyond that too much reliance should not be placed on them, for we know at present that mutation rates may differ in different strains of the same species. Such intraspecific differences were suggested already by certain data of Muller (1928b), and more recently Dubovskij (1935) has published some confirmatory evidence. Demerec (1937) found that wild-type strains of *Drosophila melanogaster* coming from different geographical sources show dissimilar spontaneous mutation rates for lethals in the X chromosome (Table 3).

The data presented in Table 3 show that the Florida, and perhaps also the Formosa and Wooster, strains are more mutable than the rest. By making appropriate crosses Demerec has ascertained that the difference between the mutation rates in the Florida and the Swedish-B strains is due to the effects of a gene, or genes, located in the second chromosome. Aside from the lethals, an unusually great number of mutations producing visible effects (yellow, forked, vermilion, lozenge, black, and others) were observed in the Florida strain, indicating that the increase of mutability in that strain is a general one.

RATE OF MUTATION OF INDIVIDUAL GENES

The spontaneous mutability of individual genes is important because it may prove to be a limiting factor in the evolutionary process. Are all genes equally mutable? In other words, is the process of mutation essentially a random matter, affecting now one and now another gene, or are some genes more predisposed to change than others? The latter answer is undoubtedly correct. In well studied organisms mutations in some genes have been observed to occur repeatedly, while other genes have mutated only once. Moreover, so-called unstable or mutable genes are known (reviews have been published by Demerec, 1935, and by Stubbe, 1933) which mutate especially frequently, in the sex-cells as well as in the somatic tissues. Somatic mutations give rise to individuals showing mosaicism, that is, a part of the tissues has a genetic constitution different from other parts. Although some writers have expressed a suspicion that the mutable genes are a phenomenon *sui generis,* and that their mutability may be caused by processes different from mutation of the "normal" (i.e., not "mutable") genes, there is no method for separating the two categories, and genes seem to form an uninterrupted series ranging from the most mutable to the most stable ones.

The maize plant is well suited for studies of mutation rates because changes in genes determining the endosperm characters can be easily observed in very large numbers of individuals. Stadler (quoted in Demerec 1933) obtained the following data for seven endosperm genes (Table 4).

TABLE 4

FREQUENCY OF SPONTANEOUS GENE MUTATIONS IN MAIZE (*after Stadler from Demerec*)

GENES	NUMBER OF GAMETES TESTED	MUTATIONS OBSERVED	MUTATION RATE PER 1,000,000 GAMETES
R (color factor)	554,786	273	492
I color inhibitor)	265,391	28	106
P₂ (purple)	647,102	7	11
Su (sugary)	1,678,736	4	2.4
Y (yellow)	1,745,280	4	2.2
Su (shrunken)	2,469,285	3	1.2
Wx (waxy)	1,503,744	0	0

Differences in mutation rates of different genes are far greater than could be accounted for by chance; the gene R for example, changes more frequently than others. No comparable data exist for Drosophila, but the general experience has been that some genes (white, cut, yellow, forked) give spontaneous mutations much more frequently than certain others. The genes with a high spontaneous mutability prove to be the most frequently mutating ones also after X-ray treatments, where observations are much easier on account of the general accentuation of the mutability.

Patterson and Muller (1930), and especially Timofeeff-Ressovsky (1929, 1931, 1933, a, b, c) have devoted a great deal of attention to the problem of mutability to and from a given allelomorph. The gene A mutates with finite frequencies to allelomorphs a^1, a^2, a^3, ... a^x. Each of these may, in turn, mutate back to A or to any other member of the series. The mutation process is potentially reversible ($A \rightleftarrows a$), but the data show that mutation pressures to and from each allelomorph are likely to be different, making the mutability in a sense directed. The gene W (normal red eye color) in *Drosophila melanogaster* changes to white (w), eosin (w^e), apricot (w^a), and to other states producing eye colors of varying intensities. In comparaable experiments, where flies received similar amounts of the X-ray treatment, Timofeeff-Ressovsky (1933b) obtained results summarized in Table 5.

The wild type (W, red), apricot (w^a), and eosin (w^e) mutate most frequently to the extreme member of the series, namely white. It is not clear, however, whether the different whites, which are indistinguishable from each other for our eyes, may not be in reality a composite group (Timofeeff-Ressovsky 1933c). The reverse mutation from white directly to red has not been observed at all, although white may mutate to eosin, and eosin derived from it may in turn

TABLE 5

MUTABILITY TO AND FROM VARIOUS ALLELOMORPHS OF THE GENE WHITE IN *Drosophila melanogaster* (*after Timofeeff-Ressovsky*)

	w	w^{bf}	w^e	w^a	w^b	w^x	W	NUMBER OF CHROMOSOMES TESTED
Wild (W)	25	1	3	1	2	5		48,500
w^{co}	1	—	—	—	—	—	—	6,000
w^b	3	—	1	—	—	—	—	12,000
w^e	1	—	—	—	—	—	—	5,000
w^a	2	—	1	—	—	—	—	11,000
w^e	13	—		—	1	2	2	39,000
w^{bf}	1	—	—	—	—	—	—	5,500
w^i	1	—	—	—	—	—	—	7,000
w		1	1	—	1	—	—	54,000

mutate to red, the reversion from white to red being thus accomplished in two steps. The mutation from the wild type to forked (forked bristles) has been observed 5 times in 19,000 chromosomes, and the reversion from forked to wild type 7 times in 29,000 chromosomes, the X-ray treatment being similar in both cases (Patterson and Muller 1930, Timofeeff-Ressovsky 1933a). Here the mutation seems to be about equally frequent in either direction.

Timofeeff-Ressovsky (1932) has reported an especially interesting case where "normal" allelomorphs of the same gene found in different strains have different mutation rates. Two strains of *Drosophila melanogaster*, one from America and the other from Russia, were given identical X-ray treatments. In the former, 55 mutations at the white locus were observed among 59,200 chromosomes; and in the latter, 40 mutations among 75,300 chromosomes. This difference

(0.093 ± 0.012 per cent vs. 0.053 ± 0.008 per cent) is statistically significant, and in addition the "Russian allelomorph" changes mostly to white and the "American" one to white and to intermediates (eosin, etc.) about equally frequently. Through special experiments Timofeeff-Ressovsky proved that the difference in the behavior of the Russian and the American strains was due to the different mutabilities of the white gene itself, and not to the influence of the genotypic environment (modifying genes of various kinds). Demerec (1929) reported that both influences may be effective. The gene miniature in *Drosophila virilis* produces mutant allelomorphs (wing characters) some of which are about as stable as the ancestral gene and some which are very unstable and revert to the wild type or change into each other with considerable frequencies. Some of the allelomorphs are mutable only in somatic cells, producing mosaic patches of normal and mutant tissues; others mutate mainly in the sex cells. In addition, Demerec finds a gene located in a different chromosome from that where miniature lies; this gene by itself produces no visible effects, but it increases the mutability of the miniature gene.

The results of Timofeeff-Ressovsky and Demerec may throw a new light on the observations made long ago by systematists and palaeontologists, namely, that the pace of evolution is not alike in all organisms. Some groups seem to possess an unlimited store of variation and evolve rapidly, while others are conservative and undergo no change during geological epochs. A classical example of the evolutionary conservatism is the brachiopod genus Lingula, which was already living in the seas of the Palaeozoic period and yet has neither changed nor become extinct in our time. Although the fossil material on Lingula gives no assurance of the unchangeability of its internal structures, the facts are still remarkable. Mutation rates in organisms like Lingula would be decidedly interesting to know.

However different may be the mutation rates of different genes and in different organisms, we are justified in concluding that the mutation process is constantly going on. If this process is allowed to progress unchecked, the species must eventually change in the direction of the greatest mutation pressure. The situation in nature is not so schematically simple however, for mechanisms that counteract the mutation pressure are known to exist. Selection is one of them; it can eliminate mutations that decrease the adaptation of the organism to its environment. Consequently the existence of mutation is in itself no conclusive evidence that evolution is caused by it. The problem of the role of mutation in evolution must be approached in a different manner.

III: MUTATION AS A BASIS FOR RACIAL AND SPECIFIC DIFFERENCES

STATEMENT OF THE PROBLEM

ORGANIC diversity may be described simply in terms of the observable morphological and physiological differences between individuals of a race, races of a species, species of a genus, etc. But a genetic analysis must ascertain first of all to what extent these differences are hereditary and what part of them is due to modifications induced by environmental influences during development. Next, the genetic basis of the hereditary differences must be analyzed to determine whether these can be resolved into genic differences. The only known method of origin of genic differences is through mutation; in fact, any change of the gene structure is a mutation by definition. It follows that any difference between individuals and populations which can be expressed as a function of gene differences is to that extent due in the last analysis to mutational changes.

The above argument can be evaded only by assuming that diverse genes and gene allelomorphs have always existed in nature and have maintained their identity more tenaciously than chemical elements. Such an assumption being granted, evolution could be conceived as a result of recombination and permutation of the gene-elements. An attempt of this sort has actually been made by Lotsy (1916 and later work), who denied that gene mutation takes place, or at any rate that it plays a role in evolution. As an alternative, he pointed out that races, species, and even genera occasionally cross with each other and produce hybrids. The hybrids, being heterozygous for all the genes in which the parental forms differ, give rise in further generations to greatly variable progenies. According to the second law of Mendel, all possible combinations of the parental genes appear in the offspring of a hybrid. Many of the new gene combinations thus produced are discordant, ill-adapted to the environment, and are destroyed by natural selection. Others, presumably a few favorable combinations, survive and become new species.

It must be admitted that Lotsy's theory is formally logical. Its difficulty is that, by assuming a *doctum ignorantium*, it impinges on one of the main principles of scientific methodology. It takes for granted the existence of a large supply of immutable gene allelomorphs. But by supposing that they have existed eternally, the problem of their origin is not solved but merely pushed back in the obscure past, and becomes a tantalizing puzzle. Creation of such a puzzle is unnecessary, since gene mutation is an established fact. Lotsy's theory falls under its own weight, although he must be credited with having correctly emphasized the significance of hybridization in evolutionary processes.

Osborn (1927) denies the importance of mutation in evolution on different grounds: "Speciation is a normal and continous process; it governs the greater part of the origin of species; it is apparently always adaptive. Mutation is an abnormal and irregular mode of origin, which while not infrequently occurring in nature is not essentially an adaptive process; it is, rather, a disturbance of the regular course of speciation." Variations of this statement have been made into professions of faith by a number of writers. The source of the difficulty here is a profound misunderstanding of the genetic conception of the mechanisms of evolution.

It is true that the formation of species is a continuous process, but its components are the numerous discontinuous steps, mutations. Mutation changes one gene at a time; simultaneous mutation of masses of genes is unknown. On the other hand, species differ from each other usually in many genes (see below); hence, a sudden origin of a species by mutation, in one thrust, would demand a simultaneous mutation of numerous genes. Assuming that two species differ in only one hundred genes and taking the mutation rate of individual genes to be as high as $1:10,000$, the probability of a sudden origin of a new species would be 1 to $10,000^{100}$. This is not unlike assuming that water in a kettle placed on a fire will freeze, an event which is, according to the new physics, not altogether impossible but improbable indeed. The integration of the mutational steps into geno-types of species is probably always an adaptive process, but to believe that every step in this process is also adaptive from the start is to believe in miracles. If the racial and specific differences cannot be resolved into mutational differences, they must be capable of resolution into something else. This is a crucial problem and fortunately it is open to an experimental attack.

MUTATIONS IN WILD POPULATIONS OF DROSOPHILA

Individuals resembling mutations obtained in the laboratory have been repeatedly found in natural populations of Drosophila (Lutz 1911, in *Drosophila melanogaster*, Sturtevant 1915, Morgan, Bridges, Sturtevant 1925, in *Drosophila repleta*, etc.). The first systematic studies on the occurrence of mutations in wild populations of *Droso-phila melanogaster* were made by Tschetwerikoff (1926, 1927) and by Timofeeff-Ressovsky (1927). These investigators concentrated their efforts not on finding in nature individuals showing visible ef-fects of mutant genes, but on a genetic analysis of the offspring of wild individuals. The theoretical basis of this method of approach is simple enough. Recessive mutant genes may be present in heterozy-gous condition without producing any perceptible effects. A popula-tion in which every individual is heterozygous for one or more reces-sives may appear perfectly homogeneous, provided the frequencies of the individual mutant genes are so low that mating of two indi-viduals heterozygous for the same mutant is unlikely to occur. To detect such concealed mutants, a study of the offspring of each indi-vidual collected in nature must be made under controlled conditions. Taking, for example, females which have already been fertilized in nature, one may see in their immediate offspring (F_1 generation) whether each particular female was heterozygous for any sex-linked mutant genes, and whether she or her mates carried autosomal domi-nants. Inbreeding F_1 individuals will permit the detection in F_2 and F_3 generations of the recessive autosomal mutants.

The results of Tschetwerikoff have been reported only briefly. In the offspring of 239 wild females, 32 different hereditary characters were found. About 50 per cent of the population (from Gelendjik, Caucasus) carried the autosomal recessive gene for polychaeta, which causes a reduplication of some bristles on the body of the fly; some wild individuals were homozygous for this gene. One of the flies proved to be heterozygous for the autosomal recessive gene arista-pedia, producing a transformation of the antennae into leglike organs. Timofeeff-Ressovsky (1927) analyzed 78 females of *Drosophila melanogaster* from Berlin; 37 of them were free of mutants, 26 were heterozygous for one, and 15 for two or more mutant genes. No less than nine different "mutants" were extracted from this population; one of them proved to be a sex-linked dominant and one a sex-linked recessive; there were two sex-linked lethals, three autosomal dominants, and two autosomal recessives.

Dubinin and his collaborators (1934) have published the most extensive data now on record on wild populations of *Drosophila melanogaster*. Their material was collected in various localities in Caucasus (the first nine localities in Table 6) and in one locality in central Russia (Tambov). These data are most interesting; not only did a large percentage of the flies prove to be "infected" by mutations, but many of them were heterozygous for mutations producing lethal effects in homozygous condition. Only one population among the ten studied seemed to be free of mutations. In the other nine populations the frequency of lethals varied from 7.7 per cent to 21.4 per cent, and the frequency of mutations inducing visible external changes, from 0 per cent to a high value of 28.6 per cent. The samples of the population taken in the same locality in successive years showed large fluctuations in the frequency of mutations. Some of the mutations thus proven to exist in the wild state were identical with previously known mutants obtained in the laboratory (the recessive genes comma, sepia, ebony, scarlet, black, purple); others were new genes, or at least new allelomorphs of the previously known ones. The mutations "extra bristles" and comma were found in a majority of the populations studied, and in some populations up to 50 per cent of the flies carried them in heterozygous condition. Other mutations were found in a single locality only, but in that locality some of them were frequent; thus, the gene ebony (body color) was detected in the population from Batum and in no others, but in Batum 7.9 per cent of the chromosomes tested carried it. As many as 3.5 per cent of the tested second chromosomes of the Vladikavkaz population contained identical lethal genes—a very high frequency for a lethal. In all, 24 mutational changes were detected more than once, and 37 mutations were found only once. In contradistinction to the high frequency of the autosomal recessives, lethal and nonlethal ones, no autosomal dominants and no sex-linked mutant genes of any kind were found in the total of 3,252 X chromosomes from all the populations studied. In this respect the data of Dubinin and his collaborators are different from those of Timofeeff-Ressovsky on the

TABLE 6

FREQUENCY OF MUTANT GENES IN SEVERAL WILD POPULATIONS OF
Drosophila melanogaster (after Dubinin and collaborators)

POPULATION FROM	YEAR	NUMBER OF CHROMOSOMES II AND III TESTED	LETHALS IN CHROMOSOME II	VISIBLE AUTOSOMAL MUTANTS
Essentuki	1931	187	20	78
Essentuki	1932	120	15	25
Piatigorsk	1931	81	8	16
Piatigorsk	1932	142	11	35
Vladikavkaz	1931	165	32	35
Vladikavkaz	1932	115	13	6
Mashuk	1931	180	19	103
Mashuk	1932	178	15	30
Erivan	1931	102	16	25
Erivan	1932	81	7	5
Kislovodsk	1931	144	13	68
Batum	1931	101	14	23
Armavir	1932	42	9	22
Delizhan	1931	92	not studied	0
Tambov	1931	120	not studied	44

Berlin population (see above), who found both dominant and sex-linked mutations in the wild state. The results of Gordon (1936) on the English Drosophila populations of the same species agree in general with those of Dubinin, but Gordon records one sex-linked mutant.

Results essentially identical with those referred to above for *Drosophila melanogaster* have been obtained also in other species of Drosophila. Balkaschina and Romaschoff (1935) made an analysis of wild populations of *D. transversa* and *vibrissina*, Gershenson (1934) of *D. obscura,* and Gordon (1936) of *D. subobscura*. Since the genetics of these species is known in far less detail than that of *D. melanogaster*, the detection of autosomal lethals was impracticable. With this exception, the populations of the various species proved to be about equally infected with mutants.

In the American species *D. pseudoobscura* an analysis of wild populations from diverse geographical localities has been made by Sturtevant (unpublished). Only two sex-linked and one doubtful autosomal dominant mutant genes were detected, but autosomal recessives, both producing visible external effects and lethals, were found in abundance. The average frequency of lethals in the third chromosome was found to be between 15 per cent and 20 per cent. If other autosomes are as likely to contain lethals as is the third chromosome (and according to preliminary data of Sturtevant this may be the case), it follows that a majority of flies found in nature contain one or more lethals in heterozygous condition. In some localities the population was found to be heterozygous for mutant genes which had been observed previously to mutate under laboratory conditions (the eye colors orange and purple).

NATURAL MUTATIONS IN ORGANISMS OTHER THAN DROSOPHILA

The detection of mutations is evidently very difficult in organisms that are unsuitable as laboratory objects, and which therefore have been only superficially studied, or not studied at all, from the point of view of genetics. Nevertheless, many scattered observations are available to show that Drosophila is at least not alone in the living world in producing mutants under natural conditions. The descriptive biological literature is full of recorded cases of appearance of aberrant individuals among masses of normal representatives of a species in nature. They are sometimes described as aberrations, phases, monstrosities, and sometimes are merely recorded without much detail. Naturalists of the old type were prone to assume without a scintilla of evidence that these unorthodox looking creatures are all due to developmental accidents, to fortuitous coincidences of some intangible external conditions, and on these grounds to discount them altogether. That developmental accidents occur in nature, and may produce teratological specimens, is not doubted by anyone. But in a number of cases it has been established with varying degrees of certainty that aberrations of this sort are in reality hereditary types, mostly recessive to the normal condition, the rare instances when recessive genes, long carried in the population in heterozygous state, emerge as homozygotes because of the occasional mating of two carriers.

Dunn (1921) was the first to apply the above genetic interpretation to the aberrant individuals in wild species of rodents (albinism, pink eyes, yellow, black, or white spotted specimens among the agouti-colored normals). The aberrations found in nature resemble well-known breeds of domestic species, rabbits, mice, and guinea pigs, in which these characteristics are inherited as Mendelian recessives.

Storer and Gregory (1934) have made a special study of exceptional individuals in the western pocket gopher, *Thomomys bottae*. They record at least seven exceptional types, all analogous to the familiar varieties of the domesticated rodents. One of them, the black fur color, is especially interesting. In California the black specimens of the pocket gopher are very rare, the normal coloration being agouti (black is recessive to agouti in genetically studied species). But in *Thomomys townsendi* from Idaho (Nampa) black individuals are no great rarity, and *Thomomys niger* from Oregon is normally black. Sumner (1932) records no less than six "mutations" in the mouse Peromyscus, which appeared after a few generations of inbreeding in the progeny of individuals taken from nature.

H. Timofeeff-Ressovsky (1935) obtained a mutation in the beetle *Epilachna chrysomelina*, which resembled a type of this species previously recorded in some localities. Mutations were detected in the Crustacean *Gammarus chevreuxi* by means of inbreeding the offspring of wild individuals (Spooner 1932). Among crickets many species are dimorphic with respect to the length of the wings, the long-winged and short-winged specimens being sometimes found in the same population and sometimes restricted to different localities. *Pteronemobius heydeni* in southern France is normally short-winged, but Chopard and Bellecroix (1928) have shown that long-wingedness exists in the population in heterozygous condition. Interesting observations on the population of white mustard (*Synapis alba*) in the vicinity of Leningrad were published by Saltykovsky and Fedorov (1936). This species is normally cross-fertilizing, and the wild population is homogeneous with the exception of rare individuals which show chlorophyll defects. If, however, wild individuals are inbred, 3.8 per cent of the offspring in the first and 6.8 per cent in the second generation manifest such defects. There can be no doubt that systematic experiments on inbreeding will show populations of most normally cross-fertilizing species to be infected with mutant genes.

SMALL MUTATIONS

A majority of the mutant types detected in the works referred to above produce rather striking phenotypical changes and more or less pronounced depressions of viability. The fact that they were found in nature almost exclusively in heterozygous condition suggests that these mutations are at present unable to become established as true racial characteristics. The great frequency of lethal genes in wild populations proves conclusively that at least some of the mutations taking place in nature are of the same "pathological" kind which is common among laboratory mutations. It is, consequently, important to inquire whether "small" mutations, of the type found by Baur (1924) and Timofeeff-Ressovsky (1934d, 1935b) to be so frequent in the cultures of Antirrhinum and Drosophila, respectively, also occur in nature. For obvious reasons, data bearing on this problem are extremely laborious to obtain, and Dubinin and collaborators (1934) are the only ones thus far to attempt such a work in *Drosophila melanogaster*. They have used three different methods for the detection of small mutations.

First, ten strains of flies obtained each from a single female caught in nature were established, and the number of certain bristles (sternopleurals) was studied in the flies from these strains. The number of bristles is a variable character, and laboratory strains of Drosophila frequently show inheritable differences in this respect. Statistically significant differences were found also between strains derived from separate wild females. Moreover, the average number of the bristles proved to be different in groups of strains coming from different localities. Second, strains of flies coming from nature were outcrossed to the dominant gene Dichaete, whose manifestation has been shown previously (Sturtevant 1918) to be sensitive to the effects of modifying genes present in various lines. Very pronounced differences in the manifestation of Dichaete were observed also in outcrosses to the wild strains from various localities in the Caucasus. Third, the manifestation of one of the commonest major mutations (extra-bristles) found in a number of wild populations was studied quantitatively, and found to be different in nearly every strain. This suggests that genes modifying the effects of this mutation are common in the population studied.

INDIVIDUAL AND RACIAL VARIABILITY

A living species is seldom a single homogeneous population. Far more frequently species are aggregates of races, each race possessing its own complex of characteristics. The term "race" is used quite loosely to designate any subdivision of species which consists of individuals having common hereditary traits. Races may be of very different orders. Formation of geographical races is probably the most usual method of differentiation of species both in the animal and the plant kingdoms (the zoological literature on this subject has been reviewed by Rensch, 1929), but a geographical race can occupy an area ranging from a few to hundreds of thousands of square miles. Geographical races are in turn subdivided into smaller secondary races (sometimes known as natio). Schmidt (1923) shows that populations of the small fish *Zoarces viviparus* inhabiting different Scandinavian fiords, or even parts of the same fiord, are distinct from each other in such characters as size, number of vertebrae, etc. The distinctions are hereditary. Turesson (1922-31) finds that many plant species are split into numerous hereditary "ecotypes" adapted to living in definite ecological stations and found wherever the proper environment exists in the distribution region of the species. Coastal, dune, forest, swamp, alpine, and other ecotypes are distinguished.

The momentous problem is whether the genetic basis of this racial variability is the same as that of the individual one, in other words whether both can be described in terms of gene differences. As we have seen, an enormous, and hitherto scarcely suspected, wealth of genic diversity has been detected in wild populations of Drosophila. The fragmentary information available suggests that the same situation obtains in other organisms as well. From a systematist's viewpoint this store of variability is more potential than actual, since it is caused mainly by recessive genes carried in heterozygous condition. But the phenotypic manifestation of recessive germinal changes is merely a question of their frequency in the population, for as soon as they become so frequent that matings of the heterozygous carriers are likely to occur on a random basis, homozygotes showing the effects of the genes involved will be produced. This is exactly what is observed. The "exceptional" individuals, the aberrations discussed above, are such outcroppings on the morphological surface of the concealed genic variability. Furthermore, all sorts of intermediate situations exist with respect to the frequency of aberrations: from a great rarity to a condition where a species is regularly polymorphic, that is, represented by two or more distinct classes of individuals.

The individual variability, in so far as it is hereditary, may be likened to a store of building materials; the process of race formation consists in arranging the materials in definite patterns. An excellent illustration of this situation is provided by the snail *Partula otaheitana* which inhabits the radially arranged valleys of the island of Tahiti (Crampton 1916). Individuals of this snail vary in many characters: the spiral of the shell may be dextral or sinistral; size, shape, and color of the shell may be different. In one of the valleys (Fautaua) a population is found in which all the characters and their combinations present in the species as a whole may be encountered in separate individuals. Other valleys harbor distinct races each restricted to a single or to a few adjacent valleys. Every race is characterized by a definite combination of traits found in the Fautaua population, but in no case are any distinctive traits added to the Fautaua complex.

Mendelian inheritance of individual and racial characteristics has been observed in so many instances that a mere enumeration is far beyond the scope of this book. A review of the work on the genic analysis of such characters in higher plants alone has been published by Matsuura (1933). Up to 1929, some 400 species belonging to 57 families had been studied, and the bibliography of the subject contains 2,077 references. Genetic investigations on animals have been restricted to a smaller number of species, undoubtedly owing to the greater technical difficulties of experimentation encountered in many groups (especially in marine forms). Nevertheless, the amount of information available is very large. No recent reviews seem to exist. It may be well to point out that Mendelian inheritance has been observed not only in higher organisms, but also in the unicellular ones (Pascher 1916), in fungi (Burgeff 1928, Dodge 1936), in mosses (Wettstein 1924), and in many others.

DISCONTINUOUS GEOGRAPHICAL VARIABILITY CAUSED BY A SINGLE GENE

Special interest is attached to studies on geographical races. Not only is the formation of geographical races very widely encountered, but such races are commonly believed to be incipient species. The phenomenology of geographical variability is rather complex, making

TABLE 7

THE FREQUENCY (IN PER CENTS) OF DEXTRAL AND SINISTRAL SHELLS
IN *Partula suturalis* IN MOOREA (*after Crampton*)

LOCALITIES	DEXTRAL	SINISTRAL	NUMBER OF INDIVIDUALS
Atimaha	100.0	0	277
Vaianai	100.0	0	369
Oio	98.7	1.3	612
Haapiti	93.0	7.0	241
Uufau	12.5	87.5	303
Moruui	0.1	99.9	537
Maraarii	0	100.0	788
Varari	0	100.0	578

it advisable to consider several examples of different kinds. The simplest situation is encountered when the variable characters are represented each by only two clearly distinct phases, so that the variation can be expressed in terms of the relative frequencies of each phase in different localities. The dextral and sinistral shells in snails are such a case. Table 7, extracted from the mass of data of this sort published by Crampton (1916, 1932), shows the frequency of the dextral and sinistral spirals in the shells of *Partula suturalis vexillum* in some valleys of the island of Moorea. Each valley has a characteristic frequency of shells of either kind; in some places the population is purely dextral or purely sinistral, in others a mixed population is found. In the water snail Limnaea the direction of coiling of the shell is determined by a single gene (Diver, Boycott, Garstang 1925, Sturtevant 1923). It is very probable that in Partula the mode of inheritance is the same as in Limnaea, and hence the variation shown in Table 7 is due to different frequencies of the gene determining the direction of the coiling of the shell in various colonies.

TABLE 8

FREQUENCIES OF THE FOUR BLOOD GROUPS AND OF THE THREE GENE ALLELOMORPHS DETERMINING THEM (IN PER CENTS) AMONG DIFFERENT PEOPLES (*after Snyder, Much Abbreviated*)

PEOPLE	AUTHORITY	BLOOD GROUP				GENE FREQUENCY			NUMBER STUDIED
		O	A	B	AB	p	q	r	
Americans	Snyder	45.0	41.0	10.0	4.0	25.9	7.3	67.0	20,000
English	Hirszfeld	46.4	43.4	7.2	3.0	26.8	5.2	68.1	500
French	Hirszfeld	43.2	42.6	11.2	3.0	26.2	7.4	65.7	500
Germans	Gundel	37.3	43.7	13.4	5.7	28.8	10.0	61.1	8,662
Swedes	Hesser	36.9	46.9	9.7	6.4	31.8	8.5	60.7	533
Italians	Mino	35.9	51.1	8.6	4.2	33.3	6.9	59.9	1,391
Rumanians	Popoviciu	36.5	40.9	14.5	7.9	28.6	12.1	60.4	2,372
Russians	Avdeieva & Grizevich	32.0	38.5	23.0	6.5	25.9	16.1	56.6	2,200
Armenians	Parr	28.3	46.7	12.6	12.4	36.1	13.4	53.2	1,536
Poles	Halber & Mydlarski	32.5	37.6	20.9	9.0	26.9	16.3	57.0	11,488
North Japanese	Miyaji	30.2	37.9	22.5	9.5	27.4	17.5	55.0	1,786
Ainu	Grove	15.8	31.3	30.9	22.0	31.7	31.4	39.7	304
Middle Japanese	Nakijima	28.7	41.7	20.2	9.4	30.1	16.1	53.5	509
Koreans	Kirihara	30.5	27.4	34.5	7.6	19.4	23.9	55.2	354
North Chinese	Liu & Wang	30.7	25.1	34.2	10.0	19.5	26.0	55.4	1,000
South Chinese	Chi-Pan	31.8	38.8	19.4	9.8	28.5	16.0	56.3	1,296
Javanese	Bais & Verhoef	39.9	25.7	29.0	5.4	17.8	19.1	63.1	1,346
Australians	Lee	60.3	31.7	6.4	1.6	18.4	4.2	77.6	377
Filipinos	Cabrera & Wade	64.7	14.7	19.6	1.0	8.2	10.9	80.4	204
Indians (India)	Hirszfeld	31.3	19.0	41.2	8.5	14.9	29.1	55.9	1,000
Madagascans	Hirszfeld	45.5	26.2	23.7	4.5	16.8	15.4	67.5	400
Senegalese	Hirszfeld	43.2	22.6	29.2	5.0	14.9	18.9	65.7	500
Moroccans	Snyder	53.6	23.1	20.8	2.3	13.8	12.5	73.2	466
Indians (No. American)	Snyder	79.1	16.4	3.4	0.9	9.2	2.3	88.9	1,104
Indians (So. American)	Mazza & Franke	82.9	12.8	4.3	0.0	6.7	2.2	91.0	94

The variation of the blood groups in man is only slightly more complicated, because the gene determining this character has three (or, according to the newest data perhaps even four) allelomorphs denoted as O, A, and B, respectively. The interaction of these allelomorphs produces four blood groups—the group O (sometimes denoted also as I) consists of individuals homozygous for O (OO); the group A (or II) has the gene A in homozygous or heterozygous condition (AA or AO); the group B (or III) carries the gene B (BB or BO), and the group AB (or IV) has both A and B (AB). The relative frequencies of the four blood groups in various populations have been determined, and from these data the relative frequencies of the three gene allelomorphs have been deduced. The frequencies of the genes A, B, and O are usually denoted as p, q, and r, respectively.* Table 8 shows the frequencies of the four blood groups and of the three gene allelomorphs in various human populations.

The differences between human races with respect to the blood groups are merely quantitative. The American Indians are at one extreme, owing to the rarity among them of the genes p and q; the tribe of the Ainu is at the opposite extreme, with high values of p and q and a relatively low value of r. On the great Eurasiatic continent the frequency of q increases rather steadily in the direction from west to east, but drops again to the south, reaching a very low value among the Australian aborigines. The gene A (p) shows a high concentration in Europe, and again in northeastern Asia, but a low concentration in southern Asia and in Africa. Many attempts have been made to correlate the blood-group situation with the classifications of the human races based on external characteristics, but almost all of them are unconvincing.

DISCONTINUOUS GEOGRAPHICAL VARIABILITY DUE TO SEVERAL GENES

The geographical variability of characters determined by interaction of several genes is more complicated, but fairly clear cases are encountered. The Asiatic beetle *Harmonia axyridis* (a lady beetle) is extremely variable with respect to the color pattern of the elytra and of the pronotum (Fig. 1). The numerous patterns known in this species can be divided into five groups, intermediates between which are absent or very rare. Individuals with yellow, or yellow spotted with black, elytra form one group, within which the varia-

---

* The formulae used for this deduction are: $p = 1 - \sqrt{O+B}$, $q = 1 - \sqrt{O+A}$, and $r = \sqrt{O}$.

tion is due to minor genetic and also to environmental factors (Fig.
1 a-c). Tan and Li (1934) have proved this type of pattern (*aa*) to
be recessive with respect to all others studied. The pattern of aulica
(Fig. 1 d) is dominant to the yellow (*AA* or *Aa*). The gene *B*, irrespective of the presence of *A*, changes the yellow into the form

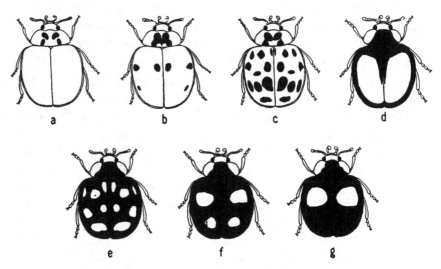

FIG. 1. Color patterns in varieties of *Harmonia axyridis*. (a) var. suc-
cinea; (b) var. frigida; (c) var. 19-signata; (d) var. aulica; (e)
var. axyridis; (f) var. spectabilis; (g) var. conspicua.

spectabilis (Fig. 1 f), which consequently can be genetically *aaBB*,
AaBB, *aABb*, *AABB*, *AABb*, or *aaBb*. Finally, gene *C* produces,
irrespective of the presence of *A* and *B*, the form conspicua (Fig.
1 g). The geographic distribution of the color classes is shown in
Table 9.

West-central Siberia (Altai, Yeniseisk) is occupied by a race mani-
festing nearly always the pattern axyridis (Fig. 1 e). In central
Siberia the yellow forms appear and rapidly displace axyridis, which
on the Pacific Coast of Siberia and in China is very rare or is absent.
Spectabilis and conspicua are found in the Far East only, the latter
apparently reaching a high frequency in Japan. Aulica is nowhere
frequent, but is found almost everywhere in the Far East (Dobzhansky 1933c). The variation in this species could be expressed in
terms of frequencies of the genes determining the various patterns

just as well as in terms of frequencies of the patterns themselves. Unfortunately, the data of Tan and Li (1934) do not include the form axyridis: the genetic basis of this color pattern being unknown, the calculation of the gene frequencies is impossible in all regions where axyridis occurs.

The case of *Harmonia axyridis* is by no means unique. Variability of the same type is known to occur in most diverse forms, animal as

TABLE 9

FREQUENCIES OF COLOR PATTERNS (IN PER CENTS) IN *Harmonia axyridis* FROM DIFFERENT REGIONS

REGION	SUCCINEA, FRIGIDA, 19-SIGNATA	AULICA	AXYRIDIS	SPECTABILIS	CONSPICUA	UNCLASSIFIED	NUMBER EXAMINED
Altai Mountains	0.05	—	99.95	—	—	—	4,013
Yeniseisk Province	0.9	—	99.1	—	—	—	116
Irkutsk Province	15.1	—	84.9	—	—	—	73
West Transbaikalia	50.8	—	49.2	—	—	—	61
Amur Province	100.0	—	—	—	—	—	41
Khabarovsk	74.5	0.3	0.2	13.4	10.7	—	597
Vladivostok	85.6	0.8	0.8	6.0	6.8	0.1	765
Korea	81.3	—	—	6.2	12.5	—	64
Manchuria	79.7	0.5	—	11.2	8.6	—	232
North China (Peiping)	83.0	0.4	—	8.8	7.3	0.5	9,676
West China (Szechwan)	42.6	2.9	0.01	28.8	25.1	0.8	1,074
East China (Soochow)	66.6	0.6	—	16.5	16.1	0.2	6,231
Japan	27.2	—	11.0	14.3	47.4	—	154

well as plant. In butterflies and moths systematists use the expression "variatio et aberratio" to denote forms which occur sporadically, among masses of differently built individuals, in some parts of the distribution area of a species and become predominant in other parts, displacing there the "typical" form. Nevertheless, cases of this sort have not received enough attention either from taxonomists or from geneticists. Quantitative data on the frequencies of the variants in different localities are only seldom to be found, and the mode of inheritance of the varying characters is mostly unknown. In several species of lady beetles (Coccinellidae) the writer (Dobzhansky 1933c) has found instructive examples of this kind of variability.

Some species are regularly polymorphic, and the frequencies of the different types composing the population are similar in all parts of the area inhabited by the species. In other species all the types are present throughout the distribution area, but their relative frequencies are not alike in different regions (the variation of the blood groups in

TABLE 10

ALTERNATIVELY VARYING CHARACTERISTICS IN GEOGRAPHIC RACES OF
Pachycephala pectoralis (after Mayr)

RACE	DISTRIBUTION	THROAT		BREAST BAND		COLOR OF BACK		FORE-HEAD		WING	
		Yellow	White	Present	Absent	Olive	Black	Yellow	Black	Colored	Black
dahli	New Britain ⎫										
chlorura	New Hebrides ⎬		+	+		+			+	+	
vitiensis	So. Fiji ⎭										
bougainvillei	No. Solomon ⎫	+		+		+			+	+	
torquata	N.E. Fiji ⎭										
melanota	S.W. Solomon	+		+			+		+		+
melanoptera	So. Solomon	+		+		+			+		+
sanfordi	N.E. Solomon	+			+	+			+	+	
ornata	Santa Cruz		+	+			+		+	+	
bella	S.W. Fiji ⎫	+		+		+		+		+	
optata	Central Fiji ⎭										
graeffii	N.W. Fiji	+			+	+		+		+	

man is a good example of this situation). In still further differentiated species the distribution area is subdivided into regions each of which is inhabited by a definite form, or a definite set of forms, which do not occur elsewhere, but mixed populations exist in the intermediate zones. The intermediate zones may be wide, or may be so narrow that mixed populations practically no longer exist.

Dr. Ernst Mayr has called my attention to the geographic races in the bird *Pachycephala pectoralis* (Table 10). It inhabits some of the South Sea islands, each island or a group of neighboring islands having a race of its own, characterized in part by alternatively varying characters (Mayr 1932). The exchange of individuals between the populations of different islands is probably very rare. The remarkable feature of this case is that at least eight combinations of the five alternative characters are known, and that the same combination may occur in races found on rather remote islands (i.e., New Britain, New Hebrides, and the southern part of Fiji, Table 10). On one island the character combination is nearly constant, the population apparently being genetically almost pure with respect to these characters. Two exceptions from this rule have been found. Three little islands of the Solomon group are inhabited by a hybrid (or mixed) population in which yellow-throated as well as white-throated individuals occur. In another case a single individual having an unusual combination of characters has been found in an otherwise homogeneous population.

CONTINUOUS GEOGRAPHICAL VARIABILITY

The common property of the geographical races discussed above is that the variability of their differentiating characters is discontinuous. In Partula the shells may be either dextral or sinistral, no intermediates occur. The blood groups in man, the five types of color patterns in Harmonia, and the alternative characteristics in Pachycephala are distinct categories, so that every individual can be placed in one or the other class without hesitation. The discontinuous variation is from the geneticist's viewpoint the simplest possible situation, because the underlying genes are represented here by allelomorphs producing distinct phenotypical effects, comparable to the effects of the major mutations dealt with in most laboratory experiments. This comparison is something more than an analogy, for the genic differences causing discontinuous variation may arise through relatively large mutational steps.

Large mutations are however less frequent than genic changes with intermediate or small visible effects, hence geographical races are encountered in which the racial differences are far less striking than those in Partula and Harmonia. In fact, continuous geographical variation is a commonly observed phenomenon. The lady beetle *Coccinella septempunctata* may serve as an example here. The geographical distribution of this species includes Europe, northern Asia, and north Africa. Throughout this area it has seven black spots on the red elytra, but the size of the spots is variable. In central and southwestern Asia a race with very small spots occurs (Fig. 2). Northern

Europe and especially the Far East are inhabited by races with large spots. In any locality the population is variable with respect to the size of the spots, but the variability forms a normal probability curve, and it is futile to attempt to classify the individual variants in any other than arbitrary classes. Likewise, there is no clear line of separation between the races with large and those with small spots.

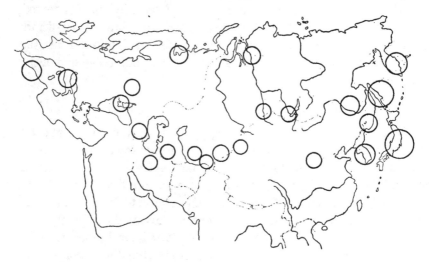

FIG. 2. Geographical variation in the size of the spots on the elytra of *Coccinella septempunctata*. The size of the circle is proportional to the diameter of the discal spot.

The populations coming from geographically remote localities are often strikingly different, but intermediate populations are found in the intervening localities, forming a continuous chain of intergradations between the extremes. The racial variability of stature, of cephalic index, and of many other characteristics used by anthropologists for the description and classification of the human races is continuous. Copious examples of continuous geographical variability have been reviewed by Rensch (1929).

The genetic analysis of continuous variability is technically difficult. The same obstacles are met with here that are encountered in studies on small mutations. Interracial crosses produce in the F_2 generation continuously varying arrays of individuals instead of a clear segregation into discrete classes. Laborious progeny tests are necessary to reveal the Mendelian segregation thus obscured by the individual variability which is mostly environmental in origin. These difficulties led some writers to a sort of a defeatist attitude; continuous variability was declared different in principle from the discontinuous one. It was said that only the latter is clearly genic, while the former was alleged to be non-Mendelian and to be due to some vague principle which assiduously escapes all attempts to define it more clearly. The sharp distinction between the continuous and the discontinuous variability is however gratuitous. All sorts of intermediate situations occur. The variation of human eye color and hair color are good examples. Moreover, and this is decisive, we are fortunate in having several published accounts of genetic analyses of continuous variation, and these accounts agree in showing that its genetic basis is similar in principle to that of the discontinuous variation. At least as a working hypothesis, and pending further studies in this rather neglected field, it is reasonable to assume that geographical variation of all kinds is caused by the inequality of the relative frequencies of different gene allelomorphs in populations inhabiting different parts of the distribution area of a species.

The classical studies of Nilsson-Ehle on the genetics of cereals first established the principle of multiple factors, which was later applied to studies on the inheritance of quantitative characters in general. An inherited difference between two individuals or races can be due to coöperation of several genes each modifying the character by a certain more or less small value. The effects produced by the different genes involved may be approximately equal, or some genes may be more effective than others. In the latter case we speak about a "main" factor and its modifiers. The heterozygotes are frequently intermediate between the two homozygotes, so that every genotype has its own phenotype, of course modified by environmental factors. The differences between the phenotypes are, however, not absolute; they are rather relative, differences of degree. The situation may be further complicated if one or both of the ancestral races are themselves not homogeneous for some of the genetic factors causing the interracial distinction. For instance, the interracial distinction may be due to five pairs of genes producing cumulative effects; one race is predominantly *AABBCCDDEE* and the other *aabbccddee*, but some individuals of the constitution *aaBBCCDDEE* are encountered in the population of the first, and some *AAbbccddee* individuals in the population of the second race.

Sumner (1923, 1924a, 1929, 1930, 1932, Sumner and Huestis 1925) has made a painstaking investigation of the geographical variability of some species of deer mice, Peromyscus. The geo-

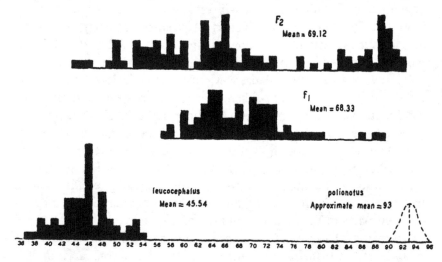

FIG. 3. Inheritance of the extent of the colored area of the pelage in the mice cross *Peromyscus polionotus* × *P. leucocephalus*. (After Sumner.)

graphical races differ here in such characters as size of body; length of tail, feet, and ears; the extent of the colored portion of the pelage; coloration of the pelage; proportionate number of different types of hairs, etc. Most of the differences are quantitative. The result of interracial crosses is usually that the F_1 generation is intermediate in its characters between the parents, and about as variable as each of the parental races themselves. In the F_2 generation the mean values of the characters are likewise intermediate, and mostly similar to those of the F_1 generation. The variability is, however, greater, so that the extreme variants fall within the range of the normal variation of the ancestral races. Backcrosses to the ancestral races cause shifts of the mean values in the direction of the backcrosses. Fig. 3 shows the inheritance of the extent of the colored area of the pelage in the cross *Peromyscus polionotus polionotus* $\times$ *Peromyscus polionotus leucocephalus*. The former is a dark animal, the extent of the colored area having a mean value of about 93. *Leucocephalus* is light, the mean value of the colored area being 45.54. The F_1 and the F_2 generations are almost exactly intermediate between the parents, mean values being 68.33 and 69.12 respectively. The variability in F_1 is however smaller (the standard deviation, $\varsigma = 6.46$) than in F_2 ($\varsigma = 13.87$).

The increase of variability in the F_2 was supposed to be evidence of the occurrence of Mendelian segregation. Taking the theoretical example of the two races differing in five genes (see above), the F_1 will consist of a single genotype (*AaBbCcDdEe*), or at most of a few genotypes (if the ancestral races are not homogeneous). But in F_2 the number of genotypes is much greater; for a character determined by five genes this number equals 3^5, or 243. Moreover, only a small fraction of the offspring will be identical in genotype with either of the parents; for the case of five genes this fraction is 4^{-5}, or $1:1,024$. Evidently, the difficulty encountered in the analysis of a character determined by multiple factors will vary depending upon the number of genes involved, the equal or unequal effectiveness of these genes in modifying the character, and the absolute size of the effect—other things being equal. Sharp differences are easier to analyse than the small ones. Crosses involving some of the racial characters in Peromyscus have given such small increases of the variability in F_2 generation that in his earlier work Sumner believed them to be non-Mendelian. Only the accumulation of further data and improvements in technique finally convinced Sumner that the multiple gene hypothesis is adequate to account for his data; in fact his work is a good example of the applicability of this hypothesis. Muller (1936) has recently pointed out that the increase of the variability in the F_2 generation of crosses is not strictly necessary if multiple genes are involved, for the variability due to the interracial hybridization may sometimes be canceled by the intraracial variability present in the parental strains. Such a situation, although theoretically possible, is probably rare in nature.

Tedin (1925) has analysed the differences between lines of the plant *Camelina sativa* from various localities in Sweden. Characters distinguishing the lines are leaf shape, hairiness, height of the plant, shape of the pods, the angle between the main axis of the inflorescence and the pod stalks, seed weight, and others. Throughout, the inheritance was found to be due to multiple genes. In some cases Tedin was able to isolate single Mendelizing factors causing the quantitative differences, in other cases the number of the factors remained unknown. As a whole, Tedin's work is one of the best examples of a successful application of the multiple gene hypothesis. Equally good results were obtained by Philiptchenko (1934) on quantitative differences between strains of the cultivated species of wheat.

A very extensive series of investigations on the continous geographic variation in the moth *Lymantria dispar* has been published by Goldschmidt. Races of this moth are often widely different in such characters as the strength of the sex-factors (Goldschmidt, Seiler, Poppelbaum 1924, Goldschmidt 1932a), color of the caterpillars (Goldschmidt 1929), duration of the winter rest of the eggs (1932c), speed of the development (1933a, b) and others (1934b). For each character Goldschmidt finds several determining genes, one of which, according to his interpretation, produces a major effect while others act as plus and minus modifiers.

The genetic basis of continuous variation is probably similar to that of discontinuous variation. The extremes of the two races (for instance, the Japanese and the central Asiatic populations of *Coc- cinella septempunctata*, Fig. 2) may be supposed to differ in a number of genes, *AABBCCDDEE* . . . and *aabbccddee* . . . respectively. The intermediate populations may be *AAbbccddee, AABBccddee AABBCCddee, AABBCCDDee,* or mixtures of these genotypes. The characteristics of the intermediate races will, then, be determined by the average number of the multiple genes present in the population in a given locality.

GENETIC CONCEPTION OF A RACE

So many diverse phenomena have been subsumed under the name "race" that the term itself has become rather ambiguous. Sometimes any subdivision of a species which is genetically different from other subdivisions is called a race. A "geographical race" has somewhat more definite implications: it is a group of individuals which inhabits a certain territory and which is genetically different from other geographically limited groups. This is the meaning of the term race as used in the taxonomic literature, with the important reservation that the hereditary nature of the racial characteristics is more often assumed than empirically known. The amount of attention which the problem of races has attracted in genetics is very small, altogether out of proportion to its theoretical and practical importance. This explains the abuses which the race concept has suffered, especially in recent years.

In classical morphology and anthropology, races are described usually in terms of the statistical averages for all the characters in which they differ from each other. Once such a system of averages is arrived at, it begins to serve as a racial standard with which individuals and groups of individuals can be compared. This simple method of racial studies is unquestionably convenient for some practical purposes. The difficulty is however that from the point of view of genetics such an attempt to determine to which race a given individual belongs is sometimes an unmitigated fallacy. The fact which is very often overlooked in making such attempts is that racial differences are more commonly due to variations in the relative frequencies of genes in different parts of the species population than to an absolute lack of certain genes in some groups and their complete homozygosis in others. Examples quoted above show that gene frequencies in different races of a species may vary from 0 per cent to 100 per cent, these being no more than limiting values. Individuals carrying or not carrying a certain gene may sometimes be found in many distinct races of a species.

The difficulty is enhanced by the genetically complex nature of most racial differences. Blood groups are certainly not the sole difference between human races; populations of Partula vary with respect to size, shape, color, and other characters, as well as in the direction of coiling of the shells; Harmonia exhibits variation in size, shape, and coloration of the body and its parts. The geographical distributions of the separate genes composing a racial difference are very frequently independent. In fact, the distribution of the blood group genes in man does not resemble that of the genes for the skin color or the cephalic index (although the related racial groups are more likely to be similar in each of these characters than the unrelated ones). Goldschmidt (1934b) has made especial efforts to correlate the geographical distributions of the various genes differentiating the Lymantria races, only to find them largely independent. The result is that with respect to some genes an individual or a population A may be more similar to B than to C, but in other genes the same individual or population A may be more like C than B. In fact, individuals of the same race may differ in more genes than individuals of different races.

The fundamental units of racial variability are populations and genes, not the complexes of characters which connote in the popular mind a racial distinction. Much confusion of thought could be avoided if all biologists would realize this fact. How important it is may be illustrated by the following analogy. Many studies on hybridization were made before Mendel, but they did not lead to the discovery of Mendel's laws. In retrospect, we see clearly where the mistake of Mendel's predecessors lay: they treated as units the complexes of characteristics of individuals, races, and species, and attempted to find rules governing the inheritance of such complexes. Mendel was first to understand that it was the inheritance of separate traits, and not of complexes of traits, which had to be studied. Some of the modern students of racial variability consistently repeat the mistakes of Mendel's predecessors.

An endless and notoriously inconclusive discussion of the "race problem" has been going on for many years in the biological, anthropological, and sociological literature. Stripped of unnecessary verbiage, the question is this: is a "race" a concrete entity existing in nature, or is it merely an abstraction with a very limited usefulness? To a geneticist it seems clear enough that all the lucubrations on the "race problem" fail to take into account that a race is not a static entity but a process. Race formation begins with the frequency with which a certain gene or genes becomes slightly different in one part of a population from what it is in other parts. If the differentiation is allowed to proceed unimpeded, most or all of the individuals of one race may come to possess certain genes which those of the other race do not. Finally, mechanisms preventing the interbreeding of races may develop, splitting what used to be a single collective genotype into two or more separate ones. When such mechanisms have developed and the prevention of interbreeding is more or less complete, we are dealing with separate species. A race becomes more and more of a "concrete entity" as this process goes on; what is essential about races is not their state of being but that of becoming. But when the separation of races is complete, we are dealing with races no longer, for what have emerged are separate species.

Racial variability must be described in terms of the frequencies of individual genes in different geographical regions or in groups of individuals occupying definite habitats. Such a description is more adequate than the usual method of finding the abstract average phenotypes of "races" because it subsumes not only an account of the present status of a population, but to a certain extent also that of its potentialities in the future (e.g., presence of cerain genes in heterozygous state which may change the phenotype of the race if increased in frequency).

The geography of the genes, not of the average phenotypes, must be studied. Only a few attempts to apply this method in practice have been made to date. The most successful one among them is that concerning the blood groups in man (Table 8). Serebrovsky (1927) and Petrov (1936) have made interesting studies in poultry, Vavilov (1928) in cultivated plants, Bernstein (1925a, b) on the human voice and certain other characters, Dubinin and his collaborators (1934) on Drosophila. The scarcity of similar data for other organisms is due less to any difficulty in obtaining such data than to a lack of appreciation of their importance. Seldom does one find in the literature accurate numerical data on the frequencies of alternatively varying characters in populations inhabiting different parts of the distribution area of a species. Clearly, here is an almost virgin field for future work.

MENDELIAN INHERITANCE IN SPECIES CROSSES

Hybrids between species, and sometimes between genera, have been obtained and studied before as well as after the rediscovery of Mendel's laws. A rather unwieldy mass of data has resulted from these studies; an attempt to account for inheritance in all species crosses in Mendelian terms is at first sight hopeless. Some of the early geneticists took it almost for granted that Mendel's laws are applicable only to intraspecific crosses, the interspecific ones being governed by some other principles. At present, however, it is sufficiently clear why attempts to resolve the differences between species into genic differences encounter serious difficulties.

The existence of genes is inferred from observations on the ratios in which the traits distinguishing the parental forms reappear in the second and following generations of hybrids. However many species can not be crossed, and among those that do cross many produce completely sterile offspring. Any pair of species that fails to give rise to an F_2 generation is therefore automatically eliminated from the ranks of cases which are amenable to analysis by direct hybridological methods. Furthermore, a prerequisite for the occurrence of a Mendelian segregation is a normal chromosome pairing and disjunction in the gametogenesis of the hybrid. Wherever this prerequisite is not fulfilled, breakdowns of the normal Mendelian ratios may be engendered by disturbances in the chromosomal mechanisms. Yet chromosome behavior is abnormal in the gametogenesis of many, if not of most, interspecific hybrids.

A brilliant analysis of the interrelation between chromosome pairing and Mendelian segregation in interspecific hybrids has been given by Federley (1913), who studied the crosses between three species of the moths Pygaera. In some of these crosses certain characteristics showed a typical segregation, while other traits followed the "constant intermediate inheritance," that is to say behaved alike in the F_1 and the F_2 generations. This constant intermediate inheritance seemed inconsistent with the Mendelian scheme and was supposed to be due to an intimate blending of the parental germ plasms, resulting in the appearance of a constant intermediate product. Federley proved, on the contrary, that in those crosses where no segregation at all was observed, no chromosome pairing had taken place in the gametogenesis of the hybrid, and segregation of traits was invariably preceded by the formation of at least some chromosome bivalents. Far from being due to a particularly intimate mixing of the ancestral germ plasms, the constant intermediate inheritance and lack of segregation are results of a failure of chromosome pairing.

The interspecific hybrids in which a straightforward genetic analysis can be made constitute, therefore, only a residue left after the elimination of the sterile and the chromosomally abnormal hybrids. The difficulties do not end here. As we have seen, even in interracial crosses the isolation and study of the individual genes is often very difficult on account of the large number of genes involved. The species differences are likely to be determined by more numerous genes than the racial ones, making the segregations in F_2 and later generations from interspecific crosses correspondingly more complex.

The results obtained by Baur (1924, 1930, 1932) and Lotsy (1911) in crosses between species of snap-dragons (Antirrhinum) may be regarded as typical for fertile hybrids between parents differing in numerous visible, partly qualitative, characteristics. In the cross *Antirrhinum majus* × *Antirrhinum molle,* the F_1 generation is in most respects intermediate between the parents. The variability of the F_1 is about as great as that of the more variable parental species, in this case *Antirrhinum molle.* The second generation obtained by selfing the F_1 is enormously variable in all characteristics. Among hundreds of plants no two identical individuals can be found, and many siblings differ in a most striking manner. A majority of plants carry various recombinations of the parental characteristics, and may be therefore loosely described as intermediates. Some individuals resemble one or the other of the parents in one or more characters, such as flower or leaf shape, color, height, manner of growth, etc., but, according to Baur, none can be mistaken for a pure *majus* or pure *molle.* Moreover, many individuals possess characteristics present neither in the parents nor in the F_1 generation, but encountered in other species of Antirrhinum or even in other genera of the family, Scrophulariaceae. These owe their origin to the formation of new combinations of the ancestral genes, which interact in such a manner as to produce apparently novel characteristics. How peculiar and striking may be such novelties appears from the fact that Lotsy described one of them as a "new species"—*Antirrhinum rhinanthoides.* This segregate possesses certain attributes of the genus Rhinanthus that do not occur at all in the known species of Antirrhinum.

The non-appearance in F_2 of individuals completely resembling the parental species indicates that the number of genes responsible for the species difference is large. The published data of Baur and Lotsy contain, however, no estimates either of the total number of genes involved or of the numbers determining separate characteristics. In this respect the data of Honing (1923, 1928) on the hybrids between *Canna indica* and *Canna glauca* are somewhat more satisfactory. These two species differ in a number of clearly defined traits, some of which Honing has succeeded in resolving into the component genic elements. For example, *Canna indica* has red flowers, and *glauca* yellow ones; this difference is due to a single pair of allelomorphs, *A-a, AA* being red and *aa* yellow. Genes *B* and *C* produce a red margin on the leaves; *D, E, F,* and *R* are intensifiers of the red flower color; *G* is a gene for anthocyanin in leaves; *K* and *L* produce a waxy covering on leaves; *M, N,* and *O* determine the presence or absence of a third staminode; *H, I,* and *P,* the color of the staminode; *J,* the coloring of the veins; and *R,* the red patches in yellow flowers. The genetic formulae of the two species may be written as follows:

Canna indica . . . *AB(C)D(E)FgHIjkl(mno)pr*
Canna glauca . . . *ab(c)DdefGhiIjKL(MNO)PpRr*

The genes that are not quite securely established are shown in the above formulae in parentheses. In *Canna glauca* the genes *D, J, P, R,* and perhaps also *M, N,* and *O,* may be either homozygous or heterozygous.

This analysis covers, however, only a portion of the differences between the two Canna species. A large number of the differentiating characters are quantitative rather than qualitative and the genes determining them were not individually isolated and studied. For example, *Canna indica* has much shorter staminodes than *Canna glauca.* The inheritance of the staminode length in the interspecific cross presents a typical picture of "blending inheritance," not unlike that encountered in the case of the racial differences in Peromyscus discussed above (Fig. 3). One may surmise that these not closely analyzed differences are due each to a joint action of several genes, and hence the total number of genes in which *Canna indica* and *Canna glauca* differ is likely to be quite large.

These crosses between species of Antirrhinum and Canna are not unique in producing an apparently boundless diversity of forms in the F_2 generation. The crosses *Dianthus armeria* × *Dianthus deltoides* (Wichler 1913), the hybrids between species of willows (Heribert Nilsson 1918), and between species of violets (J. Clausen 1931a), may be cited as further instances. An experimental garden plot with a few hundred individuals of a segregating hybrid progeny of this kind is one of the most impressive sights for any biologist who has a chance to behold it. The botanical systematic literature contains many records of similarly polymorphic hybrids observed in nature in regions where two or more distinct but closely related species of a genus overlap in their distribution and have an opportunity to cross. The "hybrid swarms" found in such regions of contact frequently represent veritable riots of genetically distinct types (Chapter X).

Segregations observed in the offspring of hybrids between species differing mainly in quantitative characteristics are, as might be expected, less spectacular than where genes producing distinct qualitative effects are concerned. Analysis of such segregations in terms of genes is extremely difficult. Nevertheless in a number of hybrids Mendelian segregation has been established beyond doubt and occasionally one or more pairs of genes have been studied individually. Hybrids between species of Mirabilis (Correns 1902), Nicotiana (East 1916, Brieger 1935), cotton (Harland 1936 and earlier works), guinea pigs (Detlefsen 1914), birds (Phillips 1915, 1921, Steiner 1935), moths (Lenz 1928), and crickets (Cousin 1934) may be quoted here (for further references regarding plant hybrids, see Renner 1929). Of course, in species crosses, as in the racial ones, there is no sharp dividing line between the blending and the clearly alternative modes of inheritance, the distinction being merely a matter of degree.

We may conclude that the total number of genes responsible for the difference between a pair of species has in no published instance been accurately determined. This is simply another way of saying that the genetic analysis of the differences between species has been in no case complete. The work on interspecific hybrids has produced so far only very general evidence showing that differences between species are due to coöperation of numerous genes, and that these genes produce characteristically complex segregations wherever the conditions of the chromosome pairing in the hybrid provide a physical basis for any segregation at all.

MATERNAL EFFECTS AND THE NON-GENIC INHERITANCE

Even a most rigidly critical appraisal of the evidence presented in the foregoing paragraphs is bound to show that at least a portion of the individual intraracial variability, as well as of the interracial and the interspecific differences, is resolvable into gene elements. Since, furthermore, the only known mode of the origin of the gene variability is through mutation, it is justifiable to suppose that the variability in question arose ultimately by occurrence of mutants in nature. It is now pertinent to inquire whether gene variation is the sole and only source of natural variability, or whether alongside with it there exist other sources of a different kind. A number of biologists, with some very eminent names included, have in fact entertained such a dualistic notion.

The genetic analysis being inexorably limited in its application to forms that cross and produce at least partly fertile F_1 hybrids, there is by the very nature of things no possibility of adducing a direct experimental proof that the differences between systematically remote organisms are also genic and mutational in origin. Although it is very doubtful that the ability of organisms to cross is strictly proportional to the closeness of their relationships, a general correlation of this sort unquestionably exists. Even intergeneric hybrids are moderately rare, and hybrids between families, orders, classes, and phyla can not be obtained except in the earliest developmental stages.

This purely negative consideration has opened the gates for contentions that genes determine only the characteristics of the lowest systematic categories (races, species, genera), while those of the higher categories are not genic. The only kind of positive evidence on which these contentions seemed to be gaining a foothold were the results of the experimental embryologists on the early developmental stages (chiefly the type of cleavage) in hybrids between remote forms. In such hybrids the effects of the foreign sperm begin to manifest themselves not immediately after the latter's entry into the egg, but only after a more or less prolonged delay, which may in some cases not pass until the hybrid embryo dies. In a number of well-known experiments, that have been for some time the subject of a lively controversy, the eggs were artificially deprived of the female pronuclei, and were fertilized by spermatozoa of unrelated animals. If such eggs develop, the development begins according to the plan characteristic for the species which has produced the eggs, not the spermatozoa. The impression is therefore gained that the development is directed by the maternal cytoplasm, and hence is non-genic.

The value of this kind of evidence for a demonstration of cytoplasmic inheritance is, however, spurious. An egg evidently can not be regarded naïvely as a mechanical sum of two independent elements—a nucleus and a cytoplasm. On the contrary, an egg is an organized system which represents the result of a long process of development in the ovary of the mother. The structure, and hence to a certain extent the potentialities, of the egg had been determined before fertilization by the genes in the nucleus of the growing oocyte, and perhaps also by the genes in the surrounding maternal tissues. The characteristics of the development which sets in after fertilization by a sperm of a foreign species need not be due to the gene complex present in the egg at that particular time; they may have been predetermined much earlier by the maternal genotype. Such a predetermination is by no means an *ad hoc* assumption invented for the refutation of cytoplasmic inheritance; on the contrary, the predetermination of the characteristics of a zygote by the maternal genotype is observed in many intra-specific, as well as inter-specific, crosses, and is usually described as "maternal effect." Three examples of the maternal effect phenomenon are given below. They differ from each other in the degree of complexity of the situation and also with respect to the duration of time during which the predetermination by the maternal genotype is noticeable in the zygote.

Toyama (1912) was the first to describe several cases of maternal effects in the silk-worm (*Bombyx mori*). Races of this moth frequently differ in the kind of eggs they produce. Some eggs are spindle-shaped and others round, some have brownish and others greenish or whitish coloration. The characteristics of the eggs produced by a given female always conform to those of the race to which she belongs, and are independent of the race of her mates. F_1 hybrid females from crosses between races differing in the kind of eggs they produce may lay eggs similar either to those of the maternal or to those of the paternal race (depending upon which gene is dominant and which is recessive), but in any case all eggs produced by an F_1 female are alike. Finally, in the F_2 generation three-quarters of the females produce eggs like the dominant, and one-quarter like the recessive race; here again all eggs produced by an individual female are alike and independent of the race of her mate. The explanation of this mode of inheritance is of course extremely simple: the appearance of an egg is due not to its own genetic constitution as established after fertilization, but to the genes of the mother in whose body the eggs develop.

The inheritance of the direction of coiling of the shell in *Limnaea peregra* (Sturtevant 1923, Diver, Boycott, Garstang 1925) is equally illuminating. Eggs of a dextral individual fertilized by the sperm of a sinistral one develop into dextral adults. These, when inbred, give again a purely dextral F_2 generation, but in the latter, three-quarters of the individuals lay eggs developing into dextrals and one-quarter lay eggs giving rise to sinistral adults. An individual of pure sinistral parentage fertilized by the sperm of a pure dextral individual produces only sinistral progeny. This progeny however produces on inbreeding a purely dextral F_2 generation. Among these F_2 individuals, three-quarters give dextral and one-quarter sinistral F_3's. These relationships, at first sight very puzzling, are easily accounted for if one assumes that the characteristics of the shell of an individual are determined by the genotype of its mother rather than by its own genotype. The gene for dextral shells (D) is dominant over that for sinistral shells (d). The results of the crossing experiments may, then, be represented by the following scheme:

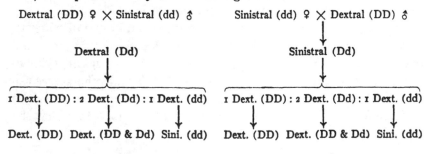

In fact, the inheritance of the direction of coiling in Limnaea is quite analogous to that of the egg characteristics in the silk worm. The direction of coiling, which is expressed in the adult mollusc in the structure of the shell, is laid down already in the egg cell before the latter is fertilized. Whether the individual developing from a given egg will be dextral or sinistral is quite clearly visible in the early cleavage stages, at which the asymmetry of the embryo is revealed by the relative position of the mitotic spindles.

The hybrids between race A and race B of *Drosophila pseudo-obscura* are different, depending upon the direction in which the cross is made. The cross B ♀ × A ♂ produces in the F_1 generation sterile males having abnormally small testes. The reciprocal cross (A ♀ × B ♂) gives rise to males that are likewise sterile, but whose testes are normal in size. An analysis of this difference between the males coming from reciprocal crosses has been made by Dobzhansky (1935c), who has shown that the development of small testes is due to an interaction between the cytoplasm of the eggs deposited by a race B mother on one hand, and the autosomes of race A brought in by the spermatozoon on the other. Autosomes of race B in eggs deposited by race A mothers may permit the development of large testes (for detail see Chapter IX).

There can be no doubt that many, if not most, of the alleged instances of cytoplasmic inheritance are actually due to maternal effects. Nevertheless, some authentic cases of inheritance through the cytoplasm have been recorded. Best known among them are those involving the transmission of the chlorophyll-bearing plastids in plants. The chloroplasts, or the cellular elements from which they arise, are endowed with a power of self-reproduction, and they sometimes undergo changes which are then transmitted independently from the genes and chromosomes. Thus, plants are produced with a piebald foliage, containing in their normal green parts normal plastids, and in the pale green or the white parts plastids more or less devoid of chlorophyll. Correns (1928 and earlier) and Renner (1934) have shown that the defective plastids are transmitted almost exclusively in the maternal line, in such a way that seed formed in flowers developing from the normal green parts of the plant give rise to green offspring, and seed from the chlorophyll-less branches are devoid of green plastids. Only in some plants in which plastids are transmitted through the pollen grains as well as through the embryo sacs can the chlorophyll defects be inherited partly in the male line also.

Less clear is the mechanism of the inheritance in certain mosses (Wettstein 1924), where the reciprocal hybrids between species sometimes differ from each other in characteristics which are not obviously due to the possession of distinct kinds of plastids. Wettstein has shown that this difference between reciprocal crosses persists for so many generations that the possibility that a maternal effect is involved seems excluded. Analogous results were obtained by a number of investigators on hybrids between species of Epilobium (cf. Michaelis 1933). The hybrid from the cross *Epilobium hirsutum* ♀ × *Epilobium luteum* ♂ is different from that obtained in the cross *Epilobium luteum* ♀ × *Epilobium hirsutum* ♂. Using both hybrids as female parents, and pollinating them by the *hirsutum* pollen in several successive generations, one should be able to obtain plants which have only *hirsutum* chromosomes in *hirsutum* cytoplasm on one hand, and in *luteum* cytoplasm on the other. If the cytoplasm does not transmit heredity independently from the chromosomes, the two kinds of plants should be alike. Michaelis, however, finds that the differences between the reciprocal crosses persist, with a gradual weakening, up to at least the seventh generation of hybrids.

Unquestionably, much remains to be done to clear up the problems connected with cytoplasmic inheritance. How widely the phenomenon is distributed, what are the mechanisms of the transmission of heredity of this kind, and above all what is the mode of origin of variations in the cytoplasm is still unclear. Nevertheless, judging from the present incomplete data, cytoplasmic inheritance is so rare relative to genic inheritance that in the general course of evolution the former can hardly play more than a very subordinate role.

IV: CHROMOSOMAL CHANGES

CHROMOSOMES AS GENE CARRIERS

AMONG the constituents of a cell, the chromosomes have attracted unquestionably a lion's share of the attention of investigators. The tremendous strides accomplished in cytology in the last forty years concern largely the chromosomes, while studies on the other cellular elements are more or less in a state of temporary eclipse, inspiring a sort of diffidence in those who find something anomalous in the transformation of cytology (study of the cell) not only into karyology (study of the nucleus), but even into chromosomology. Whatever one may think of such a procedure, its historical reason (not to say justification) lies in the fact that chromosomes have been demonstrated to be the physical carriers of heredity. This demonstration had been accomplished in the main before the rebirth of Mendelism, and is to be attributed to the nineteenth-century exponents of cytology and experimental biology. There remained to be pointed out, as Sutton did in 1903, that the transmission of genes can be understood only if they are borne in the chromosomes.

Since Sutton, an overwhelming amount of evidence has been accumulated in favor of the chromosome theory of heredity. Moreover, chiefly through inference from genetic data, the architectonics of the germ plasm and its physical carriers has been revealed in great detail. Boveri, and especially Morgan and Bridges, have proved that every chromosome is an individual in the sense that it carries a definite complex of genes. Sturtevant founded the theory of the linear arrangement of the genes within the chromosomes and it has rapidly evolved into one of the most fundamental doctrines of modern genetics and cytology. Not only the kind of genes contained in a chromosome, but even their positions with respect to each other, are constant. The more recent work of Painter, Muller, Dobzhansky, and others on the chromosomes of Drosophila changed by the influence of X-rays has given a crucial proof of the correctness of the theory of the linear arrangement of the genes in the chromosomes, and of the chromosome theory of heredity in general.

One of the corollaries of the chromosome theory of heredity is the necessity of assuming a new biological constant, the karyotype. Every individual, race, or species has a relatively constant genotype, the sum total of the hereditary factors, genes, inherited from its ancestors. The stability of the genotype is responsible for the persistence of the visible traits of the organism. But the genes are carried in the chromosomes; every organism is characterized by having a definite number of chromosomes, and each chromosome bears a definite complex of genes arranged in a fixed linear order. These stable structural features constitute the karyotype. A chromosome, as well as a gene, is a self-reproducing body, and is potentially able to give rise to an indefinitely large number of exact copies. Nevertheless, the constancy of the karyotype, like that of the genotype, is subject to certain limitations. The chromosome structure may undergo changes, and an altered chromosome, like a mutant gene, may retain the power of self-reproduction virtually *ad infinitum*.

The nature of the relationship between the genotype and the karyotype must be made clear before we can proceed further with our discussion. One may inquire whether the chromosome structure is determined ultimately by the genetic composition of the organism, or whether the karyotype develops autonomously from the genotype. In other words, is the karyotype merely a specialized province of the morphology of the body, or is it in a sense superimposed on the genotype? The answer is that the relation between the karyotype and the genotype partakes of both these possibilities.

The visible characteristics of the chromosomes are to a certain extent determined by the genes. Mann and Frost (1927) have described a strain of stocks (*Matthiola incana*) in which the meiotic chromosomes are much longer than in other strains of the same species. When the two strains are crossed, the F_1 individuals have short chromosomes, and in F_2 three-quarters of the plants have short, and one-quarter long, chromosomes. It follows that chromosome length is here determined by a single gene; short is dominant to long. Goldschmidt (1932b) has made a detailed study of the chromosome size in the geographical races of the gypsy moth, *Lymantria dispar*. Some races proved to have large chromosomes, while others were characterized by small chromosome size. As far as one can judge, all the chromosomes, which in this species number thirty-one haploid, are proportionately increased or diminished in a given race. The differences between the extreme races are quite striking, since the ratios of the chromosome volumes are as 261 : 154. The inheritance of the chromosome size in interracial crosses is not as simple as in the Matthiola strains, but it is unquestionably Mendelian and seems to depend on multiple genes. The chromosome size is positively correlated with the strength of the sex-determining factors. The same group of genes must therefore be involved in the inheritance of both characteristics.

Changes in the appearance of individual chromosomes have been recorded by Navashin (1934), McClintock (1934), and Darlington (1932b). Species of the composite Crepis frequently have every individual chromosome in a group distinguishable from the rest by its size, position of the constrictions, presence or absence of satellites, etc. These characteristics are as a rule retained in the hybrids. A hybrid between two species has every chromosome of the haploid set of the parental species present once, and preserves the morphological peculiarities which the chromosomes normally have in the pure species. But in some hybrids this rule breaks down; for instance, a satellite may be retracted into the body of the chromosome and may be no longer visible.

The physiological nature of all these changes in the appearance of chromosomes is a matter of conjecture. It must be remembered that a chromosome, as seen under the microscope, is composed of the genic materials proper, as well as of some concomitant substances, among which the nucleic acids seem most prominent in determining the staining properties of the element. It seems probable that gene changes (mutations) may modify the amount of the nucleic acids produced in the cell, and thus affect the appearance (diameter, length, degree of contraction) of the chromosomes.

Aside from the alterations in the appearance of chromosomes induced by genic changes, there exists a large group of alterations of an entirely different kind. Chromosomes may be reduplicated or lost, giving rise to individuals with an excess or a deficiency of certain genes. Blocks of genes located normally in one chromosome may become detached from their normal position and transposed to different chromosomes. The relative positions of genes within a chromosome may be changed (Fig. 4). Alterations of this sort

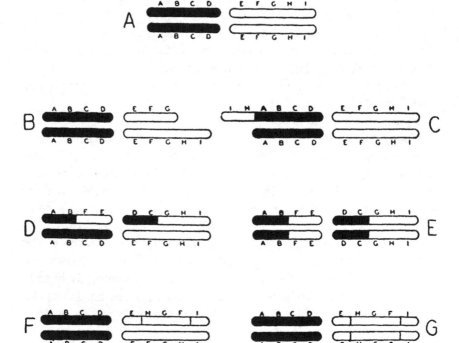

FIG. 4. (a) Normal chromosomes; (b) deficiency; (c) duplication; (d) heterozygous translocation; (e) homozygous translocation; (f) heterozygous inversion; (g) homozygous inversion. (From Dobzhansky.)

may or may not result in a visible modification of the chromosome morphology, but if they are hereditary they cause permanent transformations of the karyotype, and may acquire the role of a racial or a specific characteristic. These are chromosomal changes proper, as distinguished from changes induced in the chromosomes by genes. The chromosomal changes are classed as mutations in the wide sense of that term (see above), but they differ from gene mutations in that they represent modifications of the gross structure of chromosomes rather than of the structure of individual genes. Two organisms may possess exactly the same genes arranged in a different fashion in a different number of chromosome aggregates. It is expedient to treat the chromosomal changes as a phenomenon separate from gene mutation, although in recent years we have learned that they may not be quite as completely independent as was formerly supposed (see the discussion of position effects).

Two main classes of chromosomal changes may be distinguished. Numerical changes involve variations of the number of chromosomes, but leave the gene contents and arrangements within the chromosomes unmodified. Structural changes alter the number or the distribution of genes in separate chromosomes. The further subdivisions of these classes may be gathered from the following synopsis:

I. Numerical changes—affecting the number of chromosomes
  A. Change in the number of sets of chromosomes present in the nucleus
    a. Haploidy. Higher organisms are mostly diploid during a major part of their life cycle, that is, they possess two chromosomes of each kind in the nuclei of most cells. Gametes, and gametophytes (in plants), are haploid, and carry one chromosome of each kind. Some diploid organisms have produced under experimental conditions haploid aberrants, which have a single set of chromosomes in the tissues that are normally diploid. Such haploids are known in Datura, Crepis, Oenothera, Triticum, Nicotiana, Solanum, and other plants.
    b. Polyploidy. Normally diploid organisms may give rise to forms with more than two sets of homologous chromosomes. Such forms are known as polyploids. Triploids (three sets), tetraploids (four sets), pentaploids (five sets), hexaploids (six sets), and higher polyploids are encountered. Autopolyploids and allopolyploids are usually distinguished. The former arise from diploids that have the two members of each pair of chromosomes more or less similar in the gene contents and gene arrangement. Reduplication of the chromosome complement of a diploid hybrid, which consequently has two different sets of chromosomes, gives rise to an allopolyploid. The boundary between the auto- and the allopolyploids is not sharp, owing to the gradations which exist with respect to the similarity and the dissimilarity of chromosomes. Some species,

especially among plants, are normally polyploid, presumably because of their origin by allopolyploidy.

B. Change in the number of separate chromosomes of a set
  a. Monosomics arise through a loss of one of the chromosomes, giving
     rise to individuals which are diploid except for the lack of one of
     the chromosomes of the normal complement. Inbreeding mono-
     somics may result in appearance of some zygotes that lack one of
     the chromosomes entirely; such individuals are as a rule inviable.
     Monosomics are known as karyotypical aberrants in Drosophila,
     Datura, and in other species.
  b. Polysomics are individuals having one of the chromosomes repre-
     sented three (trisomics) or more times. They are obviously the
     reciprocal of monosomics. Monosomics and polysomics may be
     derived from polyploids as well as from diploids. In Datura, tetra-
     ploids are known in which one or more chromosomes are represented
     three, five, or six times, instead of four.

II. Structural changes—affecting the gene contents of chromosomes
  A. Changes due to a loss or a reduplication of some of the genes located
     in a chromosome
    a. Deficiency (deletion). A section containing one gene or a block
       of genes is lost from one of the chromosomes (Fig. 4B). If a normal
       chromosome carries genes *EFGHI*, the deficient chromosomes are
       EFG, GHI, etc. An organism may be heterozygous or homozygous
       for a deficiency. A short deficiency, producing no cytologically
       visible reduction of the size of a chromosome, may be mistaken
       for a gene mutation. In fact, at least some of the lethal mutations
       in Drosophila are due to deficiencies.
    b. Duplication. A section of a chromosome may be present at its
       normal location in addition to being present elsewhere (Fig. 4C).
       If normal chromosomes have genes *ABCD* and *EFGHI*, the dupli-
       cation may be *IHABCD, EFGHI* or the like. An individual may
       therefore have some genes represented three times (heterozygous
       duplication) or four times (homozygous duplication).
  B. Changes due to an alteration of the normal arrangement of genes
    a. Translocation. Two chromosomes, *ABCD* and *EFGHI*, may ex-
       change parts, giving rise to "new" chromosomes *ABFE* and
       DCGHI. This is a reciprocal or mutual translocation (Fig. 4D).
       A chromosome may also be broken, and one of the resulting frag-
       ments may become attached to another chromosome—*AD* and
       EFBCGHI (simple translocation). An individual may be hetero-
       zygous or homozygous for a translocation (Figs. 4D and 4E).
       Genes which in one species lie in the same chromosome may in
       another species lie in different ones. Such differences between

species are probably due to the occurrence of translocations in the phylogeny. Translocations have been produced experimentally in Drosophila, in maize, and elsewhere. Their genetic detection is based on the fact that individuals heterozygous for translocations produce gametes containing deficiencies and duplications for chromosome sections (see Chapter IX for more detail). Thus, at least four types of gametes may be produced in an individual having the constitution represented in Fig. 4D: (1) *ABCD, EFGHI*, (2) *ABFE, DCGHI*, (3) *ABCD, DCGHI*, and (4) *ABFE, EFGHI*. Gamete (1) is normal; gamete (2) carries the translocation but has every gene represented once; gamete (3) is deficient for *EF* but has *DC* twice; gamete (4) is deficient for *DC* but has *EF* twice. Since gametes carrying deficiencies or duplications normally die or give rise to inviable or abnormal zygotes, an apparent linkage is produced between the chromosomes involved in a translocation. The cytological methods of the detection of translocations are discussed below.

b. Inversion. The location of a block of genes within a chromosome may be changed by a rotation through 180°. The resulting chromosome carries the same genes as the original one, but the arrangement of the genes is modified from *EFGHI* to *EHGFI*, or *EGFHI*, or others (Fig. 4F). In individuals heterozygous for an inverted section (Fig. 4F), the frequency of the detected crossing over in the chromosome involved is markedly lowered, both within and outside of the limits of the inversion. Homozygosis for an inversion (Fig. 4G) restores the normal frequency of crossing over, but the linkage relations between the genes are altered.

The occurrence of structural chromosome changes in natural populations is discussed in the following paragraphs. Among the numerical changes, polyploidy presents some problems *sui generis* which will be dealt with in Chapter VII. Monosomics and polysomics as independent phenomena seem to play a very subordinate role in the evolution of most groups of organisms, except in as much as sex determination is accomplished by representatives of one sex having a certain chromosome in duplicate which the opposite sex carries only once (cf., however, Chapter X).

ORIGIN OF CHROMOSOMAL CHANGES

Chromosomal changes, like gene mutations, became known at first as spontaneous outbreaks of germinal variability. De Vries has de-

scribed spontaneous mutants in Oenothera some of which proved
to be chromosomal changes. The first deficiencies, duplications, translocations (Bridges 1917, 1919, 1923), and inversions (Sturtevant
1917, 1926) detected in *Drosophila melanogaster* were to all appearances spontaneous in origin. Spontaneous polysomics, monosomics,
polyploids, haploids, and translocations were observed in *Datura
stramonium* by Blakeslee (1922).

The discovery of the induction of gene mutations and chromosomal
changes by X-rays (Muller 1928a) has given a tremendous impetus
to the investigation of both phenomena. In genetically well studied
forms like Drosophila and maize, chromosomal aberrations can now
be obtained in virtually unlimited numbers. During less than a decade
that has elapsed since the discovery of the influence of X-rays on
the karyotype, induced chromosomal changes have been extensively
and successfully used as tools in many genetic experiments. The
chromosome theory of heredity in general and the theory of the linear
arrangement of the genes in particular have been put to a crucial
test, which has proved their coherence. Some problems, mostly bearing on the genetics of the transmission of hereditary characters (see
above), were approached and partly solved. Another class of problems, which happens to be more momentous for us in this book, has
been so far largely neglected.

Very little is known of the frequency of chromosomal changes.
Spontaneous non-disjunction of chromosomes leading to the production of monosomics and polysomics seems to be fairly common, at
least in *Drosophila melanogaster*. For the X chromosome the rate
is about 1 : 2,000, and for the small fourth chromosome it is possibly even greater. About the frequency of the spontaneous origin
of structural changes nothing definite is known, except that they are
rare. Oliver (1932) has studied the translocations and inversions
associated with the lethal mutations induced in the X chromosome
of *Drosophila melanogaster* by X-rays. Only those changes were detected that partly or wholly suppress crossing over in the X chromosome, which means that some chromosomal aberrations were overlooked. Oliver applied five different X-ray treatments, t_1 to t_{16},
which can be arranged in a series in which every following member
is twice as great as the preceding one, and the minimum dose is

equal to about 385 Röntgen units. The results are summarized in Table 11.

The doubling of the amount of the X-ray treatment doubles the frequency of the induced mutations, but the frequency of the chromosomal changes appears disproportionately great after strong treatments. It would seem to follow that the spontaneous rate of origin of chromosomal changes is below that for spontaneous mutations.

TABLE 11

FREQUENCY OF LETHAL MUTATIONS AND CHROMOSOMAL ABERRATIONS INDUCED IN THE X CHROMOSOME OF *Drosophila melanogaster* BY X-RAY TREATMENTS (*after Oliver*)

TREATMENT	PERCENTAGE OF LETHAL MUTATIONS	PERCENTAGE OF CHROMOSOMAL ABERRATIONS
t_{16}	16.09	5.52
t_8	9.87	2.43
t_4	4.90	0.35
t_2	3.23	0.40
t_1	1.24	0.075
Control	0.24	0.0

Two conjectures have been put forward regarding the mechanics of the origin of structural chromosomal changes. One of them, originated by Belling (1927) and elaborated especially by Serebrovsky (1929) and Dubinin (1930), assumes that chromosomes occasionally undergo "illegitimate crossing over" between non-homologous sections. If two different chromosomes are involved in such a process, an interchange of blocks of genes may result, and a translocation may be produced. Loop formation in a chromosome with a subsequent establishment of new connections at the point of overlapping of the chromatids may give rise to inversions, deficiencies, and duplications. The alternative viewpoint supposes that chromosomes suffer fragmentation, and that fragments are either lost or reattached in new combinations. No critical evidence that would permit a discrimination between these two alternatives is available. It is now established that reciprocal translocations in Drosophila and maize, the two organisms in which the structure of the rearranged chromosomes can be accurately ascertained, are far more frequent than simple ones. Simple translocations involving the attachment of a fragment of one chromosome to a free end of another occur very infrequently, if at all. These facts are evidently best accounted for by the illegitimate crossing over hypothesis, but they are not absolutely incompatible with the hypothesis of fragmentation and subsequent reattachment.

An important fact that has emerged from studies on chromosomal aberrations is the permanence of the spindle attachment loci in chromosomes. Some cytologists believe that a contractile fiber unites the pole of the spindle with a certain point in each chromosome and directs the movements of the latter during mitosis. Whether such fibers exist in the living cells or appear only as artifacts of fixation is an open question. There is no doubt, however, that a certain point in each chromosome, described as the spindle attachment, is especially concerned with the guidance of chromosome movements. Exchanges of sections between chromosomes may result in the formation of new chromosomes without a spindle attachment or with two attachments. Such chromosomes are, however, lacking among the fairly large sample of translocations that has been studied in Drosophila, in maize, and in several other species. A chromosome fragment that does not include the locus of the spindle attachments is either lost or is re-attached to another chromosome or a fragment which retains that locus. Dobzhansky (1930b) has consequently surmised that one and only one spindle attachment is necessary and sufficient for a normal transmission of a chromosome from one cell generation to the next. *De novo* formation of spindle attachments does not occur. All further work has borne out this supposition. From studies on the offspring of cells irradiated with X-rays (e.g., Mather and Stone, 1933) it is now known that chromosomes with two and with no spindle attachment are sometimes formed, but their behavior in cell division is so abnormal that they are ultimately eliminated. The necessity of preserving one spindle attachment per chromosome imposes a limitation upon the freedom of chromosomal rearrangements.

TRANSLOCATION AS A RACIAL CHARACTER

It has been pointed out above that the occurrence of gene mutations in laboratory experiments does not constitute a proof that evolution is caused by them. The same statement evidently applies to chromosomal changes as well. Some critics have hastened to remark that since mutations and chromosomal changes can be induced by so destructive an agent as X-rays, such changes bring about degeneration and not evolution. The logic of this criticism is certainly rather ludicrous. An acid test of the ideas concerning the rôle played by chromosomal changes in evolution must however come not from theoretical discussions but through an analysis of the differences between races, species, and other natural groups. For if these differences are similar to those observed between chromosomal aberrations obtainable under laboratory conditions, the conclusion follows that chromosomal changes are an active evolutionary agent.

An extensive series of observations on the chromosomes of the Jimson weed (*Datura stramonium*) of different geographical origins has been reported by Blakeslee (1929, 1932) and by Bergner, Satina, and Blakeslee (1933). This plant possesses twelve pairs of chromosomes, which normally form twelve bivalents at meiosis. In crosses between certain strains, one or more circles of four or six chromosomes appear, the remainder of the chromosomes forming bivalents as usual. Belling (1927) and Belling and Blakeslee (1926) have shown that the circle formation at meiosis is due to translocations involving two or more chromosomes. The chromosomes of two strains are shown in Fig. 5a and b respectively. The chromosome structure observed in one of these strains might arise from that in the other by means of a translocation; for instance, the chromosomes 1.2 and 3.4 (Fig. 5a) might exchange sections, giving rise to the chromosomes 1.3 and 2.4 (Fig. 5b). Since like parts of chromosomes come together and pair at meiosis, a cross-shaped configuration (Fig. 5d) will be formed in the hybrid between the two strains. At the metaphase of the first meiotic division, the cross-shaped figure will be transformed into a twisted circle shown in Fig. 5e. A translocation in which three different chromosomes have exchanged sections gives a circle of six chromosomes (Fig. 5j); two translocations between two different pairs of chromosomes produce two circles of four chromosomes each, etc. The formation of chromosome configurations like those shown in Fig. 5d and i has been inferred on the basis of genetic data or actually seen cytologically in spontaneous as well as in induced translocations in Datura, in maize (Burnham 1930, McClintock 1931), Drosophila (Dobzhansky and Sturtevant 1931), and in other organisms.

Each of the twelve chromosomes of *Datura stramonium* is distinguishable from the others genetically, and some of them also cytologically. One of the strains of this plant has been arbitrarily chosen as a standard of comparison, and the free ends of each chromosome in this strain designated by numbers 1, 2, 3 . . . 23, 24. The standard strain consequently has chromosomes 1.2, 3.4, 5.6 . . . 21.22, 23.24.

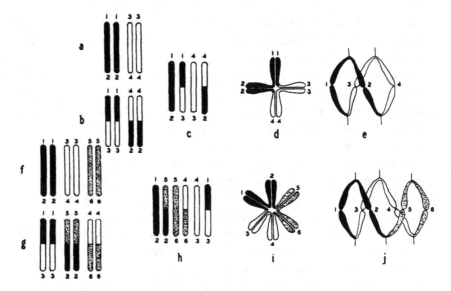

FIG. 5. Translocation between two (b-e) and between three (g-j) chromosomes. (a) and (f) Normal chromosomes; (b) and (g) translocation homozygotes; (c) and (h) translocation heterozygotes; (d) and (i) chromosome arrangement at pairing stages; (e) and (j) arrangement of chromosomes at the metaphase of the meiotic division.

By means of methods which we need not describe here in detail, the structure of the chromosomes in other strains has been ascertained. In the strains united under the name "prime type 2," chromosomes 1.18 and 2.17 are found instead of the 1.2 and 17.18 present in the standard. The origin of "prime type 2" from the standard, or vice versa, is obviously due to a translocation involving these two chromosomes; hybrids between the standard and the "prime type 2" strains display a circle of four chromosomes and ten bivalents at meiosis. The "prime type 3" strains have chromosomes 11.21 and 12.22 instead of the standard chromosomes 11.12 and 21.22; "prime type 4" has chromosomes 3.21 and 4.22 instead of 3.4 and 21.22; other "prime types" found in nature, or induced by X-rays or radium treatments have different combinations of parts of the chromosomes of the standard line.

Datura stramonium is a weed which at present is nearly cosmopolitan in distribution, owing to its involuntary transport by man with agricultural products. Its original native country is not known with certainty. In organisms so deeply influenced by man one does not generally expect to find clearly defined geographical races; even more interesting, therefore, is the fact that populations of *D. stramonium* from different geographical regions proved to be unlike in their chromosomal constitution. Plants having chromosomes apparently identical with those of the standard line have been grown from seed collected all over the United States, in the West Indies, Brazil, France, Portugal, Italy, Japan, Portuguese West Africa, and Australia. The populations from Brazil and from the United States, except along the Atlantic Seaboard, seem to have only standard chromosomes. The "prime type 2" has a wider distribution; it is very common in Central and South America (except in Brazil and Argentina), on the Atlantic Seaboard in the United States, in Europe, in Asia (except Japan), and in Africa (except the western Portuguese colonies). The "prime type 3" is restricted to Peru, Chile, and Central America, but has been found once in Spain. The "prime types" 4 and 7 occur in the eastern United States, the West Indies, the Mediterranean countries of Europe, South Africa, and Australia. The Peruvian and Chilean population seems to be homozygous for the chromosomes 1.18, 2.17, 11.21, 12.22 (combination of the "prime types" 2 and 3); the F_1 hybrids from standard and Peruvian or Chilean strains have therefore two circles of four chromosomes and eight bivalents at meiosis.

Formation of rings of chromosomes at meiosis has been observed in many other species besides Datura. Håkansson (1929a, 1931a, 1934), Sansome (1931), and Pellew and Sansome (1932) described translocations in peas (*Pisum sativum*), Gairdner and Darlington (1931) in *Campanula persicifolia*, J. Clausen (1931) in *Polemonium reptans*, Philp and Huskins (1931) in *Matthiola incana*, Håkansson

(1931b, 1933) in *Clarkia elegans* and *Salix phylicifolia*, Levan (1935a) in *Allium ammophillum*, Kattermann (1931), Müntzing (1933, 1935a), and Rancken (1934) in several species of grasses. Among animals, a perfectly clear case of a translocation has been observed in *Trimerotropis citrina* by Carothers (1931), and it seems very probable that at least some of the "multiple chromosomes" described at meiosis in the Orthopterans *Jamaicana subguttata* (Woolsey 1915), *Hesperotettix* and *Mermiria* (McClung 1917), and in the copepod *Diaptomus castor* (Heberer 1924) are due to a similar cause. In most of these investigations it remains uncertain whether the chromosome structures recorded are individual or racial characteristics. Translocations, like gene mutations, may be present in populations in varying frequencies, and there is no qualitative distinction between the individual and the racial variability in this respect. In Tradescantia chromosomal differences between individuals as well as between races have been found by Darlington (1929a), Sax and Anderson (1933), Anderson and Sax (1936), and others. Some species of this genus are remarkably inconstant in their karyotypes; polyploidy, translocation, and fragmentation of chromosomes leading to the formation of duplicating fragments are common. The extraordinary behavior of some species of Oenothera, which for years was one of the outstanding riddles in genetics, has been finally explained as due to translocations. Many Oenothera species are permanent translocation heterozygotes; translocations may involve various numbers of chromosomes, the extreme condition being one in which all the chromosomes of a plant form a large circle at the meiotic division (Darlington 1929b, Blakeslee and Cleland 1930, Cleland and Blakeslee 1931, Emerson and Sturtevant 1931).

Formation of circles of chromosomes at meiosis is not always observed in translocation heterozygotes. In Fig. 5c a translocation is shown which involves an exchange of approximately half chromosomes. Such a translocation usually gives a cross-shaped configuration (Fig. 5d) and a circle (Fig. 5e) at meiosis. A translocation may be however unequal: a very short section of one chromosome may be exchanged for a relatively long section of another, or two chromosomes may exchange sections that are short relative to the total length of the chromosomes. The arms of the cross shown in Fig. 5d may therefore be very unequal in some translocations. A genetic comparison of the behavior of various types of translocations in *Drosophila melanogaster* (Dobzhansky 1931, 1932, 1933a) has led to the conclusion that small translocated sections may fail to pair entirely, or at least may fail to cross over, as a consequence of which a chain of chromosomes will be formed instead of a ring. If the two opposite arms of a cross in Fig. 5d are much shorter in relation to the other two, no ring, but two bivalents consisting of unequal partners, will appear at meiosis. McClintock (1932) and Burnham (1932) have corroborated these conclusions through a cytological analysis of different translocations in *Zea mays*. The work of other investigators on different objects also shows that rings, or chains of chromosomes, or unequal bivalents may be present in translocation heterozygotes.

Unequal bivalents may, of course, arise through processes other than translocation. If one race differs from another by a deficiency or a duplication of a section of one of the chromosomes, an unequal bivalent is expected at meiosis in the hybrid. A duplication for a section of one of the chromosomes including the spindle attachment may be followed by a translocation of a section of the same or of another chromosome onto the "new" spindle attachment thus made available. The result will be the formation of a race having one more chromosome than the ancestral race; two chromosomes in the former will be homologous to one of the latter, and the hybrid will have a group of three chromosomes at meiosis. A loss of a spindle attachment with a subsequent translocation of the remaining portion to another chromosome will simulate a fusion of two chromosomes in a single body. The hybrid between the ancestral and the derived race will again have a group of three chromosomes at meiosis. Changes in the chromosome number a species possesses may therefore go in the plus as well as in the minus directions without a *de novo* formation of spindle attachments. Dubinin (1934, 1936) has in fact succeeded in obtaining strains of *Drosophila melanogaster* with five pairs and with three pairs of chromosomes instead of the normal four pairs. That changes of such a nature have actually taken place in the phylogeny of certain species is demonstrated by the occurrence of unequal bivalents and by the racial differences in chromosome number and chromosome morphology. In objects that are not favorable from the genetic and cytological standpoints, the exact nature of the change in the karyotype may be difficult to ascertain. Nevertheless, judging by analogy with known translocations, there is a reasonable degree of assurance that chromosomal aberrations of the translocation type are involved.

A particularly clear instance of a racial difference in the chromosome number has been described by Seiler (1925) in the moth *Phragmatobia fuliginosa.* One of the races has one chromosome pair more than the other; in the hybrid a trivalent is formed at meiosis, due to the pairing of two small chromosomes of the former race with one large chromosome of the latter. The causation of the variation in the chromosome numbers found in *Viola kitaibeliana* (J. Clausen 1927) and in *Ranunculus acris* (Sorokin 1927) is less clear than in the Phragmatobia. Translocations as well as monosomics and polysomics may be here involved. In *Galtonia candicans* (S. Nawaschin 1912, 1927) two kinds of individuals are present: one has two large satellites on a certain pair of chromosomes, and the other has one chromosome with a large and one with a small satellite. The size of the satellite is inherited. By inbreeding the strain with unequal satellites a type with two small satellites should theoretically be produced, but according to Nawashin it is not viable. The origin of the chromosome with a small satellite may be due to a deficiency or to a translocation of a section of the large satellite. Two races with a dissimilar chromosome morphology exist in *Vicia angustifolia* (Sweschnikowa 1928). Unequal bivalents have been observed in a series of species of several genera of Orthoptera (Robertson 1915, Carothers 1917, Helwig 1929, and others). Helwig *(l.c.)* finds that in *Circotettix verruculatus* populations from different localities may differ in frequencies of individuals with unequal bivalents. Here, then, a beginning of a differentiation of a species into geographical chromosomal races is witnessed.

INVERSION AND RACIAL DIFFERENTIATION

The genetic technique for the detection of inversions is based on the fact that most of them produce when heterozygous a reduction of the frequency of crossing over in the chromosome involved. Using this technique, Sturtevant (1931) has analyzed a number of strains of *Drosophila melanogaster* coming from wild progenitors collected in various localities in America and in Europe. Most strains had the standard gene arrangement, but some proved to possess various inverted sections in the second and the third chromosomes. The standard arrangement is the one encountered in most laboratory strains, especially in those in which mutants serving as chromosome markers had appeared. Sturtevant discerned no tendency for any of the inversions to occur in definite geographical regions.

The detection of inversions through the suppression of crossing over entails an expenditure of much labor. The introduction of the Drosophila salivary gland chromosome technique (Painter 1934) has enormously facilitated, and at the same time has increased the accuracy of comparing the chromosomal constitution in different strains. In the cells of the larval salivary glands in most flies (Heitz and Bauer 1933) the chromosomes increase more than a hundredfold in length as compared with their size in other tissues (gonads, nerve ganglia). They appear as cross-striped cylinders or ribbons, the cross-striation forming a constant pattern that permits the identification not only of the separate chromosomes but also of their parts. The stainable discs that form the striations may or may not correspond each to a single gene, but in any event their longitudinal seriation is known to reflect accurately the gene arrangement in the chromosome. In addition, the homologous chromosomes in the salivary gland nuclei undergo a very intimate pairing, disc by disc, thus enabling one to determine the exact positions of the homologous discs in the chromosomes present in an individual. Suppose an inversion heterozygote has two chromosomes *ABCDEF* and *AEDCBF*. The only way in which these chromosomes can pair so as to bring every homologous gene into contact with a gene in the partner chromosome is by forming a loop, shown schematically in Fig. 6 in the upper right corner. The presence of such a loop is therefore a proof of the existence of an inversion; the extent and the ends of the loop serve to discriminate between inversions in the same chromosome in different strains.

The possibilities of this method for comparisons of chromosome structure in races and species are tremendous, although its applica-

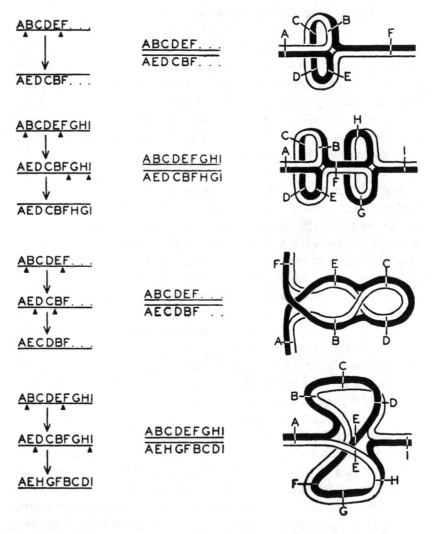

Fig. 6. Chromosome pairing in the salivary gland cells of individuals heterozygous for inversions. Upper row, a single inversion; second from the top, two independent inversions; third from the top, two included inversions; lower row, overlapping inversions.

tion is unfortunately restricted to the order of flies (Diptera) where these amazing chromosomes occur. Not only inversions but all other chromosome changes as well can be detected by this method; its accuracy is unrivaled. With inversions, it permits in certain special cases tracing the phylogeny of the chromosomal structures. Suppose the original, ancestral gene arrangement in a chromosome is *ABCD-EFHI*. An inversion of the section from *B* to *E* gives rise to an arrangement *AEDCBFGHI* (Fig. 6). A second inversion may take place in the same chromosome. The location of the second inversion may be outside the limits of the first: *AEDCBFGHI→AEDCBF-HGI*. Such inversions may be described as independent. An individual heterozygous for the chromosomes *ABCDEFGHI* and *AEDCB-FHGI* will have in the salivary gland nuclei a double loop in the chromosome affected (the configuration shown second from the top in Fig. 6). The second inversion may occur inside the first, forming included inversions: *ABCDEFGHI→AEDCBFGHI→AECDBFGHI* (the configuration second from the bottom in Fig. 6). Finally, the second inversion may have one end inside and the other end outside the limits of the first. Such inversions are termed overlapping ones: *ABCDEFGHI→AEDCBFGHI→AEHGFBCDI* (the configuration in the lower right corner in Fig. 6).

Let us consider especially the overlapping inversions. Suppose we observe in different strains the three gene arrangements (*ABCDEF-GHI, AEDCBFGHI,* and *AEHGFBCDI*). The first of these can arise from the second or give rise to the second through a single inversion. The same is true for the second and the third. But the third can arise from the first, or vice versa, only through the second arrangement, which therefore becomes a necessary intermediate step in the line of descent. A direct origin of the third from the first, or vice versa, is highly improbable, for it would require not only a simultaneous breakage of the chromosome in four places (between *A* and *B*, *D* and *E*, *E* and *F*, and *H* and *I*), but also a fortuitous reunion of the resulting fragments in such a way as to simulate the arrangement that is obtainable through the two-step process indicated above. If we find only the first and the third arrangements, it is possible to postulate with a high degree of probability that the second exists also in some unknown strains, or at least that it existed in the past. If all three are actually observed, the probability of the first and the third being related through the second becomes almost certainty. To recapitulate, the phylogenetic relationship of the three gene arrangements indicated above is $1 \to 2 \to 3$, or $3 \to 2 \to 1$, or $1 \leftarrow 2 \to 3$, but not $1 \rightleftarrows 3$. With the independent and the included inversions, no determination of the sequence of origin is possible; they are ambiguous in this respect. This theory of inversions has been applied in practice to the study of the gene arrangements found in natural population in *Drosophila pseudoobscura* (Sturtevant and Dobzhansky 1936a, also unpublished data).

Using the salivary gland method, Tan (1935) and Koller (1936) have shown that the two races A and B of *D. pseudoobscura* differ from each other in four inverted sections, two of which lie in the X chromosome, one in the second, and one in the third chromosome. These two races produce, when crossed, sterile male offspring in the F_1 generation (Lancefield 1929); the cause of the sterility and the nature of these "races" will be discussed below. Tan has also found that some strains of race B have the same gene arrangement in the third chromosome as is encountered in most of the strains of race A that he had at his disposal. The gene arrangement may therefore be different not only between races but also within a race. Sturtevant and Dobzhansky (1936b) have made a systematic study of the gene arrangements in strains of *D. pseudoobscura* coming from flies collected throughout the geographical range inhabited by the species. An almost bewildering amount of variation in the gene arrangement within the species was disclosed.

The gene arrangement is especially variable in the third chromosome, where a total of seventeen different gene sequences has been recorded. All these gene arrangements can arise from each other through inversions of some sections of the chromosome, and curiously enough nearly all of them proved to be related to each other as the overlapping inversions considered above. This fact has made it possible to construct the phylogenetic chart of the gene arrangements in the third chromosome shown in Fig. 7. Each arrangement is designated in the chart by the name of the geographical locality in which it has been first encountered. Any two arrangements connected in Fig. 7 by an arrow give a single inversion loop in the heterozygote. It may be mentioned here that some of these arrangements (Santa Cruz, Tree Line) had been postulated theoretically as the necessary "missing link" between the other arrangements, and subsequently found when more strains were examined. One of the arrangements (see Fig. 7) remains hypothetical as far as the species *D. pseudoobscura*

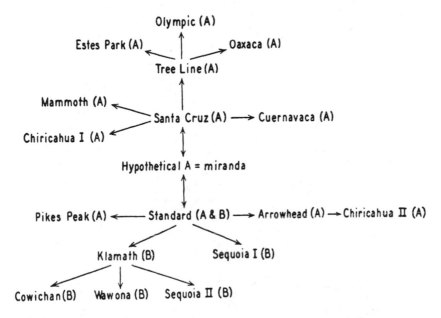

Fig. 7. A phylogenetic chart of the gene arrangements encountered in the third chromosome of race A and race B of *Drosophila pseudoobscura*.

is concerned, but an arrangement possessing the essential properties of this hypothetical one has been met with in a related species, *D. miranda*.

None of the different arrangements shown in Fig. 7 occur over the entire distribution area of the species. Some of them ("Standard," "Arrowhead") have however a rather wide distribution; others ("Pikes Peak," "Sequoia") occupy relatively restricted areas, and still others have been observed thus far in only one or a few strains from a single locality. In some localities the population is mixed, that is two or more (up to four) arrangements occur, usually with different frequencies. Some large regions contain however a popula-

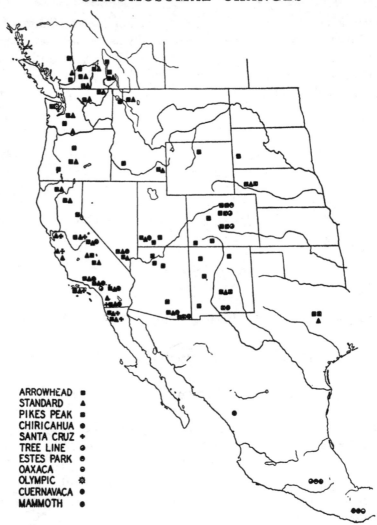

ARROWHEAD ■
STANDARD ▲
PIKES PEAK ▣
CHIRICAHUA ○
SANTA CRUZ +
TREE LINE ◔
ESTES PARK ◉
OAXACA ◎
OLYMPIC ✿
CUERNAVACA ●
MAMMOTH ●

Fig. 8. Geographical distribution of the gene arrangements encountered in the third chromosome of race A of *Drosophila pseudoobscura*.

tion that is homogeneous with respect to the gene arrangement in the third chromosome.

It is neither necessary nor desirable to discuss here the geographical distribution of each gene arrangement. Suffice it to say that all the intermediates are found between the condition where a gene arrangement may be described as an "individual variation" in flies coming from a definite locality, and one where a definite chromosome structure becomes an established racial characteristic for a certain fraction of the species (Fig. 8). Eleven of the arrangements shown in Fig. 7 occur only in race A; five others solely in race B; a single one, the "standard," occurs in both races. It is, hence, probable that the "standard" arrangement is also the ancestral one, but other hypotheses are also possible. In general it must be emphasized that the method of overlapping inversions is rigorous only as far as the establishment of the configuration of the branches of the phylogenetic tree, but not of the location of its root, is concerned. The phylogenetic tree shown in Fig. 7 may be read in various ways, depending upon which of the arrangements is chosen as the ancestral one. Even so, the overlapping inversion method is superior to other methods of phylogenetic study, for it does permit an exact description of the relationships between the different variants observed.

The gene arrangements in the chromosomes of *D. pseudoobscura* other than the third are relatively constant. Yet five arrangements are known in the second, two in the fourth, three in the right limb, and two in the left limb of the X chromosome. Some of the arrangements in the right limb of the X proved to be associated with a peculiar genetic condition that causes a male carrying it to produce mostly daughters in his progeny. This "sex-ratio" condition has been found in from 0 per cent to 30 per cent of the individuals in populations from different localities, its frequency generally increasing as one proceeds from north to south (Sturtevant and Dobzhansky 1936b). In some instances the overlapping inversion method proved to be applicable, and a phylogenetic chart has been traced for the second chromosome.

In species other than *D. pseudoobscura,* inversions in wild populations have been detected by the study of the chromosomes in the salivary glands in *D. melanogaster* and *D. funebris* (Dubinin, Sokolov, Tiniakov, 1936), in *D. repleta* and *D. sulcata* (Frolova 1936), in *D. ananassae* (Kaufmann 1936), in *D. azteca* and *D. athabasca* (Bauer and Dobzhansky, unpublished), in fact in all species in which they were looked for. Bauer (1936) found inversions in several species of flies of the family Chironomidae. It appears that inversion is an important agent leading toward diversification of the chromosome structure in subdivisions of Drosophila species. In contradistinction to inversions, translocations proved to be at least very rare in wild populations of Drosophila. In point of fact, no securely established case of translocation has been detected. The relative abundance of translocations in other organisms, especially in plants discussed above, raises the questions of the possible cause of their apparent absence in Drosophila. Only conjectures can be offered at present in attempting to answer this question. One of them might be that translocations are less frequent in wild populations of animal than of plant species. Translocation heterozygotes frequently produce gametes with an excess or a deficiency for certain chromosome sections (see above). In plants, deficiencies and duplications are usually eliminated in the gametophytic generation and thus decrease the number of functional ovules and pollen produced. In animals, the gametes with normal and abnormal chromosome contents seem equally viable, but the latter give rise to inviable zygotes when they unite with the former. Therefore, translocation heterozygotes in plants produce fewer sex cells and in animals fewer viable offspring than the homozygotes. The death of a part of the offspring may result in the elimination by natural selection of the translocations soon after their appearance. Few or no abnormal gametes are produced by most inversion heterozygotes.

So far we have discussed two methods for the detection of inversions: through studies on crossing-over frequencies, and through investigation of salivary gland chromosomes. For obvious reasons, in many organisms neither of these methods is applicable. In such organisms some inversions may be detected by still another method. As has been shown, chromosomes in the inversion heterozygotes in the salivary gland cells undergo pairing by forming loops represented schematically in Fig. 6. As shown first by McClintock (1933), such loops are observable also at the prophase of meiosis in inversion heterozygotes in *Zea mays*. If crossing over now takes place within the inverted section, the result, illustrated diagramatically in Fig. 9, will be the production of two normal chromatids, one chromatid with two spindle attachments, and one chromatid with no spindle attachment. At anaphase of the first meiotic division the chromatid with two attachments will form a "chromatin bridge" uniting the two daughter groups of chromosomes (Fig. 9e). A fiberless chromosome will be present lagging on the spindle. Such chromatin bridges and fiberless fragments were seen cytologically by McClintock (1933) in maize, by Stone (1933) in X-rayed Tulipa, by Mather (1934) in Vicia, and were postulated on genetic grounds in inversion heterozygotes in Drosophila (Sturtevant and Beadle 1936).

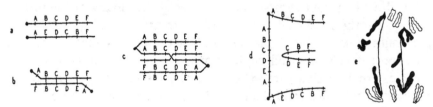

FIG. 9. Crossing over in inversion heterozygotes, leading to the formation of chromosomes with two and with no spindle attachments (d). (e) Chromatin bridges in the hybrid *Lilium martagon* × *Lilium hansonii*. (After Richardson.)

The presence of "chromatin bridges" at meiosis is therefore evidence that the individual showing them is an inversion heterozygote. Chromatin bridges were in fact observed in some strains of *Matthiola incana* (Philp and Huskins 1931), in *Trillium erectum* (Smith 1935), in rye (Lamm 1936), in *Aesculus hippocastanum* (Upcott 1936) and in Chorthippus (Darlington 1936). Unequal bivalents may also be produced by inversions, if the latter involve shifts of the location of the spindle attachment with respect to the free ends of the chromosome. It must be emphasized however that the formation of chromatin bridges depends upon the occurrence of crossing over in the inverted sections, and some inversions may not undergo crossing over. Unequal bivalents may be produced owing to translocations, to deficiencies, to duplications, as well as to inversions. A failure to detect chromatin bridges and unequal bivalents in an organism is therefore not a proof that the latter is free from inversions.

DEFICIENCIES AND DUPLICATIONS

Deficiencies and duplications involve losses or multiplications of some genes. In this respect they are basically distinct from translocations and inversions, for the latter alter only the gene arrangement but not the number of genes. A deficiency or a duplication is therefore usually accompanied by phenotypical effects, while an individual that carries a translocation or an inversion need not be different from the ancestral form (compare, however, the position effects). We have seen already that in Drosophila most deficiencies are lethal when homozygous, and that lethal "mutations" are not infrequently deficiencies for short chromosome sections. Duplications are in general less drastic in their effects than deficiencies, but many of the former are nevertheless lethal, or provoke structural and physiological changes of some kind in the carrier. The deleterious effects produced by most deficiencies and duplications may be used as an argument against their taking part in evolutionary processes; this argument would be no more valid than the similar one based on the predominantly destructive nature of some of the laboratory mutations. In either case, the critical evidence can come only from studies on the genetic nature of the variation in wild populations, which may or may not reveal the origin of this variation through a given type of change known to occur in the laboratory.

Drosophila pseudoobscura manifests a remarkable variation in the shape of its Y chromosome (Dobzhansky 1935b, 1937c). Some strains of this species have a large, V-shaped, slightly unequal-armed Y chromosome. In other strains, one or the other or both of the arms of the V are shortened, as though they have lost some of the material which is present in the type mentioned first. In still other strains the Y chromosome is J-shaped, and its length is barely half the length of the Y in other strains. In all, seven distinct types of Y chromosome are encountered in the species, each type being found only in populations that inhabit a definite part of the distribution area. Three of the seven types are rather widespread; the remaining four are each restricted to a fairly small region. For instance, one of the types is found only in southern California, another only in the highest part of the Rocky Mountains in northern Colorado, a third only around Puget Sound in the north west.

The Y chromosome of Drosophila is composed of an "inert" material, so described because it contains fewer genes per unit length or volume than other chromosomes. Few mutations of any kind occur in the Y-chromosome of any species. In the salivary gland nuclei the Y is much contracted relative to the others, and is in a condition not favorable for study. The lack of the whole of Y, or the presence of two Y's, produces no visible alteration in the fly morphology. These properties prevent a detailed analysis and comparison of the different Y chromosomes in *D. pseudoobscura,* but by the same token they also explain the relative ease with which a Y chromosome can be modified without disastrous consequences to the viability of its carrier. It is fairly certain that the seven types of the Y are all derived from a single prototype through either losses (deficiencies) or reduplications of some of the gene-contents of the latter. Variations in the Y chromosome are known also in *Drosophila ananassae* (Kaufmann 1935), and in *D. simulans* (Sturtevant 1929b, Heitz 1933).

The detailed study of the salivary gland chromosomes in *D. melanogaster* made by Bridges (1935) showed that sections having identical disc patterns may occur in two different parts of the same chromosome. Moreover, these sections, known as "repeats," manifest a certain amount of mutual attraction, and in some cells may be seen paired with each other. Since the similarity of the disc patterns in chromosome sections, and the pairing of the latter in the salivary gland nuclei, are the only criteria of the homology of blocks of genes that we have at present, Bridges concluded that the repeats are composed of similar or identical genes. If this is true, it follows that, while an ordinary diploid *D. melanogaster* carries most genes in duplicate, some genes, namely those located in the repeat sections, are present four times. The writer has observed chromosome sections that conform in their behavior to the standard established for repeats in *D. melanogaster* also in *D. pseudoobscura, D. miranda, D. athabasca,* and *D. azteca* (unpublished).

Undoubtedly much more information than is now available must be gathered before a critical evaluation of the repeats becomes possible. A study of the repeats may throw a light on the elusive problem of the formation of new genes in the phylogeny of diploid organisms, which thus far has defied all attempts to study it. Genes may be multiplied by the occurrence of duplications leading to the origin of repeats. Immediately after the repeats are formed, the "repeated" or twin sections contain identical groups of genes. The process of mutation may, however, change some of the genes in one of the twins and leave them unaffected in the other, or else mutations in the formerly identical genes may happen to go in different directions. The repeated sections may gradually become differentiated, with the result that the derived organism will have more kinds of genes than its ancestor. This hypothesis is for the time being only a speculation. But if it is corroborated by further work, the formation of repeats by duplication of some of the existing genes will have to be regarded as an important evolutionary process. Thus far no instance has been recorded in which different strains from the same wild species differ with respect to the presence or absence of repeat sections.

MORPHOLOGY OF METAPHASE CHROMOSOMES
IN DIFFERENT SPECIES

The relative constancy of the number, size, shape, and structure of chromosomes in each species is one of the long established tenets of cytology. It is equally well known that different species may differ widely in all these characteristics. The chromosome number (haploid) varies from one, two (*Ascaris megalocephala*), and three (*Crepis capillaris,* species of Crocus, *Drosophila earlei*) to several hundreds (Radiolaria). The volume of the chromosomes in a nucleus ranges from a cubic micron (certain fungi) to thousands of cubic microns (*Drosophyllum lusitanicum*). The cytological literature is filled with descriptions of chromosome numbers and sometimes of chromosome morphology in species of this or that genus. The larger part of this work is concerned with the chromosomes as they appear at the metaphase plate stage of the mitotic cycle, which is technically easiest to study.

The nature and causation of the chromosomal differences are subjects which have proved rather more elusive than the straight descriptive picture of the situation. As far as chromosome numbers are concerned, there seem to be two types of differences between related species. In many plant genera and in relatively very few animals the haploid chromosome numbers of the species of a genus form a series of simple multiples of the minimum or "basic" number (e.g., 7, 14, and 21 in species of wheats). Such differences are probably due to the reduplication of the chromosome complement (polyploidy) in the phylogenetic history, as discussed in Chapter VII. In most animals and in many plant genera no trace of polyploid series is discernible. Related species may be alike both in number and in size and morphology of chromosomes; the chromosome numbers may be similar, but the chromosomes differ in size, location of the spindle attachments and of secondary constrictions, presence or absence of satellites, and other structural characters; the chromosome numbers may be variable without formation of a clear series of multiples (e.g., 11, 13, 14, 15, 19, 21, 23, 25, 27, 28, 29, 30, 31,32, 33, 34, 37, 38, 49, 51, 56, 87 in different butterflies and moths, Beliajeff 1930); or finally both the chromosome numbers and the chromosome morphology are variable (Crepis, Fig. 10, Drosophila).

Some investigators, particularly in recent years, have displayed much interest in studies of the comparative morphology of metaphase chromosomes in related species, in the hope that when combined with the classical comparative morphological and distributional methods, these chromosome studies may become a powerful tool for the determination of the relative systematic closeness or remoteness of species of a genus and of genera of a family. Tracing phylogenetic relationships is to become at long last an exact procedure instead of an expression of opinions of the particular investigator. We have seen already that studies on overlapping inversions do give some hope for a rigorous method of establishing the phylogeny of certain chromosome structures, but so far as the application of metaphase chromosomes for the even wider purpose of tracing phylogenies of species and genera is concerned, much caution is necessary.

As a working hypothesis, most cytologists and geneticists hold that the chromosomal differences between species (except polyploidy) have arisen through structural changes of various kinds (translocation, inversion, deficiency, and duplication). This hypothesis has developed so gradually, and has been espoused by so many authors, that it appears impossible to assign credit for originating it. The instances of the control of chromosome morphology by genes (see above) show however that a cytologically visible difference in the structure of chromosomes need not necessarily be due to structural changes. This consideration may be optimistically regarded as trivial, since the rule certainly is that chromosomes retain their morphological peculiarities in the hybrid karyotype, but nevertheless it introduces an element of uncertainty into every specific instance where a comparison of chromosomes in pure species is made. Much more important is the fact that the similarity or dissimilarity of the chromosomes as seen at the metaphase plate stage is not at all necessarily proportional to the similarity of their gene arrangements.

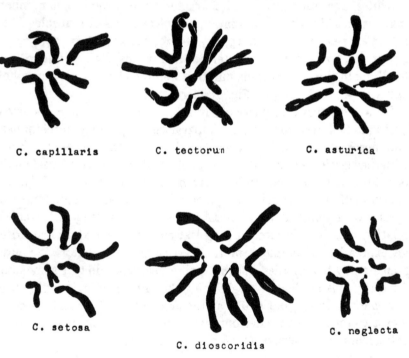

C. capillaris   C. tectorum   C. asturica

C. setosa   C. neglecta

C. dioscoridis

Fig. 10. Metaphase chromosome plates in six species of Crepis. (After Babcock and Navashin.)

Most of the chromosome types shown in Fig. 11 are encountered each in several species of Drosophila. Species possessing metaphase chromosomes that appear identical under the microscope are, however, known to be sometimes similar and sometimes sharply different in the gene arrangement (Dobzhansky and Tan 1936). The apparent identity of the metaphase chromosome configurations is therefore evidence neither of an identity nor of a dissimilarity of the internal chromosome structure. Repeated inversions may occur in a chromosome and may result in a very profound alteration in the gene arrangement, and yet the visible characteristics of this chromosome as seen at metaphase may remain unchanged, provided none

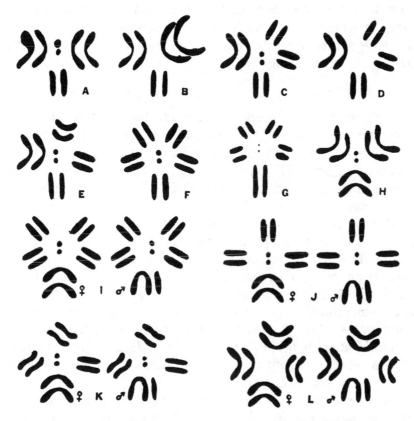

FIG. 11. Metaphase chromosome plates in several species of Drosophila. (From Metz.)

of the inversions shift the locus of the spindle attachment. Conversely, a single inversion involving the spindle attachment alters the appearance of the metaphase chromosome quite strikingly. All the metaphase configurations known in species of Drosophila may be represented as results of fusion in different combinations of the five rod-shaped and one dot-like element shown in Fig. 11F, with an occasional translocation and inversion (Metz 1914). Such a representation may have some basis in fact (Sturtevant and Tan 1937), but it is now known that a large number of inversions take place in the phylogeny that are not reflected at all in the metaphase chromosome morphology. By comparing the metaphase configurations in different species of Crepis we can not gauge how extensive were the structural changes in the chromosomes, and consequently can not legitimately decide which species are and which are not similar in gene arrangement.

TRANSLOCATIONS AND INVERSION IN SPECIES HYBRIDS

As shown first by Federley (1913) and later corroborated by a whole phalanx of investigators, chromosomes fail to form bivalents at meiosis in many interspecific hybrids, especially in those that are sterile (cf. Chapter IX). This rule has, however, many exceptions. In some hybrids the chromosomes of one of the parents find their homologues among those of the other, and the formation of the bivalents proceeds more or less normally. Such species hybrids are similar to the interracial ones in that the same cytological methods for the detection of translocations and inversions are applicable to both. If an interspecific hybrid shows circles or chains of chromosomes at meiosis, it is reasonable to infer that genes located together in the same chromosome in one of the species lie in different chromosomes in the other. This condition is probably due to the occurrence of translocations in the phylogeny. Similarly, formation of chromatin bridges between the separating chromosomes at the meiotic divisions indicates the presence of inversions.

Blakeslee (1932), Bergner and Blakeslee (1932, 1935), and Bergner, Satina, and Blakeslee (1933) have studied the chromosome behavior in hybrids between species of Datura. Species of that genus, despite the sometimes extensive morphological differences between them, all have twelve chromosomes in the haploid, twenty-four chromosomes in the diploid. Within some of the species, races exist that must have arisen by translocation; hybrids between such races show circles or chains of chromosomes at meiosis in addition to bivalents (see p. 83). Analogous conditions obtain also in the hybrids between species. Thus, the hybrid between the standard line of D. stramonium and D. leichardtii has two circles of four chromo-

somes and eight bivalents; the *D. leichardtii* × *D. meteloides* hybrid
has a circle of eight, three circles of four, and two bivalents; *D.
leichardtii* × *D. innoxia*, a circle of eight, two circles of four, and
four bivalents; *D. innoxia* × *D. meteloides*, a circle of six, two circles of four, and five bivalent chromosomes.

Blakeslee and his collaborators have described the arrangement
of the chromosome parts in species of Datura in terms of that observed in the standard strain of *D. stramonium*, the form best known
both cytologically and genetically. Each of the twelve chromosomes
of *D. stramonium* has its ends indicated by a number, so that the
chromosome "formula" of this species may be written thus: 1.2, 3.4,
5.6, 7.8 . . . 19.20, 21.22, 23.24. Among the chromosomes of *D. discolor*, seven appear similar and five are different from *D. stramonium*, The chromosomes that are thus characteristic for *D. discolor* are 1.11, 2.17, 12.22, 15.21, and 16.18. In the *D. stramonium*
× *D. discolor* hybrid seven bivalents and a circle or a chain of ten
chromosomes appear. The structure of the circle is as follows:

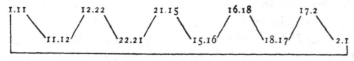

In this scheme the *D. discolor* chromosomes are shown in the upper line and those of *D. stramonium* in the lower. In *D. quercifolia*,
six chromosomes are similar and six are different from *D. stramonium*. The chromosomes that differentiate *D. quercifolia* are 1.18,
2.17, 12.22, 11.21, 7.20, and 8.19. The *D. stramonium* × *D. quercifolia* hybrid has six bivalents and three circles of four chromosomes
each. The structure of the circles is as follows (writing the *D. quercifolia* chromosomes in the upper and those of *D. stramonium* in the
lower line):

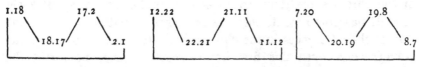

These are the chromosome configurations found in crosses between
the standard line of *D. stramonium* on one hand and *D. discolor* and
D. quercifolia on the other. But we have seen that within the species

D. stramonium there exist also races that differ from each other in the chromosome structure. Thus, in Peru and Chile the population of *D. stramonium* has chromosomes 1.18, 2.17, 11.21, and 12.22 instead of the chromosomes 1.2, 11.12, 17.18, and 21.22 present in the standard. It may be noticed however that all the chromosomes peculiar for the population of *D. stramonium* from Peru and Chile exist also in *D. quercifolia,* and two of them (2.17 and 12.22) also in *D. discolor.* If, then, one compares the chilean *D. stramonium* with *D. quercifolia,* they prove to differ in only two chromosomes (7.20 and 8.19 in *D. quercifolia* instead of 7.8 and 19.20 in *D. stramonium*), and give a single circle of four chromosomes and ten bivalents at meiosis. The difference between the two species, *D. stramonium* and *D. quercifolia,* seems to be no greater than the racial differences within *D. stramonium.*

Both *D. discolor* and *D. quercifolia* have the chromosome 12.22. The cross *D. discolor* × *D. quercifolia* shows however that this chromosome forms in the hybrid a bivalent consisting of two unequal halves. The inequality of the bivalent discloses the fact that although the numbered ends 12 and 22 are present in both *D. discolor* and *D. quercifolia* in the same chromosome, this chromosome is nevertheless not similarly constructed in the two species. This brings us face to face with the inadequacy of the observations on circle formation in hybrids as a method for studying chromosome structure. As pointed out above, translocations result in circle or chain formation at meiosis if they involve exchanges of large sections of chromosomes. If only small sections are exchanged, unequal bivalents may be produced, but the inequality is easily overlooked. The presence of circles including the same chromosomes in two different hybrids does not necessarily prove that the translocation in the two cases has involved exchanges of identical sections. The comparison of the chromosome structure in species of Datura is, as Blakeslee himself has pointed out, only the first approximation to an elucidation of the differences in the chromosome structure between species. Nevertheless, even this first approximation leaves no doubt that translocation is an active agent in the evolutionary process.

Formation of circles and chains of chromosomes has been recorded in many interspecific hybrids besides those between the Datura species discussed above. Kihara and Nishiyama (1930), Kihara and Lilienfeld (1932, 1935), Lilienfeld and Kihara (1934), Mather (1935), and others have observed them in hybrids between species of wheat (Triticum) and Aegilops. J. Clausen (1931a, b) found them in hybrids between species of Polemonium, and also between species of Viola; Håkansson (1931b, 1933, 1935) in the *Godetia amoena* × *Godetia whitneyi* hybrid, in a hybrid between species of willows (Salix), and in *Pisum humile* × *Pisum arvense;* Levan (1935a) in *Allium cepa* × *Allium fistulosum*. In the hybrid between the two moths *Orgyia antiqua* and *Orgyia thyellina*, Cretschmar (1928) saw indications that parts of the same chromosome of one species pair with parts of two different chromosomes of the other; at the metaphase of the meiotic division, aggregates of several chromosomes are present, suggesting circle or chain formation. The silk-worm (*Bombyx mori*), with twenty-eight chromosomes haploid, differs from the related species *Bombyx mandarina*, which has twenty-seven chromosomes. In the hybrid between them a trivalent is present, apparently due to the pairing of two chromosomes of *B. mori* with one chromosome of *B. mandarina* (Kawaguchi 1928). Similarly, the hybrid between *Vicia sativa* (six chromosomes haploid) and *V. amphicarpa* (five chromosomes) shows at meiosis either a trivalent, or an unequal bivalent and a free univalent (Sveshnikova and Belekhova 1936). Uniqual bivalents were also observed by Meurman (1928) in hybrid Ribes, and by Liljefors (1936) in wheat × rye hybrids.

Chromatin bridges between the disjoining halves of bivalents at the anaphase and telophase of the meiotic divisions in hybrids indicate, as we know, that the parents differ in having inverted sections in some of the chromosomes. Müntzing (1934) seems to have been the first to describe such chromatin bridges in an interspecific hybrid, *Crepis divaricata* × *C. dioscoridis*, and to recognize their significance. These two species of Crepis have four as the haploid number of chromosomes. At meiosis in the hybrid, from none to four bivalents appear; at the anaphase some chromatin bridges and fiberless chromosome fragments are found. The number of the latter is variable, ranging from none to five per cell, one being the average number. Similar observations were made by Müntzing (1935) in *Nicotiana bonariensis* × *N. langsdorfii*, by Richardson (1936) in *Lilium marta-gon* × *Lilium hansonii*, by Liljefors (1936) in *Triticum turgidum* × *Secale cereale*. The drawings of Sipkov (1936) representing meiosis in the *Triticum durum* × *Agropyrum glaucum* hybrid might also be interpreted as showing chromatin bridges and fragments resulting from crossing over between chromosomes containing inversions.

GENE ARRANGEMENT IN SPECIES OF DROSOPHILA

For years the gene arrangement in species of Drosophila has been studied by means of finding the linkage relationships between mutant genes and by constructing genetic chromosome maps. Despite the enormous amount of labor involved in the process, this method is by no means obsolete now, since it is in many ways more adaptable to the peculiarities of the material than the far simpler method of chromosome comparison in the salivary gland cells. Sturtevant and Plunkett (1926) and Sturtevant (1929) found that the arrange-ments of the loci of the mutant genes on the genetic maps of the chromosomes of *Drosophila melanogaster* and *D. simulans* are alike, except for a single long inversion in the third chromosome. Pätau (1935) and Kerkis (1936) have reëxamined the problem by observ-ing the pairing of the chromosomes in the salivary gland cells of the hybrids. Their results agree with Sturtevant's, and in addition a slight alteration of the gene seriation in the free end of the X chromosome has been detected. Although *D. melanogaster* and *D. simulans* are very similar externally, it is rather surprising to find two "good species" so little different in gene arrangement. Indeed, within both *D. melanogaster* and *D. simulans* laboratory strains may be obtained that are more different chromosomally than are the two species.

Crew and Lamy (1935), Donald (1936), and Sturtevant and Tan (1937) have compared the genetic maps of *D. melanogaster* with those of *D. pseudoobscura* and other species. The metaphase chromosome configuration in *D. melanogaster* is that shown in Fig. 11A and in *D. pseudoobscura* in Fig. 11J. The former species has a rod-shaped X chromosome and two V-shaped autosomes, and the latter has a V-shaped X and three rod-like autosomes. Each has a pair of dot-like autosomes. The mutants in one of the two limbs of the X of *D. pseudoobscura* proved similar to those in the X of *D. melanogaster*. The other limb of the X of *D. pseudoobscura* seems to contain the genes that lie in the left limb of the V-shaped third chromosome of *D. melanogaster*. The rod-shaped second chromosome of *D. pseudoobscura* is apparently homologous to the right limb of the V-shaped third of *D. melanogaster*, the third of *D. pseudoobscura* to the right limb of the second of *D. melanogaster*, the fourth of *D. pseudoobscura* to the left limb of the second of *D. melanogaster;* the dot-like autosomes seem to be homologous. If, however, one turns to the arrangement of the genes relative to each other in the homologous chromosomes in the two species, these arrangements are clearly very different. Sturtevant and Tan conclude that the evolution of the chromosomal apparatus in the section of the genus Drosophila to which *D. melanogaster* and *D. pseudoobscura* belong has proceeded mostly by inversion of chromosome segments, and that relatively few translocations have taken place. It must be noted that the establishment of the homology of the mutant genes in *D. melanogaster* and *D. pseudoobscura* (which is critical for the inference of homology of the chromosomes themselves) can be made only on the basis of the similarity of the external effects they produce; a direct test, by crossing the different mutants and observing their effects in the interspecific hybrid, is here impracticable, since *D. melanogaster* and *D. pseudoobscura* fail to cross. Despite this element of uncertainty, the results obtained by Sturtevant and Tan are so self-consistent, that their conclusions are convincing.

Observations on the arrangement of the stainable discs in the salivary gland chromosomes in different species are, as already indicated above, the easiest method of comparison of the gene arrangements in species of Drosophila that cross and produce at least mature hybrid larvae. Chromosome sections that carry homologous genes may be expected occasionally at least to pair in the hybrids. The sections thus seen paired are recorded, and by accumulating enough such records the structure of the chromosomes in the species under investigation can be gradually unraveled. This method has been used by Dobzhansky and Tan (1936) for a comparison of the gene arrangement in *D. pseudoobscura* and *D. miranda*, two species that resemble each other externally to about the same extent as *D. melano-*

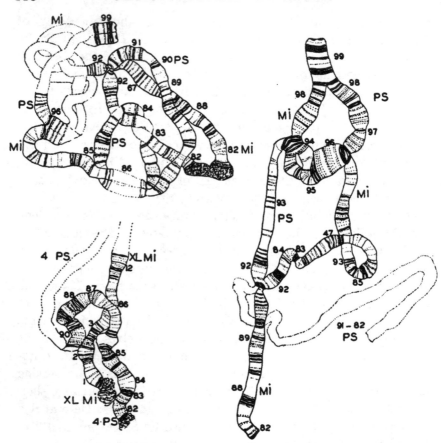

FIG. 12. Chromosome pairing in the salivary gland cells in the hybrid *Drosophila pseudoobscura* × *D. miranda*. Upper left, the fourth chromosomes; lower left, the fourth chromosome of *D. pseudoobscura* partly paired with the left limb of the X-chromosome of *D. miranda*; right, the fourth chromosomes. (After Dobzhansky and Tan.)

gaster resembles *D. simulans*. The metaphase chromosome groups of *D. pseudoobscura* and *D. miranda* females are identical. And yet the gene arrangement in the chromosomes of the two latter species proved to be most strikingly different.

The differences in the gene arrangement between *D. pseudoobscura* and *D. miranda* are so profound that the chromosomes either fail to pair in the salivary gland cells of the hybrid, or else form extremely complex pairing configurations, examples of which are shown in Fig. 12. Genes that in one species lie adjacent, in the other species may be far apart in the same chromosome. This indicates

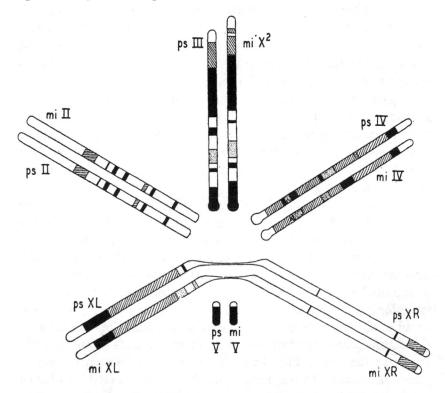

FIG. 13. A comparison of the gene arrangements in *Drosophila pseudoobscura* and *D. miranda*. Sections having the same gene arrangements in the two species are white; inverted sections, cross-hatched; translocations, stippled; sections of which the homologues are not detectable in the other species, black.

that inversions have taken place in the phylogenetic development of *D. pseudoobscura* and *D. miranda* from their putative common ancestor. Some genes which in one species are located in the same chromosome are borne in different chromosomes in the other. Such differences are apparently due to the occurrence of translocations. Finally, homologues of certain chromosome sections in *D. pseudoobscura* have not been detected at all in *D. miranda,* and vice versa.

The nature of such sections, which are present in one species only, is a rather perplexing problem. It is conceivable that the common ancestor has had more genes than either of its surviving descendants, and that losses of genic materials have taken place in evolution. A vastly more probable explanation is however that some chromosome sections have been so thoroughly rebuilt by repeated inversions and translocations that their disc patterns in the salivary gland chromosomes no longer resemble each other, and no pairing of the homologous genes takes place. This explanation is supported by some genetic evidence. The *D. pseudoobscura* strains carrying various recessive mutant genes have been outcrossed to wild-type *D. miranda,* and a supression of the mutants has invariably been observed in the hybrids. It follows that *D. miranda* possesses wild-type allelomorphs of all the mutant genes so tested.

A comparison of the chromosome structure in the two species under consideration is shown diagrammatically in Fig. 13. If the chromosome sections that are present in one species only are temporarily disregarded, one may conclude that the chromosomal differences between *D. pseudoobscura* and *D. miranda* are due chiefly to repeated inversions. The translocations are less numerous than inversions, and wherever the former do occur they involve only relatively small blocks of genes. It is easy to visualize the derivation of the chromosome structure observed in one species from that observed in the other. The chromosomes must be broken in several places, dismantled to fragments, and then reconstructed by placing the fragments in a new linear series. Evidently, the dismemberment need not be thought of as due to a sort of a catastrophic explosion, but rather to a gradual rebuilding by inversions and translocations. Dobzhansky and Tan have estimated that a minimum of forty-nine breakages is necessary for such a process. The sections whose homologues are not identifiable in the other species suggest however that this minimum estimate falls short of the actual number, which can be surmised to be at least twice as large as this minimum.

The different chromosomes have been rebuilt to a different extent in *Drosophila pseudoobscura* and *D. miranda.* The right limbs of the X and the second chromosomes are relatively similar. On the contrary, the third and the fifth chromosomes are very thoroughly altered (Fig. 13). It is interesting that the third chromosome is the one that is most variable in the gene arrangement also within the species *D. pseudoobscura*. The overlapping inversion method of tracing the phylogeny of chromosome structures (see above) is applicable to *D. pseudoobscura* and *D. miranda*. "Race A" and "race B" of the former species differ in five inversions in four chromosomes (Tan 1935). The chromosome structure in *D. miranda* is widely distinct from either race of *D. pseudoobscura*. Yet, a careful comparison shows that the gene sequence of *D. miranda* is less different from "race A" than from "race B" of *D. pseudoobscura*. The evidence derived from a consideration of different chromosomes is consistent throughout: it takes one less inversion step to derive the *D. miranda* arrangement from "race A" than from "race B." The possibilities that may be opened by the application of the overlapping inversion method, when the chromosome structure in other species of Drosophila becomes better known, are difficult to gauge at present.

No more than preliminary exploration work has been done in comparing the salivary gland chromosomes in other pairs of Drosophila species. The work of Pätau and Kerkis on *D. melanogaster* and *D. simulans* has been referred to already. *D. azteca* and *D. athabasca*, two species that are very similar externally but that produce sterile F_1 hybrids, have been examined by Bauer and Dobzhansky (unpublished). They differ from each other in the gene arrangement even more than do *D. pseudoobscura* and *D. miranda*, and no similarity whatever seems to be detectable between the chromosomes of the first and the second pair of species, although they are not too widely remote systematically. A like negative result has been met with in an attempt to compare *D. melanogaster* with *D. pseudoobscura*: no chromosome sections seem to be recognizably alike (Dobzhansky and Tan 1936). This comparison has been made not in a hybrid between these species, for they do not cross, but in pure species themselves. There are some objections against such comparisons, but the negative result obtained is not contradictory to those of Sturtevant and Tan (1937) who homologize the chromosomes of *D. pseudoobscura* with those of *D. melanogaster*. If inversions are the main factor in the chromosome differentiation in species of Drosophila, the chromosomes of two species may carry identical genes and

yet the order of the latter may be so greatly changed that cytologically the chromosomes appear totally dissimilar.

It has been shown that individuals, races, and species may differ
in chromosome structure. The differences are sometimes small and
sometimes almost bewilderingly great, but everywhere they are explicable on the supposition that inversions, translocations, and probably also deficiencies and duplications occur in the phylogeny. It is
undeniable that recent investigations have bridged the gap between
the chromosomal aberrations observed in the laboratory and the
variations of the chromosome structure found in the wild state. It
looks as though in some organisms, e.g., in Datura, the interspecific
differences in the chromosomes may be not appreciably greater than
the interracial ones, and therefore no correlation between the systematic rank and the degree of chromosomal differentiation obtains.
This conclusion can not be securely established before the technical difficulties in the detection of all classes of chromosomal changes
are surmounted. Be that as it may, there can be little doubt that
chromosomal changes are one of the mainsprings of evolution.

It may appear paradoxical that inversions and translocations are
so important as evolutionary agents. Indeed, they change only the
gene order but not the quantity or quality of the genes, and therefore need not produce any effects on the phenotype. Admittedly, the
precise rôle of inversions and translocations in evolution is for the
time being not sufficiently understood. We shall attempt however to
show in Chapter IX that they are effective in the genesis of some
forms of hybrid sterility, the evolutionary importance of which is
beyond doubt. Here it may be pointed out that the discovery of the
position effects on genes shows that phenotypical effects may be produced not only by gene mutations and by reduplications and losses
of genes, but also by alterations in the location of the genes within
a chromosome.

Classical genetics conceives the hereditary materials, the germ
plasm, as absolutely discontinuous. The germ plasm is a sum total
of discrete particles, the genes. The genes are pictured as independent of each other, both in inheritance and in evolution. Each of

them can undergo mutational changes, or can separate from its neighbors by crossing over and by chromosome breakage, without affecting the adjacent genes. A chromosome is, then, a sort of sausage stuffed with a definite number of layers of genes, arranged in a definite but fortuitous linear order. A mere rearrangement of the layers within the sausage need not have any effects on the properties either of the layers or of their container.

The discontinuous or particulate nature of the germ plasm is of course beyond doubt. It is a corollary to Mendel's laws, and the existence of the elementary hereditary particles, genes, is as well established as that of molecules or atoms. A rapidly growing amount of evidence indicates, however, that the genes are not quite so impregnable and impervious to the influence of their neighbors as has been thought. The effect of a gene on development is a function of its own structure as well as of its position in the chromosome. A change of the linear order of the genes in a chromosome may then leave the quantity of the gene unaffected, and yet the functioning of the genes may be changed. Only a brief summary of the evidence on which the viewpoint just stated is based can be given here. For more detailed reviews of the present status of this problem, see Dobzhansky 1936c, d.

The mutant gene Bar in *Drosophila melanogaster* produces rather frequently "mutations" to a more extreme allelomorph, known as double-Bar, or to the wild type. Sturtevant (1925) has shown that these "mutations" are in fact due to a reduplication or a loss of a certain gene or of a group of genes, and Bridges (1936) and Muller, Prokofyeva, and Kossikov (1936) have recently shown that the original mutation from the wild type to Bar was also a duplication for a short chromosome section, presumably containing several genes. A wild-type chromosome consequently has these genes represented once, Bar twice, and double-Bar three times. In Bar and double-Bar the homologous sections in the chromosome follow each other in tandem. Bar and double-Bar decrease the number of ommatidia, and consequently the eye size as compared to the wild type. As mentioned above, double-Bar is more extreme than Bar. Sturtevant (1925) compared the number of ommatidia in flies homozygous for Bar with that in flies heterozygous for double-Bar and wild type, and found that the former have larger eyes than the latter. Now a fly homozygous for Bar has a total of four Bar genes, two in each chromosome; a fly heterozygous for double-Bar has also four Bar genes, three in the double-Bar chromosome and one in the wild-type chromosome. The two kinds of flies have the same number of Bar genes and differ only in the position of the latter in the chromosomes. Hence the fact that their phenotypes are not identical can be due only to a position effect. Bar genes lying in the same chromosome reinforce each other's action more than the same number of Bar genes when located in different chromosomes.

A study of the translocations, inversions, and other aberrations involving a rearrangement of genes in the chromosomes of Drosophila has shown that they often produce unexpected effects. Translocations and inversions are frequently lethal when homozygous, or else they may produce visible changes in the morphology of the flies (Bridges, Muller, Dobzhansky, Patterson, and others). The behavior of translocations and inversions is often such as would be expected if their origin were accompanied by mutation in some genes. The remarkable fact is that these apparent "mutations" occur in the immediate vicinity of the places where the chromosomes were broken or reattached in the process leading to the production of the respective translocations or inversions. For example, Dobzhansky has observed a "mutation" from the wild type to an allelomorph of Bar, and this "mutation" arose simultaneously with a breakage of the chromosome near the Bar locus. It is now believed that in cases of this sort we are dealing not with gene mutations at all, that is not with changes in the structure of the genes themselves, but with position effects. Suppose the chromosomes $ABCD$ and $EFGHI$ are involved in a translocation which gives rise to the chromosomes $ABFE$ and $DCGHI$ (Fig. 4). The genes B and C, and F and G, were adjacent in the original chromosomes but in the new chromosomes they are far apart. The genes B and F, and C and G, were not adjacent in the original chromosomes, but they are in the new ones. If the functioning of a gene is influenced by its neighbors, the removal of the old and the substitution of the new neighbors may induce position effects, and the gene will appear changed, as though its own structure has been altered by a mutation.

There is no simple criterion to decide which of the changes associated with chromosome reconstructions are due to real mutations and which to position effects. That some of them are due to the latter cause has been proved in a series of ingenious experiments of Dubinin and Sidorov (1935). In a translocation in *Drosophila melanogaster* one of the genes lying in the chromosome near the breakage point appeared changed. By crossing over, a portion of a normal chromosome was substituted for the translocation chromosome containing the seemingly changed gene. A fresh gene so introduced into the broken chromosome immediately assumed the properties of the gene which was there originally.

Next to Drosophila, *Zea mays* is the organism in which the greatest number of translocations and inversions has been studied. Curiously enough, no position effects have been thus far recorded in Zea, contrasting with rather numerous instances in Drosophila. No satisfactory explanation of this has been given, although position effects that are lethal are obviously difficult to detect in a plant, where they are eliminated in the haploid generation. To what extent the differences between such species as *Drosophila pseudoobscura* and *D. miranda* are due to position effects is also a matter of speculation; the greatly different gene arrangements in these species may be responsible for many alterations in the morphological and physiological properties of their carriers. In any event, position effects show that gene mutations and chromosomal changes are not necessarily as fundamentally distinct phenomena as they at first appear.

V: VARIATION IN NATURAL POPULATIONS

PREMISES

AS POINTED OUT by Darwin, any coherent attempt to understand the mechanisms of evolution must start with an investigation of the sources of hereditary variation. Toward the accumulation of evidence bearing on this problem Darwin bent his principal efforts. He was able to satisfy himself that hereditary variations are always present, in wild as well as in domesticated species, somewhat less abundantly in the former, more on the average in the latter. But the mode of their origin, and the cause of their appearance remained obscure to Darwin, and he was not afraid to confess his ignorance on this point. The conjecture that the increase of the food supply causes an accentuation of the variability in the domesticated forms was a purely provisional hypothesis, the inadequacy of which soon became apparent, prompting Darwin to fall back on the Lamarckian principle of direct adaptation.

A solution, though a partial one only, of the problem of the origin of hereditary variation has been arrived at in the present century, due to the application of the methods of modern genetics. It is now clear that gene mutations and structural and numerical chromosome changes are the principal sources of variation. Studies of these phenomena have been of necessity confined mainly to the laboratory and to organisms that are satisfactory as laboratory objects. Nevertheless, there can be no reasonable doubt that the same agencies have supplied the materials for the actual historical process of evolution. This is attested by the fact that the organic diversity existing in nature, the differences between individuals, races, and species, are experimentally resolvable into genic and chromosomal elements, which resemble in all respects the mutations and the chromosomal changes that arise in the laboratory. It may perhaps be objected that despite this resemblance, the mode of origin of laboratory and natural variations may be different. There is nothing to be said against such a criticism, except that an unnecessary multiplication of unknowns is contrary to accepted scientific procedure.

There is no contradiction between the foregoing statements and the acknowledgment which must be made of our ignorance of the exact nature of mutational and chromosomal changes. For despite the recent successes in inducing these changes by X-rays and perhaps by other agents, the nature of the difference between an ancestral and a mutated gene and the mechanics of the production of such differences and of chromosomal aberrations remain obscure. In a sense we are, then, in the same position in which Darwin was: the intimate nature of the hereditary variation is still unknown. But in another respect we are in a much better position compared with Darwin, since at least some of the attributes of the mutation process in the wide sense of that term are no longer a mystery.

The origin of hereditary variations is, however, only a part of the mechanism of evolution. If we possessed a complete knowledge of the physiological causes producing gene mutations and chromosomal changes, as well as a knowledge of the rates with which these changes arise, there would still remain much to be learned about evolution. These variations may be compared with building materials, but the presence of an unlimited supply of materials does not in itself give assurance that a building is going to be constructed. The impact of mutation tends to increase variability. Mutations and chromosomal changes are constantly arising at a finite rate, presumably in all organisms. But in nature we do not find a single greatly variable population of living beings which becomes more and more variable as time goes on; instead, the organic world is segregated into more than a million separate species, each of which possesses its own limited supply of variability which it does not share with the others. A change of the species from one state to the other, or a differentiation of a single variable population into separate ones, the origin of species in the strict sense of the word, constitutes a problem which is logically distinct from that of the origin of hereditary variation.

The origin of variation is a purely physiological, and in the last analysis physico-chemical, problem which is at the root of any theory that tries to account for the origin of species. Nevertheless, when the hereditary variation is already produced and injected into the population, it enters into the field of action of factors which are on a different level from those producing the mutations. These factors, natural and artificial selection, the manner of breeding characteristic for the particular organism, its relation to the secular environment and to other organisms coexisting in the same medium, are ultimately, physiological, physical, and chemical, and yet their interactions obey rules *sui generis,* rules of the physiology of populations, not those of the physiology of individuals. The former are determined by the latter only in the same sense in which the structure of a human state may be said to depend upon the physiology of its members.

No end of misconceptions are caused by the failure to grasp the significance of this distinction. The preposterous accusation that genetics denies the importance of the environment in evolution belongs here. Genetics does assert that there is no evidence of a direct adaptation in the process of mutation, that the organism is not endowed with a providential ability to respond to the requirements of the environment by producing hereditary changes consonant with these requirements. At the individual level, the changes produced are determined primarily by the structure of the organism itself, which is of course the result of a historical process in which the environment has played a part. But the historical process (cf. Dubinin 1931), the molding of the hereditary variation into racial, specific, generic, and other complexes, is due to action of the environment through natural selection and other channels to be discussed below.

It must be stated at once that the physiology of populations is for the time being the most neglected, although very probably the most essential, part of the theory of evolution. The processes which occur in free living populations after the mutations and chromosomal changes have been produced and have become an integral part of the population genotype remained totally obscure for a long time. Only in recent years, a number of investigators, among whom Professor Sewall Wright of Chicago should be mentioned most prominently, have undertaken a mathematical analysis of these processes, deducing their regularities from the known properties of the Mendelian mechanism of inheritance. The experimental work that should test these mathematical deductions is still in the future, and the data that are necessary for the determination of even the most important constants in this field are wholly lacking. Nonetheless, the results of the mathematical work are highly important, since they have helped to state clearly the problems that must be attacked experimentally if progress is to be made.

THE PARTICULATE THEORY OF HEREDITY

The most fundamental difference between the conception of heredity which Darwin had to rely upon and the one which is at the basis of all the modern views on the mechanisms of evolution can be characterized as the antithesis between the particulate and the blending theories of inheritance (Tschetwerikoff 1926, Fisher 1930). It is known now that the germ plasm consists of a finite number of discrete particles, the genes, which maintain their properties in the process of hereditary transmission. According to the first law of Mendel, the crossing of purple peas with white peas results in the reappearance in the F_2 generation of purple and white flowered plants of the same shades as their ancestors. The difference between the genes for purple and for white flowers is not encroached upon in any way by the hybridization; no pink or cream-colored flowers, for example, are produced. Even where the hybrid, or heterozygote, is intermediate between the parents, no contamination of the genes takes place, and the homozygotes recovered from the hybrids are like the parental races. The discovery of the position effects has shown that genes are not quite as discrete and separate from each other as they were formerly supposed to be, but this does not impinge on the thesis that the differences between genes are not swamped by hybridization.

The concept of blending inheritance, which was held universally in Darwin's time and which still persists in some recesses of biology, was based on the assumption that the germ plasms of the parents undergo a sort of amalgamation in the hybrid. The difference between the ancestral germ plasms was supposed to be either lost entirely or at least impaired by passage through the hybrid organism. As an analogy, the theory of blending inheritance may be said to assume that germ plasms can mix as a water-soluble dye mixes with water. Interbreeding of the genotypically distinct forms causes, then, an eventually uniform dissolution of the dye in the germ plasm of the resulting hybrid population. The particulate inheritance theory assumes, by contrast, that genes are more like blocks of solid material; these blocks are placed side by side in the hybrid, but they separate uncontaminated in the processes of germ cell formation.

The corollaries of the two theories are strikingly different. If the germ plasms can combine as a dye combines with water, then the amount of variation present in a sexually reproducing random breeding population must be halved in every generation. Given a population which exhibits a large variability at the start, we are bound to observe a progressive, rapid, and irretrievable decay of the variability, until a complete homogeneity is reached. The only escape from this conclusion is to suppose that the variability is constantly being developed *de novo*, at a rate which is at least equal to the rate of its loss due to crossing. This is exactly the conclusion to which Darwin and his immediate followers were inexorably driven by their initial assumption of blending inheritance. In modern language this would mean that new mutations must occur with a prodigious frequency, far in excess of anything ever observed in any experiment.

No such difficulty is encountered if the germ plasm is particulate, for in this case the variability is automatically maintained on an approximately constant level despite the interbreeding, although it will be shown below that a relatively slow decay of the variability is expected even with a particulate inheritance. Very much lower mutation rates will suffice to increase the variability and to furnish the materials for evolution, since the maintenance of the variability requires practically no fresh supply of mutational changes. This deduction is applicable both to genic variation and to that in the chromosome structure. Only in the obscure realm of cytoplasmic inheritance a situation approaching blending may obtain, although too little is known about it to make any conclusion secure. With the obsolescence of the blending inheritance theory, one of the greatest impediments to the progress of evolutionary thought was removed.

A distinction must obviously be drawn between the assumption of the blending of hereditary factors and the assumption that discrete groups of forms differing in many genes will lose their discreteness if allowed to interbreed freely. Owing to the free assortment of genes (the second law of Mendel), an interbreeding of two such groups will result in a single population in which all the possible recombinations of the ancestral traits will be present in separate individuals. The amount of variation is not reduced but actually increased by crossing, but the difference between the formerly discrete groups as groups is obliterated (cf. Chapter VIII).

HARDY'S FORMULA AND THE GENETIC EQUILIBRIUM

Imagine that equal numbers of homozygous individuals differing in a single gene (*AA* and *aa*) are brought together in some sort of unoccupied territory (an island, for example), and are left there to breed for an indefinite number of generations without the admission of fresh immigrants. Provided the mating is random, that is to say, individuals of the genotype *AA* are equally likely to mate with *AA* or with *aa* individuals, then in the next generation one-half of the population will consist of heterozygotes (*Aa*), and each of the two homozygotes will be represented by one-quarter of the population. The frequencies of the different genotypes in the population will be, hence, $1AA : 2Aa : 1aa$. If none of the three genotypes has any advantage with respect to the survival over the other two, and the mating continues to be random, the same ratio $1AA : 2Aa : 1aa$ will be repeated in the second and in all following generations.

Hardy (1908) has shown that the relative frequencies of various genes in a population remain constant irrespective of the absolute values of their initial frequencies. If the *AA* and *aa* individuals are mixed in proportions q and $1-q$ respectively, the population in the second and all following generations will be:

$$q^2 \text{ AA} : 2q(1-q) \text{ Aa} : (1-q)^2 \text{ aa}$$

The formula describes the equilibrium condition in a sexually reproducing random-breeding population where the component genotypes are equivalent with respect to selection. If breeding is not random, the relative frequencies of the homozygotes and heterozygotes will be affected, but the gene frequencies $qA : (1-q)a$ will remain constant. For instance, if self-fertilization takes place, or if the carriers of each genotype exhibit a preference toward mating with their like, the relative frequencies of the homozygotes *AA* and *aa* will increase, and the frequency of the heterozygotes (Aa) will decrease. A selection favoring the survival and breeding of some genotypes over others will, on the other hand, cause alterations in the values of q and $1-q$. The direction, extent, and speed with which the alterations will take place are dependent upon the intensity and the manner of action of selection. These problems have been treated mathematically by Haldane (1924-1932), Fisher (1930), and others.

Hardy's formula is important because it shows that the hereditary variability once gained by a population is automatically maintained on a constant level, instead of being eroded and finally leveled off by crossing. As mentioned already, this is a corollary to the particulate as contrasted with the blending type of inheritance. The maintenance of the genetic equilibrium is evidently a conservative and not a progressive factor. Evolution is essentially a modification of this equilibrium. We shall proceed now to show that agents that tend to modify the equilibrium actually exist in nature. A significant fact is that each of such agents is counteracted by another opposite in sign, which tends to restore the equilibrium. A living population is constantly under the stress of the opposing forces; evolution results when one group of them is temporarily gaining the upper hand over the other group.

MUTATION AND THE GENETIC EQUILIBRIUM

The value of q in Hardy's formula, e.g., the frequency of a gene or a chromosome structure in a population, can be modified by mutation pressure in the wide sense of the term, that is by gene mutations and chromosomal changes. If the change from the gene or the chromosome structure A to the state a takes place at a finite rate, the frequencies q and $1-q$ must change accordingly. Let the mutation in the direction $A \to a$ have a rate equal to u; the change in the frequency of A in the population will be $\Delta q = -uq$, where q is the frequency of A. If the mutation in the direction $A \to a$ is unopposed by any other factor, the population will eventually reach homozygosis for a.

Wherever the mutation is reversible, the change in the direction $A \to a$ is opposed by the change $a \to A$. With the rate of reverse mutation equal to v, the frequency of A will change as $\Delta q = -uq$

$+ v(1-q)$. An equilibrium will be reached obviously when the change $\Delta q = 0$. The equilibrium value of q determined by the two mutually opposed mutation rates is therefore $q = v / u + v$. Taking, for example, the rate of the mutation $A \to a$ to be equal to one in a million gametes per generation ($u = 0.000001$), and the rate of the mutation $a \to A$ equal $v = 0.0000005$, the equilibrium value for q will be 0.33, which means that 33% of the chromosomes will carry the gene A and 67% the gene a. If the mutation rates to and from a given gene are alike ($v = u$), the equilibrium values for both q and $1-q$ are evidently 0.5, or equal. Starting from an initially homogenous population, the mutation pressure will tend to increase the hereditary variability until the equilibrium values determined by the opposing mutation rates are reached for every gene. In an indefinitely large population this level of variability will be preserved (as shown by Hardy's formula) until some agent disturbs it.

It is important to realize that an increase of variability through mutation will take place even in case the mutational changes are unfavorable to the organism, and thus opposed by natural selection. This is especially true when the mutational changes are recessive to the ancestral condition, which is actually the case for a majority of mutations arising in many species. If the heterozygote Aa (a being a mutant gene decreasing the viability) is as viable as the ancestral homozygote AA, the frequency of the gene a will be allowed to increase until the Aa individuals become so frequent in the population that their mating together is likely to take place, and the homozygotes aa are then produced. The aa condition being unfavorable, the aa individuals will be eliminated, and this will impose a check on the further spread of the mutant gene a in the population.

That this is not only a theoretical possibility but an actual process going on in nature is demonstrated by the results of the genetic analyses of wild Drosophila populations (Chapters III and IV). Despite their external uniformity, the free living populations of. Drosophila carry a great mass of chromosomal variations as well as of recessive mutant genes. While the former, at least in the *Drosophila pseudoobscura* strains studied by Sturtevant and Dobzhansky, seem to be indifferent for viability, many of the latter are lethal when homozygous. Among those mutants that are not outright lethals, a majority, at least the more striking ones, are clearly unfavorable. Their unfavorable effects are demonstrable under laboratory conditions, and the rarity in nature of the individuals homozygous for the mutant genes justifies the inference that they are unfavorable in the wild state as well. And yet the species is, according to the succinct metaphor of Tschetwerikoff (1926), "like a sponge," absorbing both the mutations and the chromosomal changes that arise and gradually accumulating a great store of variability, mostly concealed in heterozygous condition.

It is not an easy matter to evaluate the significance of the accumulation of germinal changes in the population genotypes. Judged superficially, a progressive saturation of the germ plasm of a species with mutant genes a majority of which are deleterious in their effects is a destructive process, a sort of deterioration of the genotype which threatens the very existence of the species and can finally lead only to its extinction. The eugenical Jeremiahs keep constantly before our eyes the nightmare of human populations accumulating recessive genes that produce pathological effects when homozygous. These prophets of doom seem to be unaware of the fact that wild species in the state of nature fare in this respect no better than man does with all the artificiality of his surroundings, and yet life has not come to an end on this planet. The eschatological cries proclaiming the failure of natural selection to operate in human populations have more to do with political beliefs than with scientific findings.

Looked at from another angle, the accumulation of germinal changes in the population genotypes is, in the long run, a necessity if the species is to preserve its evolutionary plasticity. The process of adaptation can be understood only as a continuous series of conflicts between the organism and its environment. The environment is in a state of a constant flux, and its changes, whether slow or catastrophic, make the genotypes of the past generations no longer fit for survival. The ensuing contradiction can be resolved either through the extinction of the species, or through a genotypical reorganization. A genotypical change means, however, the occurrence of a mutation or of mutations. But nature has not been kind enough to endow the organism with ability to react purposefully to the needs of the changing environment by producing only beneficial mutations where and when needed. Mutations are random changes. Hence the necessity for the species to possess at all times a store of concealed, potential, variability. This store will presumably contain variants which under no conditions will be useful, other variants which might be useful under a set of circumstances which may never be realized in practice, and still other variants which were neutral or harmful at the time when they were produced but which will prove useful later on.

It has already been pointed out (Chapter II) that mutational changes that are unfavorable under a given set of conditions may be desirable in a changed environment. Mutations that decrease viability when taken separately may have the opposite effect when combined. It should be kept in mind that selection deals not with separate mutations and separate genes, but with gene constellations, genotypes, and the phenotypes produced by them (Wright 1931a, 1932). A species perfectly adapted to its environment may be destroyed by a change in the latter if no hereditary variability is available in the hour of need. Evolutionary plasticity can be purchased only at the ruthlessly dear price of continuously sacrificing some individuals to death from unfavorable mutations. Bemoaning this imperfection of nature has, however, no place in a scientific treatment of this subject.

SCATTERING OF THE VARIABILITY

The supply of hereditary variation is constantly being augmented by the occurrence of mutations and chromosomal changes. In an ideal infinitely large population which is unaffected by selection and by changes in the environment, the accumulation of variability would ultimately result in every gene's being represented by several allelomorphs, and every chromosome by several structural variants. The equilibrium frequency for each allelomorph or variant would be determined solely by the mutation rates, according to the formula $q = v / u + v$ derived above. Ideal populations of this sort actually do not exist in nature. If they did exist, evolution would come to a standstill as soon as the genetic equillibria for every gene and chromosome structure were reached. In reality, natural selection tends to purge the population genotype of all the unfavorable variants. For an understanding of the dynamics of free living populations it is essential, however, not to disregard the existence of another agent besides selection, which tends to diminish the supply of hereditary variation irrespective of the adaptive value of the latter, and thus to counteract the mutation pressure. This agent is the scattering of the variability in the process of reproduction. Its significance has been almost entirely overlooked by most evolutionists.

For the sake of clarity, it is best to start the discussion of the scattering of the variability by examining the conditions that obtain in a certain ideal population. Following Dubinin and Romaschoff (1932), let us imagine a population which remains numerically stationary from generation to generation, and in which every pair of parents produce only two offspring, all of which survive to maturity and replace the parents before the next breeding season begins. A mutation from the gene A to a produces a single individual Aa. This individual must mate to a normal one, $Aa \times AA$. The offspring expected from this cross is $1Aa : 1AA$, but owing to chance, the two individuals actually produced will in 25 per cent of the cases both be $AA;$ in another 25 per cent, both Aa; and in 50 per cent, one AA and the other Aa. According to Hardy's formula, the frequency in the population of the newly arisen gene a is expected to remain constant. But actually this is not necessarily the case because in 25 per cent of cases no Aa individual is present in the next generation, and hence the mutant gene a is irretrievably lost. If no such loss takes place in the first generation, the mutation is again exposed to the risk of being extinguished in the next and in following generations. The same mechanism that causes the loss of some mutants produces a doubling of the frequency of others. But since the loss of a mutation is an irreversible process, the conclusion follows that some mutants, both harmful and beneficial ones, never become established in the population, and are lost to the species.

The same process of loss of the hereditary variability may be demonstrated in another way. Imagine a popualtion in which every gene is represented only once, that is to say, every individual carries two different allelomorphs neither of which is present in any other individual in the same population. If the breeding system is the same as in the above example (i.e., only two offspring are produced by each pair of parents), 25 per cent of the variety of allelomorphs will be lost in the very next generation, and 25 per cent of the allelomorphs will be each represented twice, hence doubled in frequncy. The loss of some allelomorphs and the increase of the frequency of others will go on in further generations as well, until the population will ultimately become homozygous for one of the allelomorphs, all others being lost.

As a model of such a population, Dubinin and Romaschoff (1932) used a bowl containing 100 marbles, each marble with a different number. To imitate the results which the process of reproduction would entail in such a population, 25 marbles taken at random were withdrawn from the bowl and discarded. Another set of 25 marbles was taken out, and in place of each of these two new marbles carrying the same numbers were returned to the bowl, so that the total number of marbles (100) was restored. This operation represented one generation of breeding, and resulted in elimination of 25 "allelomorphs," and in doubling the frequency of 25 others. Repeating the same operation several times, Dubinin and Romaschoff observed a progressive decay of the variability in the "population." Fewer and fewer different numbers remained in the bowl, although some of those which were left became represented by many marbles each. Finally, all the marbles had the same number, that is, the "population" had become homogeneous. In ten separate experiments of this sort Dubinin and Romaschoff observed the attainment of a complete "homozygosis" after from 108 to 465 "generations," the victorious allelomorph each time having, as might be expected, a different number.

These simple experiments illustrate the idea, first advanced apparently by Hagedoorn and Hagedoorn (1921), and later elaborated largely independently by Fisher (1928, 1930, 1931), Dubinin (1931), Romaschoff (1931), Dubinin and Romaschoff (1932), and especially by Wright (1921, 1930, 1931a, b, 1932, 1934, 1935). Hardy's formula is strictly applicable only to ideal infinitely large populations, and only to gene frequencies q and $1-q$ that are not too close to zero or to unity. In reality, no living population consists of an infinitely large number of random breeding individuals, and gene frequencies may be very small and very large. Hardy's formula describes, therefore, only an ideal situation, and the gene frequencies in finite populations will vary at random. But as soon as these random variations reach the values $q = 0$ or $q = 1$, a certain gene allelomorph is lost from the population genotype. A finite population left to its own devices must, therefore, suffer a progressive decay of its hereditary variability and sooner or later must reach a complete

TABLE 12

PROBABILITY OF EXTINCTION AND OF SURVIVAL OF A MUTATION APPEARING IN A SINGLE INDIVIDUAL (*after Fisher*)

GENERATION	PROBABILITY OF EXTINCTION		PROBABILITY OF SURVIVAL	
	No Advantage	1% Advantage	No Advantage	1% Advantage
1	0.3,679	0.3,642	0.6,321	0.6,358
3	0.6,259	0.6,197	0.3,741	0.3,803
7	0.7,905	0.7,825	0.2,095	0.2,175
15	0.8,873	0.8,783	0.1,127	0.1,217
31	0.9,411	0.9,313	0.0,589	0.0,687
63	0.9,698	0.9,591	0.0,302	0.0,409
127	0.9,847	0.9,729	0.0,153	0.0,271
Limit	1.0,000	0.9,803	0.0,000	0.0,197

genetic uniformity. This, of course, disregards the occurrence of new mutations.

The too artificial models of populations devised by Dubinin and Romaschoff must be modified to conform more closely to the conditions that actually obtain in natural populations. First of all, the number of offspring produced by a pair of parents reproducing sexually is always larger than two. If the population size remains approximately constant from generation to generation, it follows that a major part of the offspring die before they reach the breeding stage themselves. The problem of the retention or loss of mutants which appear singly (and almost all mutants do appear in single individuals) in such populations has been treated mathematically by Fisher (1930) and Wright (1931a). The results are summarized by Fisher in the following table (Table 12).

If the mutational change confers no selective advantage on its carrier, only 153 out of 10,000 mutants avoid extinction after 127 generations of breeding. With a 1 per cent selective advantage, the favorable mutations will suffer much the same fate, but still 271 instead of 153 out of every 10,000 will be retained. Clearly, a majority of mutations turning up in natural populations are lost within a few generations after their origin, and this irrespective of whether they are neutral, harmful, or useful to the organism. The numerous mutations which persist are the "lucky" remainder which may be increased in frequency instead of being lost. And yet, we know from experiments that a majority of these mutations are decidedly unfavorable for the organism under the conditions of the environment in which the species is normally living.

Wright has made a most searching analysis of a more general aspect of the same problem. Hardy's ratio, $q^2 : 2q(1-q) : (1-q)^2$, describes only the average state of affairs in a population. If two allelomorphs of a gene, A and a, are present in the initial population in equal numbers ($q = 1-q = 0.5$), their frequencies in the following generations will fluctuate up and down, the average, of course, tending to remain close to 0.5. Wright shows that these fluctuations, which looked at superficially represent a disconnected series of haphazard events, may lead to most significant results. Indeed, so long as the values of q remain between 0 and 1, the fluctuations are essentially reversible, but as soon as the latter values are reached an irreversible change has taken place: one of the two allelomorphs is lost and the other has reached a state of fixation, that is, the population is now homozygous for it. Thus both fixation and loss of a gene in a population may occur without the participation of selection, due merely to the properties inherent in the mechanism of Mendelian inheritance. In fact, as shown below, these events may take place even in spite of natural selection; genes which produce slight adverse effects may reach fixation and more advantageous genes may be lost.

The deviations of the actual results observed in a living population from the constancy predicted by Hardy's formula are inversely correlated with the absolute number of breeding individuals in the populations, with its "population number" denoted by the symbol N. Wright (1931, pp. 110-111) defines the population number as follows: "The conception is that of two random samples of gametes, N sperms and N eggs, drawn from the total gametes produced by the generation in question. . . . Obviously N applies only to the breeding population and not to the total number of individuals of all ages. If the population fluctuates greatly, the effective N is much closer to the minimum number than to the maximum number. If there is a great difference between the number of mature males and females, it is closer to the smaller number than to the larger."

The greater the population size, the more closely the results of breeding will approach the constancy of the gene frequencies postulated by Hardy's formula. Conversely, the smaller the population size, the more rapid is the scattering of the variability and the eventual attainment of genetic uniformity. The great significance of the population number, N, comes from the fact, emphasized by Wright, that in a population of N breeding individuals, $1 / 2N$ genes either reach fixation or are lost in every generation. Suppose that in a population many genes are represented each by two allelomorphs which are equivalent with respect to selection, and that the initial frequency of each allelomorph is 50 per cent ($q = 0.5$). With no mutations taking place, the frequencies of the different allelomorphs will in the following generations fluctuate up and down from 50 per cent, some becoming more and some less frequent owing to chance. Sooner or later a condition will obtain when gene frequencies from 0 per cent to 100 per cent will become equally numerous, and $1/4N$ of the genes will reach fixation and $1/4N$ will be lost in every generation, as indicated above (Fig. 15). The progressive depletion of the supply of hereditary variability is inevitable. Wright's formula describing this process is

$$L_T = L_0 \, e^{-T/2N}$$

where L_0 and L_T are the numbers of the unfixed genes in the initial and in the T generation of breeding respectively, N is the population number, and e the base of natural logarithms.

Dubinin and Romaschoff (1932) have furnished an illustration of the significance of the population number by making their experiments with marbles in two different ways. In some experiments they used bowls containing one hundred marbles, and in others bowls with but ten marbles in each. In the former series of experiments it took many more "generations" to attain a complete uniformity of the "population" than in the latter (from 108 to 465 "generations" with one hundred marbles, and from 14 to 51 "generations" with ten marbles).

Mutation pressure and the scattering of the variability are two mutually opposed processes: the former tends to increase genetic diversity, and the latter uniformity in populations. For the sake of clearness, we were forced to consider the action of each of these processes separately, by studying the properties of abstract populations where one or the other process was assumed to be non-existent. In reality mutations do occur, presumably in all organisms, although probably at different rates. The interactions of the mutation pressure with the scattering of the variability present some rather involved mathematical problems which have been treated by Wright (1931a, 1932).

In very large populations the scattering of the variability is evidently ineffective, and the gene frequencies are determined, in the absence of selection, by the opposing mutation rates to and from a given allelomorph. Each allelomorph eventually reaches its equilibrium frequency, which is maintained constant with insignificant variations. Conversely, in very small populations the gene frequencies are largely independent of the mutation rates, unless the latter are very large. This means that one of the allelomorphs of each gene displaces, owing to mere chance, all others, and reaches the state of approximate fixation, practically the entire population being homozygous for it. In populations of intermediate size the greatest freedom obtains. Some gene allelomorphs are lost or reach fixation at random; others fluctuate in frequency over a wide range of values, the modal value approaching that determined by the mutation rates.

Wright considers the situation that may present itself in a species whose population is subdivided into numerous isolated colonies of different size, with the exchange of individuals between the colonies prevented by some natural barriers or other agents. As we shall attempt to show below, such a situation is by no means imaginary; on the contrary, it is very frequently encountered in nature. In the course of time such a species will become differentiated into numerous local races which will differ from each other in the frequencies of various genes. The populations of the separate colonies at the time of their isolation may be assumed to be all alike in their genetic constitution. Some colonies, especially the large ones, will preserve the constitution of the ancestral population approximately unchanged. Each of the colonies with very small breeding populations will soon become genetically uniform owing to the depletion of the store of the hereditary variability they once possessed. It is important to realize that in different colonies different genes will be lost and fixed, the loss or fixation being due, as we have seen, simply to chance. Hence, at least some of the colonies will become genetically distinct from others, giving rise to local races. Finally, colonies with populations of an intermediate size will likewise become distinct from each other and from the ancestral population, but the supply of the hereditary variability present in them will not be depleted so drastically as in the very small ones. That is to say, the colonies of intermediate size will preserve a certain amount of the evolutionary plasticity which will be lost in the small ones. Wright shows that the intermediate population size which permits a reasonably rapid differentiation of the local races without too much loss of variability is the one where the products $4Nu$ and $4Nv$ are close to unity. With mutation rates of the order 1 : 10,000 or 1 : 100,000, this means breeding populations of thousands or tens of thousands of individuals.

The conclusion arrived at is an important one: the differentiation of a species into local or other races may take place without the action of natural selection. A subdivision of the species into isolated populations, plus time to allow a sufficient number of generations to elapse (the number of generations being a function of the population size), is all that is necessary for race formation. This statement is not to be construed to imply a denial of the importance of selection. It means only that racial differentiation need not necessarily or in every case be due to the effects of selection. The rôles of selection and migration in conjunction with that of isolation will be discussed in the following chapter. Here it remains only to consider some facts that illustrate the genetic rôle of isolation as such.

Descriptive biological literature is replete with observations that might be used to illustrate the importance of isolation for racial differentiation. A few examples will suffice here. Species of the land snails, owing to their limited means of locomotion, are particularly apt to be subdivided into colonies the members of which seldom pass over the barriers that separate one colony from another. On the volcanic islands that are fairly numerous in the South Seas, conditions that are especially favorable for a strict isolation of

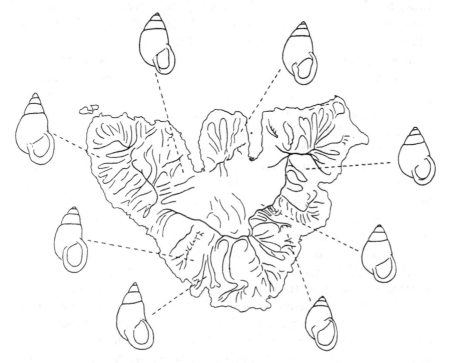

FIG. 14. Shape of the shell in *Partula taeniata* inhabiting different valleys of the island of Moorea. (After Crampton.)

colonies are sometimes created. Most of such islands have a high volcanic cone in the center, the steep slopes of which are deeply crevassed by radiating valleys that run to the sea and are separated by more or less narrow ridges. The snail species inhabiting the valleys, such as the representatives of the genus Achatinella on the Hawaiian islands studied by Gulick (1905) and of Partula on Tahiti and Moorea studied by Crampton (1916, 1932), show an extreme diversity of local races. Only species that have a very narrow distribution and are restricted to a single valley or a part of a valley are not subdivided into races. An example of such a species is *Partula tohiveana,* which is fairly common in an area about 350 meters wide and a few kilometers long in one of the valleys of the island of Moorea (Crampton 1932). In more widely distributed species, each group of adjacent valleys or each valley or even each part of a valley is inhabited by a race of its own, distinct from other races in a combination of characters such as color, size and shape of the shell, dextrality or sinistrality, and so forth (Fig. 14).

It seems impossible to establish any connection between the characteristics of the race inhabiting a given valley and the environment predominant in the valley. It is likewise impossible to ascribe any adaptive significance to the peculiarities distinguishing the races from one another. Both Gulick and Crampton are sceptical about the possibility that natural selection might be responsible for the formation of these local races. The difficulty of proving that a given trait has not and never could have had an adaptive significance is admittedly great; nevertheless, the facts at hand are explicable, without stretching any logical point, on the assumption that racial differentiation is due to mutations and to random variations of the gene frequencies in isolated populations. Crampton (1916, p. 247) emphasizes that the average dimensions and proportions of the shells may vary from valley to valley without any consistent relation either to the geographical sequence or to the peculiarities of the environment in the latter. Adjacent valleys may harbor populations that are more different in some characters than populations from remote valleys (Fig. 14). Different species co-inhabiting the same territory may vary in different directions. Thus, the shells of *Partula taeniata* in the northwestern part of the island of Moorea are smaller and stouter, and those of *Partula suturalis* are longer and more slender than in other parts of the island, and yet the two species are found on the same food plant and to all appearances are ecologically identical. *Partula mooreana* and *Partula suturalis* have each a giant race, but in the two species the giant races are found in quite different valleys, and other species in the same valleys show no tendency toward gigantism (Crampton 1932). Clearly, it would be fantastic to imagine that each combination of the alternative traits distinguishing the races of *Pachycephala pectoralis* (Table 10) is best adapted to the environment existing on the particular island they inhabit.

Kinsey (1936, 1937) reports some interesting comparisons between the "insular" and the "continental" species of the gall wasps, Cynipidae. Each species of these gall wasps is attached to a single food plant, mostly a species of oak. On the plains and hills of the eastern United States the same kind of oak may grow without major interruptions for many thousands of square miles, and frequently the same species of gall wasp occurs on it throughout. Such species Kinsey calls continental. By contrast, in the southwestern United States and in Mexico, oak forests are confined to high elevations and consequently occur mainly on mountain ranges. These island forests are separated by great expanses of arid desert lands which support no population of Cynipidae of any kind. Many species of gall wasps inhabit a single forest or a group of adjacent island forests, and are designated insular. The racial or specific diversity among the gall wasps is characterized by a predominance of the insular (225) over the continental (64) forms. The average distribution area in continental species is evidently much larger (about 300,000 square miles) than in the insular (4,600 square miles) ones. The population size in a continental species is moreover a hundred times greater on the average than in an insular species. Kinsey classifies his species into very constant, fairly constant, variable, highly variable, and most extreme. The continental and insular species show a suggestive difference in their variability. Among the continental, 74 per cent are variable, 13 per cent are fairly constant and 13 per cent are highly variable. Among the insular, a majority (82 per cent) are very constant, 12 per cent fairly constant, 4 per cent variable, and 2 per cent extremely variable. The very constant and the extremely variable forms are encountered, therefore, among the insular but not among the continental species. In large populations a moderate variability may be expected to be the rule, and in small ones any condition, from a great uniformity to a great variability, may occur.

Pronounced variations in the genetic composition of the same population from generation to generation have been observed in *Drosophila melanogaster* by Dubinin and his collaborators (1934).

As shown in Table 6, the population from Mashuk in 1931 contained 33.1 per cent of autosomal mutant genes producing visible effects, and in 1932 only 8.8 per cent. The Erivan population in 1931 had 15.6 per cent and in 1932 only 5.6 per cent of autosomal lethals. Although the probable errors of these frequencies are rather large, the differences seem statistically significant. The methods of taking samples of the populations are said to have been alike in both years. Here, then, we have a change which is difficult to account for in any other way than by assuming the random variations of the gene frequencies postulated by Wright and others. *Drosophila melanogaster* populations are fairly large during the summer months, but in winter they are probably reduced to a small fraction of their former size. The minimum effective population size is consequently likely to be small enough to permit large fluctuations of the gene frequencies. Admittedly more data are necessary to establish these conclusions securely.

A peculiar feature of the process of the racial differentiation through a scattering of the variability is its apparent indeterminacy. Given a species segregated in colonies with a definite population size, it may be possible to calculate, with the aid of Wright's formulae, the average expected result in the whole collection of such colonies, but not in the individual ones.

POPULATION SIZE

The foregoing discussion shows how important is the effective size of the breeding population of a species for its evolutionary perspectives. If population sizes in most species tend to be small on the average, the scattering of the variability and the random variations of the gene frequencies will loom large as evolutionary agents. If, on the other hand, the population sizes are usually so large that they may be regarded for all practical purposes as infinite, the evolutionary role of these agents is negligible. It is no exaggeration to say that the conclusions which eventually may be reached on the dynamics of the evolutionary process will depend in no small degree on the information bearing on the problem of the population numbers. The dearth of pertinent data in the existing biological literature is, therefore, most unfortunate. Anything like accurate statistics even on the total number of the existing individuals of a species is a rarity, and such data are far from being what one would desire to know for the determination of population numbers.

As stated above, the population number N refers not to the total number of individuals of a species, but to the number of individuals in the breeding population. This is, in the simplest case, the number of individuals of the previous generation which were parents of the generation now living, or the number of the living individuals which are going to be the actual progenitors of the following generation. Nor is this all. Any species distributed over a large territory is not a random breeding population, for individuals living, for example, in Europe will breed among themselves infinitely more frequently than they will breed with individuals of the same species living somewhere in China. Most species live in colonies of individuals, there being more or less intermixture with other colonies. It is the size of the breeding populations in such colonies, together with data on the degree of interbreeding between the colonies, which we need to know. Strange as it may seem, information of this sort is lacking even for the most talked about species, *Homo sapiens*. In the face of this situation, the discussion of the population sizes can not but be restricted to bare generalities. We shall have to be content if we can assemble some data from which inferences may be drawn that could serve as provisional working hypotheses for future investigations.

Elton (1927) has correctly emphasized that total numbers of individuals in many species are staggeringly great, far greater than is usually realized. Owing to the smallness of many creatures, or to their habit of hiding themselves from view, a great majority of individuals of a species are always overlooked. Even diregarding such phenomena as the "water-bloom," which is caused by the presence of stupendous numbers of unicellular organisms in the waters of ponds, lakes, and seas, accumulations of much larger organisms often attain colossal proportions. A school of herring approaching the sea shore for egg laying may weigh up to 100,000 tons and more. The human population, being close to two billions, is probably more than equalled by many higher animals and large plants (e.g., trees), and is certainly far exceeded in many species of lower animals and

small plants. Only in species which are in a state of temporary
eclipse, being reduced to a fraction of their former size, or in relic
and dying-out species, are the total numbers likely to be relatively
small. The Monterey cypress (*Cupressus macrocarpa*) occurs naturally in an area of a few square kilometers on the coast of the Monterey peninsula in California; the number of individuals is probably
less than one thousand. Other relic species are even less numerous.

Nevertheless, it would be a mistake to conclude that the effective
breeding populations of most organisms reach semi-astronomical
figures. Only in relic species in which all the individuals subsist on
a very small territory, may they be regarded as constituting a single
interbreeding community. In species that are spread over large areas,
the crossing of individuals born in different localities is unlikely or
altogether impossible. The size of the distribution area is evidently
a consideration the meaning of which is contingent on the propen-
sities of the particular organism for movement. For a snail, a journey
from one corner of a field to the other may be more unlikely than
commuting from one continent to the other would be for a migratory
bird. Species that contain large numbers of individuals may be subdivided into very many breeding communities. The population sizes
in such communities, rather than in species as wholes, are the primary
concern for a genticist.

The problem is not whether an individual is able to move from one
end of the species area to the other, but whether such movements
occur in fact. A body of data is accumulating that suggests that even
animals with excellent means of locomotion tend to breed within
a narrow range from the place of their birth. No better examples of
this could be wished for than these actually provided by the work
on marking migratory birds (cf. the review of this subject by Wachs,
1926). The stork (*Ciconia ciconia*) is a bird nesting in Europe and
hibernating in Africa. The storks from each part of Europe have
fairly definite routes for their yearly migrations in both directions;
those from eastern Europe go along the west coast of the Black Sea,
Asia Minor, Syria, and the Nile valley; the western ones take the
route to Gibraltar and the Sahara. The birds marked by rings at
their nesting places have been observed to return year after year to
the same nest. Moreover, and this is especially important, in no less
than 83 cases the marked young have returned and have built their

nests in the neighborhood of their birth place. Evidently, only a minority of the marked storks are ever seen again, but this need not mean that a majority change their habitation place. The yearly mortality, especially during migrations, is known to be very high. If the birds did not tend to return to their old homes, the probability of recovering any marked bird in the vicinity of its old nest would be negligibly small. Wachs (1926) quotes only two instances when a marked stork was seen several hundred kilometers from the original place. The old belief that swallows return every year to the same locality has been scientifically proved. The nesting places of swallows and of their offspring are found usually within two kilometers from the old nest.

In some species of fish, given to large-scale migrations, a differentiation of the population into local subgroups had been demonstrated. For herring (*Clupea harengus*) a penetrating analysis of this problem was made first by Heincke (1898), and since then has been corroborated by newer investigators (Scheuring 1929-30, Schnakenbeck 1931, and others). The herring is one of the fish that comes, for purposes of reproduction, to shore waters, while the young lead a pelagic life in the open sea. Among the herrings of the North Atlantic, North Sea, and the Baltic, there exist separate strains differing from each other in the place and the season of the breeding, the paths of the yearly migration, and also in morphological characters. The latter differences are usually small, the variation limits for the different strains overlap, but the averages are distinct. Heincke has shown that if proper statistical methods (the least-square method) are applied, the strains are distinguishable even in single individuals. Every strain is, then, a separate breeding community, and deserves the name of "elementary race" suggested by Heincke. The herrings of the White Sea are divided, according to Rabinerson (1925) and Averinzev (1930), into at least six races, as follows:

RACE	BREEDING PLACE	BODY SIZE	BREEDING SEASON
I	Kandalaksha Bay	small	April, under ice
II	Onega Bay	small	May, no ice
III	Archangel Bay	small	May, no ice
IV	Kandalaksha Bay	large	July, no ice
V	Archangel Bay	large	April-May
VI (pelagic)	Outside the White Sea	?	?

The Salmonidae which for breeding enter the rivers of the Pacific Coast of North America do so at a definite season, the seasons however, being different for different rivers. Moreover, the characteristics of the population entering a given river (e.g., the percentage of individuals with pink and with white meat) are more or less constant from year to year (Scheuring 1929-30). It is most probable that the situation in the Salmonidae is like that which obtains among the herrings.

Populations of the periodical cicadas (*Cicada septendecim* and *C. tredecim*) are subdivided into separate breeding communities by a different method. As their species names suggest, the development of these insects takes seventeen and thirteen years respectively. The nymphs live underground feeding on the roots of plants, while the relatively very short-lived adults come to the surface, breed, deposit their eggs, and die. *C. septendecim* occurs practically everywhere in the eastern half of the United States except along the coast of the Gulf of Mexico; *C. tredecim* is restricted to more southern localities, especially to the southeastern states. The ranges of the two species broadly overlap. Great numbers of the cicadas are known to have apeared in many localities, but such mass appearances are nowhere a yearly occurrence, and, moreover, different localities have them in different years. Marlatt (1907) was able to show the existence of seventeen broods of *C. septendecim* and ten of *C. tredecim*, and to map their distributions. Thus, brood I of *C. septendecim* has appeared in masses in 1893, 1910, and 1927 in Maryland, Virginia, West Virginia, Southern Pennsylvania, part of North Carolina, and also in some places in Illinois, Indiana, Missouri, and Kansas. Brood II appeared in 1894, 1911, and 1928 in the region mostly east of that of brood I, namely, Connecticut, southern New York, New Jersey, eastern Pennsylvania, Maryland, eastern Virginia, North Carolina, with a scattering in Indiana and Michigan. Brood III (1895, 1912, 1929) is restricted to Iowa, Missouri, and a part of Illinois.

In localities which have no large brood of Cicada scheduled to emerge in a given year, only a few scattered individuals usually appear. In years immediately preceding or following the ones with large broods, the number of these scattered individuals is likely to be larger than in the middle of the period between the large broods. This leads Marlatt to suppose that the scattering is due to variations in the time of the development—occasionally an individual takes sixteen or eighteen years to complete its cycle, giving rise to small secondary broods filling the gaps between the large ones. Here, then, we have two species divided each into breeding communities that are more or less isolated from each other by the time of emergence from the ground, as well as by geographical factors. Some of the communities have very large, and others, at least by comparison, very small population numbers.

Where the life cycle of an organism does not involve large migrations as in some birds and fish, the formation of partially isolated breeding communities is evidently even more probable. Any species which is bound to a definite kind of soil, or which preys upon a definite other organism, or, in general, is connected with a definite ecological station, is not likely to be omnipresent in the whole area of its distribution. Geographical maps showing the areas of species in terms of continents or large subdivisions thereof are abstractions, for the actual distribution areas are mostly patchworks of small stretches of territory where the species is found, alternating with stretches where it does not occur. Each inhabited stretch may be regarded as a separate colony. The discontinuity of the distribution has been especially emphasized by Anderson (1936) for Iris. This author published a map of a representative region of 50 square miles in southern Michigan showing the areas occupied and unoccupied by *Iris virginica*. The former constitute only a small fraction of the total area. The total number of the separate colonies, each consisting of from one to several thousand individuals, is enormous. Anderson estimates the average frequency of colonies for one hundred square miles to be 120 in northern Michigan, 350 in southern Michigan, 170 in northern Illinois, 30 in southern Missouri, and 5 in Alabama and Mississippi. With the aid of statistical methods, Anderson detects small but significant differences between the populations composing the separate colonies, a result that immediately reminds one of the local races of land snails formed in the separate valleys on South Sea islands (see p. 135). As far as the writer is aware, nobody else has attempted an investigation along lines similar to Anderson's.

Precisely how much isolation the breaking up of a species into colonies entails in any given case depends upon how frequently the individuals migrate from one colony to the next. Modern ecology has little accurate information on this subject, partly because it is difficult to approach experimentally, but perhaps largely because the significance of the problem involved has not been appreciated either by ecologists or by geneticists. Whatever data exist are ostensibly contradictory. The presence in many organisms of "adaptations" that engender a wide distribution of the offspring of an individual has been repeatedly emphasized, especially in the older literature. On the other hand, some data show that many species endowed with good means of locomotion are remarkably sedentary. The amazing speed with which insect pests and plant weeds conquer new countries is well known, and in some instances has been followed from year to year. The same is true for the reoccupation by autochthonous organisms of territories where the flora and fauna had been exterminated by some cause. The best known example of this is the island of Krakatau where life was destroyed by a volcanic eruption in 1883. In three years thereafter 22 species of flowering plants were found on the island, and the number of both species and of individuals was rapidly growing with time. Similar data exist also for apparently spontaneous expansions of the distribution areas of some species. The plant *Senecio vernalis,* which is native in Russia, was recorded in 1756 in East Prussia, in 1826 on the lower Vistula, in 1835 on the Oder, in 1860 near Berlin, in 1886 near Hamburg and on the middle Rhine, and in the twentieth century on the lower Rhine, in Switzerland, Austria, and southern Denmark (Wulff 1932).

An extrapolation of the data just quoted to prove that there is no isolation between populations in separate colonies formed by the species is vitiated by the following simple consideration. The migrants arriving in a territory unoccupied by their own species may find no opposition, and spread as fast as their reproductive potential and their means of distribution will permit. On the other hand, a migrant which appears in a location already occupied to the saturation limit by representatives of the same species has to contend for a place in the sun with the latter. Provided the migrant has no advantage over the local inhabitants, it has only a scanty chance to establish its progeny in the new place. The exchange of genes between the colonies will, then, be very limited, unless the number of migrants per generation is large in relation to the population size in an average colony. How small the number of migrants may be even in forms which, judged superficially, have excellent distribution means, is shown by the studies of Noble (1934) on the lizard *Sceleropus undulatus*. A number of lizards of both sexes were tagged in spring in their natural area of habitation, and the movements of the tagged individuals were watched during the following summer. Out of 226 tagged lizards, 155 were seen again, a large number of them many times. It appears that the individuals in a locality subdivide the suitable terrain among themselves, each individual patrols its own hunting grounds, does not admit intruders therein, and only seldom attempts to interfere with the neighbors. The furthest point a lizard was seen from its original place was about 250 feet. Moreover, when a lizard was artificially transferred 790 feet from its home territory, it worked its way back in eighteen days; a lizard transferred 420 feet came back in seven days. An astonishing degree of attachment to a piece of territory has been observed also in some species of bats (Eisentraut 1934).

In summarizing, one may say that the available data, meager as they are, tend to show that the effective population sizes may prove to be small, at least in some species. In any case, to suppose that the breeding population equals the total number of individuals in a species is erroneous. It is probable that in some species the population numbers tend to be small and in others large; in some parts of the distribution area of a species large, and other parts small, colonies may be formed. Theoretically it is evident that such peculiarities of the separate species are bound to exert a determinative influence on their evolutionary patterns; the mode of evolution need not be the same in all organisms, not even in related ones. The latter idea has been expressed repeatedly by some systematists and morphologists; a reëxamination of the evidence in an attempt to correlate it with the possible influence of population size would be a very promising undertaking.

MICROGEOGRAPHIC RACES

The differentiation of the population of a species into numerous small colonies, some of which acquire distinctive morphological characteristics, has received an entirely insufficient amount of attention. That this phenomenon actually exists is a fact more or less familiar probably to most biologists who have had experience in field work. In many widely distributed species, may be found populations restricted to a single or to very few localities, which strike the eye as being somehow unorthodox for representatives of the given species or race. Sometimes such local variations are sufficiently pronounced to be noticed at once by a collector. Nevertheless, observations of this kind are on the whole rather seldom recorded in the literature, or at best mentioned in passing. The reason for this is not their scarcity, but rather a lack of appreciation of their probable significance.

A majority of field biologists unfortunately still adhere to obsolete notions, according to which geographical variation is merely a persistent modification induced by the environment, and hence they are prone to ascribe the small local variations to the effects of some accidental and elusive external cause, which for no assignable reasons was operative in one narrowly circumscribed locality and not elsewhere. The very strict localization usually prevents these variations from being described as geographical races, for the latter term is applied in most groups to changes embracing populations of larger parts of the species area. It seems advisable to introduce the expression "microgeographic race" to designate the variations restricted to small fractions of the distribution region of the species. Needless to say, there is probably no sharp dividing line between the microgeographic and the major geographic races; the two merge into each other, and perhaps the former might be regarded as a stage in the development of the latter (cf. Chapter VI). The microgeographic races are important from a geneticist's viewpoint because they may provide evidence of the effectiveness of the random changes in the gene frequencies in the local subgroups of a species.

We have already had occasion to quote observations that are strongly reminiscent of microgeographic races. The slight but significant differences in the averages for many characters in local populations of Iris (Anderson 1936) belong here. Varieties of Partula

(Crampton 1916, 1932) might be classed as microgeographic races if it were not for the fact that the total distribution regions of the species are in this case so small that the area of each race constitutes a fairly large portion of the total. Schmidt's (1917, 1923) results with the races of *Zoarces viviparus* and those of Turesson (1922-31) with the local variations in plants belong to a separate category, since these variations are said to recur wherever a definite kind of environment is represented in the distribution area of the species.

A clear instance of the formation of microgeographic races has been observed by the writer (unpublished) in the European ladybird beetle, *Sospita vigintiguttata*. This beetle is represented by two clearly distinct and undoubtedly hereditary forms, one of which has black and the other yellow elytra with ten white spots on each. The species is not very common in central and in eastern Europe, and occurs only locally, on trees, especially on alders (Alnus), growing on marshy soils. It is apparently not given to migrations from place to place like some other species of the same family, and hence its distribution is broken up into a large number of small patches. In the neighborhood of Kiev (Russia) and elsewhere, several dozen colonies of this species were observed, a majority of which contained both black and yellow specimens in ratios approaching equality. Two colonies were, however, different, because one of them had only blacks and the other only yellows. This condition persisted for at least three consecutive years. The only reasonable interpretation of these observations is that the allelomorph producing the black coloration has been lost in one colony and has reached fixation in the other. The intercolonial migrations are apparently so infrequent that the colonies were at least temporarily inhabited by populations different in their genetic composition from the rest of the species. A less clearly marked microgeographic race has been observed also in an American species, *Coccinella transversoguttata*, between Rapid City and Deadwood, Black Hills, South Dakota.

The butterfly *Anthocharis cethura*, living in the foothills of the mountain ranges in southern California, has two distinct color forms among the females. The so-called typical form of the species has a yellow, and the form deserti a white, tip on the forewing. Mr. Charles Rudkin kindly informs me that in most localities where the species has been collected a large majority of individuals are yellow, and the whites constitute no more than 5 per cent of the population. A single locality is known however, a valley of a few square kilometers in area, in which the frequency of whites approaches 50 per cent. On either side of this exceptional valley, populations with normal frequencies of white and yellow individuals have been recorded.

As pointed out above, to prove that natural selection could not be responsible for the formation of any one race is an impossibility in practice. No amount of negative data would suffice for this purpose. Nevertheless, the supposition that selection has produced the microgeographic races in Sospita and Anthocharis involves assumptions so far-fetched that most biologists would prefer some other alternative. Indeed, the localities where the exceptional populations are found do not seem to be appreciably different from localities which are inhabited by "normal" populations. Random variations of the gene frequencies are a much more probable source of the microgeographic races. With the present status of our knowledge, the supposition that the restriction of population size through the formation of numerous semi-isolated colonies is an important evolutionary agent seems to be a fruitful working hypothesis. The action of selection interwoven with and following after that of isolation becomes, as we shall attempt to show in the next chapter, more effective than selection alone is likely to be.

VI: SELECTION

ADUMBRATED already by some writers of classical antiquity, the principle of natural selection was raised to the status of a scientific theory by Darwin. With consummate mastery Darwin shows natural selection to be a direct consequence of the appallingly great reproductive powers of living beings. A single individual of the fungus *Lycoperdon bovista* produces 7×10^{11} spores; *Sisymbrium sophia* and *Nicotiana tabacum,* respectively, 730,000 and 360,000 seed; salmon, 28,000,000 eggs per season; and the American oyster up to 114,000,000 eggs in a single spawning. Even the slowest breeding forms produce more offspring than can survive if the population is to remain numerically stationary. Death and destruction of a majority of the individuals produced undoubtedly takes place. If, then, the population is composed of a mixture of hereditary types, some of which are more and others less well adapted to the environment, a greater proportion of the former than of the latter would be expected to survive. In modern language this means that, among the survivors, a greater frequency of carriers of certain genes or chromosome structures would be present than among the ancestors, and consequently the values q and (1-q) will alter from generation to generation.

During the seventy-seven years that have elapsed since the publication of the theory of natural selection, it has been the subject of unceasing debate. The most serious objection that has been raised against it is that it takes for granted the existence, and does not explain the origin of the hereditary variations with which selection can work. Those who advance this objection fail however to notice that in so doing they commit an act of supererogation: the origin of variation is a problem entirely separate from that of the action of selection. The theory of natural selection is concerned with the fate of variations already present, and the merits and demerits of the theory must be assessed accordingly. In the beginning of the present century, real progress was made when Johannsen showed that selection is effective in genetically mixed populations but inoperative in genetically uniform ones. Johannsen's work was preceded by De Vries's discovery of the origin of hereditary variations through the occurrence of mutations. Following this discovery, some writers contended that De Vries and Johannsen had disproved Darwin's theory of evolution by natural selection and had supplanted it by a theory of evolution by mutation. The polemics that ensued around this weird contention both in popular and in scientific literature seems in retrospect to have been a sort of a modern confusion of tongues. It is hardly necessary to reiterate that the theory of mutation relates to a different level of the evolutionary process than that on which selection is supposed to operate, and therefore the two theories can not be conceived as conflicting alternatives. On the other hand, the discovery of the origin of hereditary variation through mutation may account for the presence in natural populations of the materials without which selection is known to be ineffective. The greatest difficulty in Darwin's general theory of evolution, of the existence of which Darwin himself was well aware, is hereby mitigated or removed.

In its essence, the theory of natural selection is primarily an attempt to give an account of the probable mechanism of the origin of the adaptations of the organisms to their environment, and only secondarily an attempt to explain evolution at large. Some modern biologists seem to believe that the word "adaptation" has teleological connotations, and should therefore be expunged from the scientific lexicon. With this we must emphatically disagree. That adaptations exist is so evident as to be almost a truism, although this need not mean that ours is the best of all possible worlds. A biologist has no right to close his eyes to the fact that the precarious balance between a living being and its environment must be preserved by some mechanism or mechanisms if life is to endure. No coherent attempts to account for the origin of adaptations other than the theory of natural selection and the theory of the inheritance of acquired characteristics have ever been proposed. Whether or not these theories are adequate for the purpose just stated is a real issue.

Whether the theory of natural selection explains not only adaptation but evolution as well is quite another matter. The answer here would depend in part on the conclusion we may arrive at on the problem of the relation between the two phenomena. No agreement on this issue has been reached as yet. Fisher (1936), who is probably one of the most extreme among the modern selectionists, has expressed his opinion very concisely as follows: "For these two theories (Lamarckism and selectionism) evolution is progressive adaptation and consists of nothing else. The production of differences recognizable by systematists is a secondary by-product; produced incidentally in the process of becoming better adapted." And further: "For rational systems of evolution, that is for theories which make at least the most familiar facts intelligible to the reason, we must turn to those that make progressive adaptation the driving force of the process." A good contrast to this is provided by the conclusions reached by Robson and Richards (1936): "We do not believe that natural selection can be disregarded as a possible factor in evolution. Nevertheless, there is so little positive evidence in its favor . . . that we have no right to assign to it the main causative rôle in evolution." And: "There are many things about living organisms that are much more difficult to explain than some of their supposed 'adaptations.' "

EXPERIMENTAL STUDY OF ADAPTATION

The action of natural selection can be studied experimentally only in exceptionally favorable objects and under favorable circumstances. No major evolutionary change is noticeable in most species of organisms within a human lifetime, hence the supposition that species have become what they are now through evolution by natural selection can be at best no more than a very probable inference. It is obviously impossible to reproduce in the laboratory under controlled conditions the evolution of, for example, the horse tribe or of the anthropoid apes. Moreover the work on natural selection is of necessity confined mainly to experiments in which the environment of the organism is modified artificially, and the resulting changes in the genetic make-up of the populations are recorded. A modification of the environment is sometimes brought about unintentionally by man; the effects of such a modification on the free-living organisms may furnish valuable, though mostly indirect, information on selection. Skeptics may contend that if the change in the environment is wrought directly or indirectly by man, the resulting selection is no longer "natural." Anyone who is prepared to reject the evidence on these grounds has no choice but to do so; a similar objection is applicable to any experimental work.

The experiments of Sukatschew (1928) on the relative viability of different strains of dandelions (*Taraxacum officinale*) are very instructive. Three plants were collected on the same meadow near Leningrad, and the strains, denoted as A, B, and C, respectively, were established from them; the strains proved to be morphologically recognizable. Seedlings were planted on experimental plots in two densities, namely, at a distance of three and of eighteen centimeters from each other, respectively. On some plots, representatives of a single strain, and on others of all three strains, were planted. After a lapse of two years the number of surviving individuals was counted. The percentage of individuals that had died on plots with a pure stand of a single strain was found as follows:

DENSITY	STRAIN A	STRAIN B	STRAIN C
Low	22.9	31.1	10.3
High	73.2	51.2	75.9

Strain C is most viable at a low and B at a high density. But in mixed stands, where the three strains grow side by side and compete with each other, the percentages of the individuals that die prove to be different, namely:

DENSITY	STRAIN A	STRAIN B	STRAIN C
Low	16.5	22.1	5.5
High	72.4	77.6	42.8

Strain C is distinctly superior to A and B at both densities employed. Besides the survival of individual plants, the numbers of flowers per plant was recorded. This has a bearing on the reproductive values of the different strains under the different conditions. At the low density in pure stands, the number of flowers was highest in strain B and lowest in C, but in mixed stands C was highest and A lowest. At the high density in pure stands, strain C had least and A and B most flowers, but in mixed stands the relation was reversed. It is clear that the flower production in a strain is not always correlated with its viability, and the relative survival value of a strain is a function of the environment in which it is placed.

In a second series of experiments, Sukatschew compared the viability of strains of different geographical origin grown under the climatic conditions of Leningrad. The two local strains, B and C, were used, and also a strain X from the extreme north (Archangel), Y from the northeast (Vologda), and Z from the south (Askania-Nova, north of Crimea). In pure stands the strains gave the following percentages of mortality:

DENSITY	B	C	X	Y	Z
Low	31.1	10.3	39.6	22.9	73.0
High	51.2	75.9	63.0	71.6	82.0

The southern strain (Z) is a failure in the new environment, but the Vologda strain did at least as well as one of the local ones (B) at the low, and the Archangel strain at the high, density. For technical reasons it was not convenient to grow all the five strains in mixed cultures, but two experiments were carried on, each with four strains in a mixture. The results are shown in Table 13.

TABLE 13

PERCENTAGE OF DEAD INDIVIDUALS IN MIXED CULTURES OF DIFFERENT STRAINS OF *Taraxacum officinale* (*after Sukatschew*)

DENSITY	FIRST EXPERIMENT				SECOND EXPERIMENT			
	B	C	X	Y	B	X	Y	Z
Low	66.3	37.5	66.6	50.0	4.2	12.5	8.3	41.6
High	99.5	96.3	56.0	49.2	89.0	29.3	72.8	99.0

The results obtained may seem paradoxical: the Archangel and the Vologda strains (X and Y) show in dense stands definitely greater survival values than the local strains (A and C). One may surmise that a competition between the local and the introduced strains would lead to an elimination of the former. At the lower density, however, the local strains at least hold their own in a competition with the immigrants. Such facts as these provide an excellent illustration of the correctness of the argument employed by us in discussing the survival values of mutants (Chapter II): unless the characteristics of the environment are known in detail, a judgment regarding the survival value of a given genotype is meaningless.

A very ingenious piece of experimental work has been done by Timofeeff-Ressovsky (1933d, 1935a) in comparing the survival values of strains of *Drosophila funebris* of different geographical origin. The geographical races in this, as well as in most other, species of Drosophila are morphologically indistinguishable. To surmount the difficulty arising from this fact, the following technique was employed. A known number (150) of eggs of a given strain of *D. funebris* was placed together with the same number (150) of eggs of *D. melanogaster* in a small culture bottle with a standard amount of food. The amount of food was deliberately made insufficient for the optimal development of the 300 larvae that hatched from these eggs. Due to the crowding and the scarcity of food, a part of the larvae died, and the number of the adult flies that hatched in a bottle was below, and frequently much below, 300. By counting the numbers of the *D. funebris* and the *D. melanogaster* adults that did hatch, it is possible to gauge the relative viability of the two species under the conditions of the experiment. In different cultures, different strains of *D. funebris* were thus compared with the same strain of *D. melanogaster*, and their viability was expressed in percentages of that of the latter (that is to say in each experiment the number of the adult *D. funebris* was expressed in per cents of the number of the adult *D. melanogaster* obtained). The experiments were done at three different temperatures, namely 15°, 22°, and 29°C. The left half of Table 14 shows the data obtained.

It may be noted that the viability of *D. funebris* is in general lower than that of *D. melanogaster*. The difference is relatively slight at the lower temperature (15°), more pronounced at the intermediate one (22°), and striking at the high one (29°). This conclusion derived from the experiment agrees well with the known difference in the geographical distribution of the two species. *D. melanogaster* is probably a native of the tropics and has been introduced to the temperate zone by man, while *D. funebris* does not occur in the tropics and can live further north than *D. melanogaster*. A more detailed analysis of the data discloses characteristic differences in the behavior of the strains of *D. funebris*. The strains from the Mediterranean countries (the second group from the top in Table 14) are distinctly inferior to *D. melanogaster* at 15°, while the strains from central and northern Russia (the second group from below) are almost as viable at that temperature as *D. melanogaster*. At 29° the strains from southern Russia suffer less than others.

TABLE 14

THE RELATIVE VIABILITY OF THE STRAINS OF *Drosophila funebris* OF DIFFERENT GEOGRAPHICAL ORIGIN (*from Timofeeff-Ressovsky*)

STRAINS OF D. FUNEBRIS	VIABILITY IN % OF THAT OF DROSOPHILA MELANOGASTER			VIABILITY IN % OF THAT OF THE BERLIN STRAIN OF D. FUNEBRIS		
	15°	22°	29°	15°	22°	29°
Berlin	81	42	18	100	100	100
Sweden	88	40	21	108.6	95.2	116.6
Norway	80	41	21	98.7	97.6	116.6
Denmark	79	44	22	97.5	104.7	122.2
Scotland	84	43	20	103.7	102.4	111.1
England	78	42	21	96.3	100.0	116.6
France	80	44	25	98.7	104.7	138.8
Portugal	71	45	28	87.6	107.1	155.5
Spain	69	48	30	85.2	114.3	166.6
Italy	78	43	25	96.3	102.4	138.8
Gallipoli	75	44	26	92.6	104.7	144.4
Tripoli	64	47	31	79.0	111.9	172.2
Egypt	68	46	30	83.9	109.5	166.6
Leningrad	90	43	22	111.1	102.4	122.2
Kiev	91	44	28	112.3	104.7	155.5
Moscow	101	43	28	124.7	102.4	155.5
Saratov	92	42	30	113.6	100.0	166.6
Perm	98	41	26	121.0	97.6	144.4
Tomsk	96	42	28	118.5	100.0	155.5
Crimea	87	42	28	107.4	100.0	155.5
Caucasus I	89	43	31	109.9	102.4	172.2
Caucasus II	86	45	32	106.2	107.1	177.7
Turkestan	90	44	34	111.1	104.7	188.8
Semirechje	92	46	36	113.6	109.5	200.0

A comparison of the strains of *Drosophila funebris* with each other is facilitated if their viabilities are expressed in percentages of that of some one arbitrarily chosen strain. The right half of Table 14 shows the data recalculated on the assumption that the viability of the Berlin strain equals 100 at every temperature. It can be seen that at 15° the viability of the strains from the Mediterranean region is consistently lower, and that of the strains from Russia higher, than that of the strains from western, central, and northern Europe. At 22° the viability of all the strains is about the same, but at 29° the strains from Russia and from the Mediterranean countries are definitely more viable than the western European ones. Timofeeff-Ressovsky points out that these properties of the strains of *D. funebris* are consonant with the climatic characteristics of the countries of their origin. Indeed, western and northern Europe have a relatively mild climate both in summer and in winter; the climate of Russia and Siberia is more rigorous, with a hot summer and a cold winter; the Mediterranean countries enjoy a mild winter but have a hot summer. Correlated with this, stands the sensitivity of the western European strains both to warmth and to cold, the greater adaptability of the Mediterranean ones to heat than to cold, and the relative hardiness of the strains from Russia and Middle Asia in both extremes.

A drawback to the experiments just discussed is that they furnish no evidence as to the particular physiological process or property of the experimental strains that render the latter more or less resistant at a given set of conditions. In the experiments on Taraxacum this drawback is particularly acute, although the significance of the results for a demonstration of the different adaptive values of the strains is not thereby impaired. An analysis of the geographical variation in the gypsy moth (*Lymantria dispar*) made by Goldschmidt (1932c, 1933a, 1934b) serves to elucidate the methods of adaptation in more detail. One of the racial differences in this insect involves the length of the incubation period of the eggs. The embryonic development is here arrested during the coldest period of winter, but it takes a different sum of temperatures (= the incubation time) to bring out the young caterpillars to hatching in spring. The incubation time for strains coming from Europe (except the Mediterranean region) and from northern Asia (including the northern island of Japan) is very short, which fact is correlated with the long winter and the rapid onset and progress of spring. The Mediterranean strains have a much longer incubation time. Goldschmidt points out that with a mild winter characteristic for that region, a short incubation time would bring the caterpillars out before the appearance of the foliage on the food plants. In southern and central Japan the incubation time generally increases from southwest to northeast, which Goldschmidt correlates with a peculiarity of the Japanese climate that makes for a relatively late (compared to Europe) development of the foliage with a similar sum of temperatures. The race from Manchuria has the longest incubation time of all of those studied; the adaptational significance of this remains however unclear.

The variations in length of larval development (from the hatching of the caterpillar to pupation) show a slightly different geographical regularity than that exhibited by the incubation time. The shortest larval development is observed in strains from the northern island of Japan, the northern tip of the main island, and the northern part of the Eurasiatic continent, which may bear a relation to the short vegetation period characteristic of these countries. The Mediterranean region, Turkestan, and central Japan have a longer vegetation period, and races with a longer larval development occur there. In southern Japan the larval development is again shorter, which Goldschmidt regards as an adaptation that permits the insect to complete its larval stages before the advent of the very high summer temperatures that occur in that region. The strains from mountain localities in Japan have shorter developmental times than the strains from the adjacent subtropical coastal region (cf. also the races of Carabus, Krumbiegel 1932).

In *Drosophila pseudoobscura*, race B is restricted in its distribution to the Pacific Coast of the United States and Canada, and the Sierra Nevada and Cascade Mountains. Race A, morphologically indistinguishable from B, does not live on the coast north of San Francisco Bay, but it does occur, together with B, further inland and extends its distribution much further east and south, up to and including the Rocky Mountains and Mexico. It may be noted that the distribution region of race B is characterized by a mild, and that of race A by a hot, summer. Dobzhansky (1935d) has observed a relation between the fecundity of different strains and temperature that is very suggestive in connection with the above difference in the geographical distribution of the two races. The total egg production (the average number of eggs deposited by a female in her lifetime) in the strains studied is shown in Table 15.

At all temperatures studied, race A reaches the maximum productivity sooner after the hatching from the pupa than race B. At 25°C, the productivity of race A remains higher than that of race B throughout the lifetime, which accounts for the greater total egg output in race A. At 19°, and still more at 14°, race B reaches the peak of productivity later than race A, but retains the productivity much longer than the latter. At 27½° race B is mostly sterile, but race A still deposits some eggs. At 9° the oviposition is very irregu-

TABLE 15

TOTAL EGG PRODUCTION IN DIFFERENT STRAINS OF *Drosophila pseudo-obscura* (*from Dobzhansky*)

RACE	STRAIN	9°	14°	19°	25°	27°
A	LaGrande	347.55	984.13	1,144.57	395.63	14.20
A	Texas	182.69	560.04	1,130.27	331.80	0.87
B	Humboldt	217.33	1,257.81	1,697.47	138.08	...
B	Seattle	104.19	1,093.49	1,591.65	75.01	2.38
B	Sequoia	415.06	1,178.79	1,229.76	35.89	...

lar. Obviously, the optimum temperature for the egg production is lower in race B than in race A.

As a rule, the strains of *D. pseudoobscura* developed in the laboratory at temperatures exceeding 27½° C fail to produce viable adults. Two exceptional strains of this species (both of race A) have been found, however, not far from Yuma, Arizona, which in the laboratory have developed for three consecutive generations at 28½-29° and still produced some viable and fertile adults, although the larval mortality was high (Dobzhansky, unpublished). The region of Yuma is notorious for its hot summer; moreover, *D. pseudoobscura* is not as a rule found in arid and hot localities. The connection between the exceptional heat hardiness of the Yuma strains and the climatic conditions of that place seems to require no further comment.

HISTORICAL CHANGES IN THE COMPOSITION OF POPULATIONS

The evidence just presented provides an experimental demonstration of the effectiveness of natural selection. It likewise shows that the differences between strains, races, and species are in some cases of such a kind that their origin through natural selection is the sole reasonable inference. It does not follow of course that any interracial or interspecific difference is necessarily of a similar kind. The number of instances where the origin of such differences through natural selection is virtually certain could be greatly increased, but there exist also other instances where the molding action of selection is by no means so evident. Books have been written to disprove the theory of natural selection by stressing the latter group of facts—an amazing procedure which seems to assume that natural selection may be either omnipotent or absolutely impotent, and that no third choice exists.

We may turn now to a consideration of evidence of another kind, where a change in the genetic composition of a free living population has been observed to take place within historical time. The occurrence of such changes is *per se* not a sufficient proof that the transformation is due to selection. To make such an assumption would mean taking for granted what one pretends to be trying to prove. Nevertheless, in some instances the participation of selection in the process seems reasonably clear. The observations of Weldon (1899) on the crab *Carcinus maenas* are perhaps most often quoted. The building of a breakwater in a certain sound in England resulted in an increase of the amount of silt in the water. Measurements of the dimensions of the carapace of the crab population of the sound showed that during the period from 1893 to 1898 covered by the observations, the mean frontal width of the carapace (a ratio between the length and width) was decreasing. Weldon attributes this change to the elimination of individuals having a high frontal width, because of the accumulation of the silt in the gill chamber of the animals. This conclusion has been checked by an experiment. Crabs were placed in aquaria containing silty water, and it was found that the individuals that died had a larger mean frontal width of the carapace than the survivors. Weldon's results have been criticized (cf. Robson and Richards 1936), mostly on the ground that the accumulation of silt in the gill chamber was not directly observed, and that the size of the entrance to the gill chamber rather than that of the carapace ought to have been measured. It may be granted that the observations are not as complete as might be desired; nevertheless, Weldon's explanation remains the most plausible one.

The appearance and the spread of melanic forms in several species of moths is probably the clearest instance of an evolutionary change scientifically recorded in natural populations (Harrison 1920, Hasebroek 1934, further references in the latter work). Variants more darkly pigmented than the norm appear in the population, and in the course of several decades become more frequent than the original form, finally supplanting the latter. In a number of instances the melanic forms have been shown experimentally to differ from the paler relatives in a single or in a few genes. The apparently well authenticated fact is that the first appearance of the melanic forms is always recorded in the vicinity of large industrial cities, and that the spread of the new variants goes hand in hand with the industrialization of the countries in question. The first records of the increased melanism were secured in the industrial districts of England in the middle of the nineteenth century. Thus, the melanic form of *Amphidasys betularia* was observed originally in Manchester in 1850 and in the twentieth century has superseded the normal form. In Germany the development of the "industrial melanism" was observed somewhat later, beginning with the Rhine district and the environs of Hamburg. Still later, analogous phenomena appeared in France and in other countries.

Unfortunately the work on the industrial melanism has been restricted mainly to collecting the records of the happenings as they occur, and the causal analysis of the problem has lagged far behind, except for more or less gratuitous speculations. Harrison and Garrett (1926) have described an artificial production of the melanic mutants by feeding the caterpillars on foliage containing salts of heavy metals, but for several reasons their results are not regarded as critical. If they are right, the origin of the melanics might be accounted for by an increase of the mutation rates of some genes owing to the contamination of the plants in the industrial districts by certain chemical waste products. It may be noted that the mutation rates that need to be assumed to account for the process are staggeringly great compared to what is known in other organisms. The possibility that a selection favoring the melanic mutants may be operative is at least not excluded, although the attempts to ascribe to the darker coloration a protective significance are certainly not convincing. Comparative physiological studies on the dark and the light forms of the same species would seem to be the most hopeful source of information on the subject.

The changes in the populations of at least three species of scale insects (Coccidae) observed in California are most interesting. These species of scales are pests attacking the groves of citrus trees; to control them, the trees are fumigated at definite time intervals in special tents filled with hydrocyanic gas. The concentration of the gas sufficient to kill approximately 100 per cent of individual scales is determined and applied regularly. Professor H. J. Quayle permits me to quote the following data from his manuscript dealing with observations on the development of strains resistant to the hydrocyanic fumigation. The red scale, *Aspidiotus* (*Aonidiella*) *aurantii* is widespread in Southern California. In 1914 it was noted that in orchards near Corona the regular fumigation was insufficient to destroy the red scale, although in previous years no difficulty was encountered. This condition has persisted in the same locality ever since, while in other localities the non-resistant populations continue to exist. Prior to 1912 a certain gas concentration gave uniformly satisfactory results in destroying the black scale, *Saissetia oleae*, but on that year it was first noted that in orchards near Charter Oak, Los Angeles County, dosages greatly in excess of that normally required failed to accomplish the desired end. In 1925 the area occupied by the resistant race extended from Charter Oak over a solidly planted citrus belt for a distance of over thirty miles. Outside of this zone the non-resistant race continued to occur. The appearance of a resistant race in *Coccus pseudomagnoliarum* was first recorded in 1925 in a very limited area near Riverside, but in the course of four years this area extended itself very rapidly, and came to include most of the groves infected with this species of scale in southern California. During the winter of 1933-34, this species, however, disappeared almost completely; the cause of this is unknown.

Careful laboratory experiments have established beyond a reasonable doubt that the difference between the cyanid-resistant and the non-resistant strains is a real one. It will probably remain forever obscure whether the original infestation of the California orchards consisted of a mixture of non-resistant and resistant strains, or whether the latter have appeared by mutations. In any event, the spread of the resistant strains constitutes probably the best proof of the effectiveness of natural selection yet obtained.

PROTECTIVE AND WARNING COLORATIONS AND RESEMBLANCES

Besides the general evidence on the efficacy of natural selection, there exist in biological literature some observations of a more special kind which furnish additional testimony in favor of the selection theory. The doctrine of the protective and warning colorations and resemblances constitutes one such topic. The essentials of this doctrine are so well known that they do not need to be explained in detail. Two of the ways in which organisms are supposed to become adapted to their environment are by becoming as inconspicuous as possible in their normal surroundings, or else by acquiring a similarity to some particular object which is dangerous or distasteful to their natural enemies. By the first method the organism escapes being noticed by its predators, or, in the case of the predators themselves, is able to approach the prey without being prematurely noticed by the latter. The second method leads the organism, contrariwise, to become as conspicuous as possible to advertise its presence and to warn its potential enemies of its obnoxiousness. A special kind of warning coloration and form is *mimicry,* whereby a harmless creature resembles some other which is in fact obnoxious (Batesian mimicry). The different forms protected by being dangerous or unpalatable and having warning colorations, may become even more thoroughly protected if they resemble each other, so that their enemies need learn a single sign of distastefulness instead of many (Müllerian mimicry).

The theories of protective resemblance and of mimicry were developed and widely used by the early Darwinists as illustrations of the action of natural selection and of evolution in general. The greatest effervescence of these theories was observed in the late nineteenth and the early twentieth century. Undoubtedly much uncritical and valueless speculation in this field has been indulged in by some writers, bringing only disrepute to the whole theory. The dangers of assuming that a given coloration is protective or mimetic are admittedly great. What to a human eye may seem to be a far reaching similarity between an organism and some inanimate object or another organism need not be such to the eyes of a predator against whom the protection is supposed to operate. Our own opinions of the dangers or distastefulness of a creature should not without further proof be imputed to the enemies of the latter. Some of the alleged cases of protection and warning have been in fact shown to be merely armchair protection and museum mimicry. A heavy barrage of criticism has recently been laid against the whole theory by Heikertinger (1933-36), who seems to reject it in its entirety. Nevertheless, the theory has survived both criticism and the damage done by its over-enthusiastic supporters, and in fact shows signs of a resurgence.

It may be noted that the process of development of protective and warning characteristics has not been observed in a species either in the laboratory or in nature (color changes occurring in certain fish, amphibians, reptiles, and insects within the lifetime of an individual obviously do not count). The concealing and mimetic resemblances that we record in nature are the end products of the historical processes that have taken place, and it only remains for us to infer whether their origin through natural selection is or is not probable. Perhaps the sole exception to the above statement is the observation of Harrison (1920) on two neighboring colonies of the moth *Oporabia autumnata,* one of which inhabits a coniferous wood and the other a birch wood. The population of moths in the former has been found to consist of darker individuals than in the latter. Harrison points out that dark coloration is probably protective in a coniferous, and light in a birch, forest, and regards it as likely that the two populations have diverged very recently, owing to the elimination of the more conspicuously colored types by birds and bats. As an evidence of the reality of this process, he records that among the fifteen pairs of wings of Oporabia found on the ground in the coniferous wood a majority were pale, while the frequencies of the dark and the pale types in the population from the same wood relate as 25 : 1 in favor of the dark. Assuming that the wings found on the ground are the remains of the moths attacked by the enemies, the case is a fairly strong one.

The best evidence in favor of the theories of protective coloration and mimicry might come from observations on the food preferences of the enemies of the supposedly protected and unprotected organisms. If, for example, an insect species is protected by an obnoxious taste and a warning coloration, another species is not obnoxious but mimics the first, and a third is neither obnoxious nor a mimic, one might expect that the former two will be eaten by the predators relatively less frequently than the third. Observations on this subject have been conducted on an unrivaled scale by McAtee (1932), who examined the contents of the stomachs of about 80,000 birds from the United States, and recorded the insects found therein. The results show that insects which are supposedly poisonous, distasteful, and protected by concealing or warning colorations are destroyed by birds in great numbers. McAtee concludes that the theory of protection is a myth. The finality of this conclusion is, however, weakened by the lack of assurance that the supposedly protected and unprotected insects are destroyed with frequencies exactly proportional to the frequency of their occurrence in the environment in which the birds live. A protective resemblance need not be absolutely effective for natural selection to foster an evolutionary change in that direction. If a "protected" form of species is destroyed by its enemies only slightly less frequently than an unprotected one, the change in the genetic constitution of the species may still be rapid.

The experiments of di Cesnola (1904), Beljajeff (1927), and Jones (1932), were conducted by exposing known numbers of different insects to birds and recording the number of insects destroyed. Here some preferences for and aversions against definite species of insects were found, but the relevance of the data to the conditions that obtain in nature has been questioned by some critics with a certain amount of justification. Taken as a whole, an unprejudiced observer must, I think, conclude that an experimental foundation for the theory of protective resemblance is practically non-existent. The theory still rests on very indirect evidence, which, however, has not been disposed of by those who wish to relegate it to the limbo of scientific delusions. As pointed out already by Poulton (1908), protective and mimic resemblances frequently involve a multiplicity of modifications of various parts of the body that combine to produce the resemblance in question, while the parts that are concealed from view and consequently can not aid in carrying the resemblance still further are not modified. This is especially evident in the mimetic circles of some butterflies belonging to different families, where a striking superficial similarity is not correlated with any convergence in the characteristics that a casual observer usually overlooks, but which are used by systematists for classification purposes. Furthermore, in different groups of animals the resemblance to the same model is often attained by quite different means. For example, Poulton finds four distinct methods used by unrelated insects to attain a resemblance to ants. Fisher (1930) points out that "this characteristic, analogous to opportunism in human devices, seems to deserve more attention than it has generally received," since natural selection appears to be the only known mechanism which can bring about such phenomena.

REGULARITIES IN GEOGRAPHICAL VARIATION

The extensive descriptive studies of the last few decades on geographical variation have disclosed a number of most interesting regularities which seem to govern this at first sight haphazard process. Some of these regularities had, in fact, been noted already by the systematists of the nineteenth century, but the more recent observations have served to test the early generalizations on a much more abundant material. It appears that representatives of various families, orders, and even classes inhabiting a given geographical region often undergo there convergent changes, while in another region the same groups seem to vary in a direction which is again similar for all of them but different from the direction observed in the first region. Rensch (1929, 1936) has recently summarized the available evidence, especially for the higher vertebrates where such phenomena have been most frequently and thoroughly observed. The following "rules" seem to be the best established ones.

I. Gloger's rule (usually named Allen's rule in the American literature) states that in mammals and birds, races inhabiting warm and humid regions have more melanin pigmentation than races of the same species in cooler or drier regions; the arid desert regions are characterized by races with an accumulation of yellow and reddish-brown (phaeomelanin) pigmentation. Rensch (1936) compared sixteen species from the bird families Paridae and Sittidae, each species having from two to four races inhabiting different climatic regions, and found the rule to hold in fifteen and to be broken in one case. Among the palaearctic Alaudidae the rule does not hold in 12 per cent of cases (among 140 examined), and among the western European mammals only 11.5 per cent (among 34) of the comparisons are exceptions to the rule.

Among insects the pigmentation increases under humid and cool conditions and decreases in the dry and hot regions, the humidity being apparently more effective than temperature. Dobzhansky (1933c) has shown this rule to apply to the ladybird beetles (Coccinellidae) of the North Temperate Zone, and data are available to show that it holds for some other insects as well (Zimmermann 1931). For the ladybirds, eastern Asia (northeastern Siberia, Japan) is the center of heavily pigmented races; going southwest and southeast from there, we encounter lighter and lighter races of the species, until centers of very pale races are reached in southern California in the Western, and in Turkestan in the Eastern Hemisphere. It is interesting that although the darkest and the lightest races, respectively, are usually confined to the "centers" just defined, the exact location of the "center" is somewhat variable for different species. Thus, in some of the Old World species the maximum depigmentation is observed in Russian Turkestan, in others in Chinese Turkestan (Tarim depression), and in still others in Persia.

II. Bergmann's rule, originally formulated for mammals and birds, but applicable to at least some invertebrates as well, states that races living in cooler climates are larger in body size than races of the same species in warmer climates. Bergmann's rule has also some exceptions, but not enough to invalidate it. According to Rensch (1936), among the palaearctic non-migratory birds, only 8 per cent, among those from the Small Sunda islands 12.5 per cent, and among the North American ones 26 per cent are exceptions. The rule fares less well among European mammals (40 per cent of exceptions), and better among those of North America (19 per cent of exceptions). The rule is extended to cover not only the horizontal, but also the vertical distribution: in mountain countries races from higher elevations are larger than those from the lower ones.

III. Allen's rule states that the races of mammals inhabiting cooler regions have relatively shorter tails, legs, and ears than races of the same species from warmer regions. Among birds the same is true for the relative lengths of beak, legs, and wings. An impressive list of examples of this rule has been given by Rensch (1936), who has examined the information available in the literature on various families. The rule has from 10 per cent to 36 per cent of exceptions.

IV. The rule for the wing shape in birds states that races inhabiting cooler regions have relatively narrower and more acuminate wings than races of the same species from warmer regions. In a fairly adequate sample of species inhabiting the temperate as well as the subtropical and tropical regions, Rensch (1936) found only 3 per cent of exceptions to this rule.

V. The rule for the pelage in mammals states that races from warmer countries have shorter but relatively coarser hair and less down than races from cooler countries. Rensch (1936) quotes among others the examples given in Table 16.

TABLE 16

THE LENGTH (IN MM) AND THE WIDTH (IN MICRA) OF THE HAIR IN RACES OF DIFFERENT CLIMATIC REGIONS (*after Rensch*)

SPECIES	ORIGIN	LENGTH	WIDTH
Felis concolor	Mexico	31.3	78.5
" "	Amazon	11.5	80.9
Felis pardalis	Tibet	20.5	87.8
" "	Zambesi	10.1	84.7
" "	Kongo	12.9	72.4
Lutra lutra	East Germany	22.9	46.2
" "	Ceylon	15.4	95.5
Canis vulpes	East Germany	45.7	77.0
" "	Algeria	39.4	77.0
Capreolus capreolus	Altai	47.5	66.2
" "	Germany	27.8	41.6

The adaptations of the plant species to the climatic and other conditions of their habitat are too well known to necessitate any extended discussion here. The parallel or convergent alterations that representatives of most diverse plant families undergo in a desert environment are at present school examples of adaptations, and are described in many treatises on plant biology. Zoology is far outdistanced by botany in the formulation of rules describing the interrelations between the organism and its surroundings. It should be noted that the formation of races, or "ecotypes," with analogous characteristics clearly consonant with the environment has been observed in many plant species often remotely related to each other (cf. especially Turesson 1922 and later work).

Strange as it may seem, the correlations between race formation and the environment revealed by the "rules" such as those just discussed have been repeatedly quoted as arguments against the natural selection theory. This amazing confusion of thought was due in the past to the almost universal acceptance among biologists of the belief in inheritance of acquired characteristics. Racial differentiation has been considered a result of the modification of the phenotype by the environment, perpetuated by a gradual change of the germ plasm in the direction of the phenotypic change. This interpretation was regarded as established almost beyond doubt by the results of experiments which showed that a change analogous to that observed in a particular geographical race can be brought about in other races of the same species as well by an exposure to certain environmental agents. Thus, in the classical experiments of Standfuss, butterflies resembling the varieties known from Syria and southern Italy were obtained from the pupae of the central European races of the same species exposed to heat treatment. On the other hand, treatment of the central European race with cold resulted in a resemblance to the form from northern Scandinavia. An exposure of a mammal to cold or heat may produce respectively an increase or a decrease of the hair length, a change that is quite analogous to that distinguishing the geographical races from the high and the low latitudes (cf. Table 16). Bergmann's, Allen's, and also Gloger's rules (see above) have their counterparts among the changes that can be induced in an animal by an application of appropriate environmental stimuli, namely, temperature and humidity or a combination thereof. The same is true for plants, where the phenotypic changes wrought by treatments with external agents may simulate the characteristics of races and species existing in nature.

Goldschmidt (1935) points out that theoretically any change of the phenotype caused by an alteration of the germ plasm may be reproduced as a purely phenotypic modification if a suitable experimental technique is evolved. Indeed, Goldschmidt (1935) describes special temperature treatments that induce in *Drosophila melanogas-*

ter phenotypic variations ("phenocopies") that more or less resemble some of the well-known mutant types of that insect. Friesen (1936) secured similar results by treating the developmental stages of Drosophila with X-rays. It is nevertheless obvious that there exists a fundamental difference between a mutant and a phenocopy. A mutant is a genotypic change and a phenocopy is a modification solely of the phenotype. The offspring of a phenocopy is indistinguishable from the original type, unless it develops in the same environment which has induced the phenocopy to begin with; the offspring of a mutant is a mutant. In discussing this subject one must constantly keep in mind the elementary consideration which is all too frequently lost sight of in the writings of some biologists; what is inherited in a living being is not this or that morphological character, but a definite norm of reaction to environmental stimuli. The norm of reaction of the "normal" *Drosophila melanogaster* is such that in the usual, standard, environment the development results in the appearance of a fly with "normal" or "wild type" characteristics, but with environment modified in a definite fashion a phenocopy is produced. The norm of reaction in a mutant is such that certain aberrant characteristics appear in individuals that have developed in the standard environment. In other words, a mutation changes the norm of reaction, but in a phenocopy the norm of reaction remains unaltered.

Evidence showing that the formation of geographical races involves hereditary, and not merely phenotypic, changes has been already discussed above. It may be reiterated here that the work of the botanical school of experimental taxonomy (Turesson, J. Clausen, and others) has brought to light the fact that even the small local variations in plants are frequently hereditary. Representatives of the same species from different habitats when brought together in an experimental garden do not become identical, although they frequently undergo noticeable and sometimes striking phenotypical changes. Few systematic studies of this kind have been made in animals, but evidence to show that the geographical races are mostly hereditary is nevertheless not lacking. In particular, the "rules" of the geographical variation discussed above pertain to genotypic changes and not merely to phenotypic modifications, as attested by the experimental data reviewed by Rensch (1936). To suppose that geographical races were originally modifications that have subsequently been fixed by heredity is contrary to the whole sum of our knowledge, just as it would be absurd to assume that mutations are phenocopies that have become hereditary. Therefore, the regularities observed in the process of geographical race formation can not be due to direct effects of the secular environment, albeit the results of this process can in part be imitated by some phenotypic modifications. In the light of these considerations, the parallelism between the genotypic and the phenotypic variability acquires a new and profound significance.

Adaptation, that is, a harmony between the organism and its environment, may be arrived at by two distinct methods. First, the norm of reaction of the genotype to the external stimuli may be adjusted to the effects of the environmental agents that are likely to be encountered under the "natural" conditions. In other words, the organism may respond to the impact of these agents by producing a phenotype that is most likely to survive in the particular environment. An ideal genotype would be capable of producing an optimal response to any environment. It appears, however, that no organism has evolved such a paragon of adaptability. A second method of becoming adapted has been resorted to, namely a genotypic specialization. A change in the genotype alters the reaction norm, and some of the alterations may enable the new genotype to produce a harmonious response where the ancestral has become a failure. Thus, the normal genotype of *Daphnia longispina* enables the organism to survive at 20° but not at 27°, while the mutant found by Banta and Wood (1927) survives at 30° but not at 20°C. Selection deals not with the genotype as such, but with its dynamic properties, its reaction norm, which is the sole criterion of fitness in the struggle for existence.

One of the principal tenets of Lamarckism is that the response of the genotype to the environmental stimuli that are encountered in the "normal" milieu is often, perhaps even as a rule, definitely adaptive. Modern genetics may, I think, accept this as an accurate statement of the observed facts, with the important reservation that the genotype responds not by changing itself but only by begetting an altered phenotype. The adaptive value of the development of a longer pelage and a greater amount of wool in a cool climate is indeed obvious. Once this is admitted, the conclusion follows that in a cold climate natural selection must favor the genotypes which, other conditions being equal, produce a warmer pelage. In a warm climate the sign of the selective value of a genotype may be reversed. The fact that the races of mammals inhabiting cold countries usually have longer hair and more down than races of the same species from hot countries is consequently evidence for and not against the effectiveness of natural selection.

The interpretation of the other rules governing the geographical variability (some of which have been discussed above) is admittedly more difficult than is the case with the variations of the pelage. The problem involved is essentially a physiological one, but the comparative physiology of geographical races is in its infancy. As a working hypothesis, one may assume however that Bergmann's rule (large body size in the cool and small size in the warm climates) is concerned with the temperature regulation of the animal. A large body size is correlated with a relatively smaller body surface, and consequently with a more limited loss of heat. A similar explanation would appear to apply to Allen's rule as well. The protruding body parts, the extremities, tails, and ears, are especially subject to a rapid loss of heat. The increase of the body surface in just these regions is therefore unfavorable in the cold, and may prove desirable in the warm, climates. Gloger's rule for the vertebrates resembles the sun tanning reaction of the human skin, and may have an analogous significance (Rensch 1936).

ORIGIN OF DOMINANCE

In a series of articles, Fisher (1928, 1930, 1931, 1932) shows that one of the longest known among the purely genetic phenomena, namely, that of the dominance of one allelomorph over another, may be interpreted as an evidence of the effectiveness of natural selection in free living populations. The point of departure in Fisher's line of reasoning is the well-established fact that mutations, whether spontaneous or induced by treatments with X-ray, are as a rule recessive to the ancestral or wild type. Among the mutations recorded in *Drosophila melanogaster*, more than two hundred are recessive and only thirteen are dominant. These figures do not include the large class of the recessive lethals, since their counterpart, the dominant lethals, are difficult to detect, and consequently the relative frequencies of the two kinds can not be accurately estimated. Of course, it must be kept in mind that dominance versus recessiveness is a matter of degree. Some of the mutant genes classed as recessives produce a slight effect in the heterozygous condition, and none of the dominants suppress the action of the wild type allelomorphs completely. Nevertheless, the greater frequency of the recessive mutants in *Drosophila melanogaster,* as well as in most other species in which mutants have been observed, is clear enough. Yet, if the degree of the dominance of a mutant were due to chance, one might expect that a majority of the mutants would be neither dominant nor recessive but intermediate in the heterozygotes. The more nearly dominant and the more nearly recessive mutants should be about equally frequent and on the whole rare.

Fisher makes the well justified assumption that most of the mutations that we observe occurring in the laboratory or in nature at our time level have repeatedly taken place in the history of the species, and proceeds to consider the consequences of this assumption. It is known (cf. Chapter II) that a majority of the mutations exert when homozygous an unfavorable influence on the viability of the carrier. Owing to their recessiveness, the heterozygotes are, however, little if at all inferior to the ancestral type. In wild populations the frequency of the heterozygotes for most mutants is very much greater than the frequency of the corresponding homozygotes. Now, a "new" mutation (that is a mutation that has not frequently taken place in the past) may be supposed to be mostly neither dominant nor recessive, and therefore the heterozygotes should be intermediate between the wild type and the homozygous mutant, both in the morphological characters and in the viability. This would place the heterozygotes at a disadvantage in a competition with the wild type. If a mutation takes place at a rate $1 : 1,000,000$, and the viability of the heterozygote is 99 per cent of that of the ancestral type, the equilibrium frequency of such mutation in a population is $1 : 10,000$ ($q = 0.0001$). With a viability of 90 per cent, the equilibrium frequency is $q = 0.000,01$ and at 50 per cent it is $q = 0.000,002$. In many species whose total population is large, this would still mean absolutely large numbers of the heterozygous carriers.

The viability of the heterozygous carriers might be improved if they had a gene or genes that suppressed the effects of the heterozygous mutation, in other words made the mutation recessive. Provided that such modifying genes produce themselves no undesirable effects on the wild type, they must, then, have a selective advantage in a population in which a given mutation is occurring, and the species might in the course of time become homozygous for the modifiers. Fisher's conception is that the dominance of the wild type over the mutant allelomorphs is due not to the intrinsic properties of the former, but rather to the presence of a system of modifying genes which make a majority of the frequently occurring mutations recessive. The evidence in favor of this conception may be summarized briefly as follows.

Some of the genes in *Drosophila melanogaster* have been observed to mutate repeatedly and to produce series of allelomorphs with distinguishable properties (multiple allelomorphs). Thus, the gene "white" in Drosophila has allelomorphs producing white, ivory, eosin, apricot, and other colors of the eyes, instead of the normal red. Multiple allelomorphs are known also in other organisms, both in animals and in plants; their behavior has been studied in detail, especially in rodents (Wright 1925). A rule applicable to most of the known multiple allelomorphic series is that a majority or all of the mutant allelomorphs are recessive to the wild type gene (that is, to the allelomorph found in the normal, free living, representatives of the species). The mutant allelomorphs display however no dominance or recessiveness with respect to each other, the heterozygotes being intermediate between the two homozygous mutants. This is what one might expect on the basis of Fisher's theory: individuals heterozygous for two mutant allelomorphs must have been very rare in natural populations, and the species could not acquire modifiers that would regulate so exceptional a condition. A similar argument accounts also for another fact: a mutant may be completely recessive to its wild type allelomorph when tested on an otherwise wild type background, but the recessiveness may be weakened in the presence of other non-allelomorphic mutant genes. Among the mutant genes having pronounced manifold effects (cf. Chapter II), the degree of dominance may be different for different characters. In *Drosophila melanogaster* the genes white, yellow, ebony, and others produce not only striking external changes (eye and body colors, etc.) but also what seem minor changes in the internal organs, such as the spermatheca. The external effects are recessive to the wild type condition, but the internal organs are more or less intermediate in the heterozygotes (Dobzhansky 1927). According to Ford (1930) and Fisher (1932) selection is operative upon the external changes but the internal ones are probably indifferent. That being the case, dominance modifiers would develop for the former but not for the latter. For some exceptional species in which many mutations are dominant to the wild type condition (e.g., in domestic poultry) Fisher finds a special explanation which seems to reconcile these cases with the theory.

Wright (1929, 1930, 1934) and Haldane (1930) believe however that the selection rates for the modifying genes producing dominance must be so negligibly small that the phylogenetic process postulated by Fisher could take place only very rarely, if ever. They put forward an alternative hypothesis that likewise involves natural selection, but in a way very different from Fisher's. If gene action is similar to that of enzymes (which is a notion favored by most investigators at present) then each wild type gene must produce a quantity of enzyme sufficient to accomplish the development of what we designate as the wild-type or "normal" phenotype. It may well be that threshold reactions are frequently involved, and a doubling or even a multiplication of the effectiveness of the gene in question produces no appreciable modification of the normal course of the development, but a halving of the activity leads to a retardation or an arrest of the whole chain of reactions. Under such conditions, wild type allelomorphs possessing a "factor of safety," that is having an activity well above the necessary minimum, will be of advantage to the organism. A mutation that curtails the activity of such a gene even to zero would then be recessive to the normal allelomorph.

Experimental data that would permit a discrimination between Fisher's and Wright-Haldane's hypotheses are scanty. Harland (1932-36) has made, however, a series of investigations on the dominance relationships of certain genes in different species of cotton (Gossypium) the results of which are very suggestive. A mutation "crinkled dwarf" producing a crinkling of the leaves has been repeatedly observed to occur in the Sea Island cotton, a variety of the species *G. barbadense*. In that variety the mutation is recessive to the normal condition. Another species of cotton, *G. hirsutum*, is not known ever to have produced the crinkled mutant. In the hybrids between certain strains of *G. barbadense* and *G. hirsutum* the F_1 generation, heterozygous for crinkled, shows, however, an incomplete dominance of the normal over the crinkled allelomorph, and in the F_2 generation a whole series of types appear, in some of which crinkled is recessive, in others intermediate, and in still others more nearly dominant. This is due to the presence in *G. barbadense* of modifiers that make crinkled recessive, these modifiers being absent in *G. hirsutum*. So far, then, the results are in an agreement with the Fisher theory. Crinkled dwarf is recessive in the form in which this mutation is known to have occurred, and the recessiveness is impaired in hybrids with a form in which it is not known.

Harland has, however, carried the analysis of the situation much further. In a series of experiments the crinkled mutant of *Gossypium barbadense* has been outcrossed to *G. hirsutum*. By means of repeated backcrosses of the hybrids to *G. hirsutum*, strains were obtained that had the genotype of the latter species except for a chromosome or a part thereof carrying the crinkled gene. Using different strains of *G. hirsutum*, Harland found that the dominance of the wild type over crinkled is due not only to special modifying genes but also to the presence of a variety of wild type allelomorphs. Some strains carry wild type allelomorphs that, other conditions being equal (that is in the presence of the same system of the dominance modifiers), exhibit a nearly complete dominance over crinkled; in other strains crinkled proves to be semi-dominant. Moreover, some of the *G. hirsutum* strains in which the crinkled mutation has never been observed possess a system of modifying genes that make the mutation recessive, though these modifiers are different from those in the Sea Island variety of *G. barbadense*. Similar results can be accomplished by different means. As a consequence, crinkled is recessive when placed on the genotype background of any of these strains; when, however, the strains are outcrossed, the systems of the dominance modifiers are broken down owing to Mendelian recombination. The original results obtained by Harland in the G. *barbadense* × G. *hirsutum* cross (see above) were due probably to such a breakdown of the modifier systems rather than to the absence of the dominance modifiers in G. *hirsutum*. The weight of the evidence seems to militate against the theory of Fisher, and to be rather in favor of that of Wright and Haldane.

SELECTION RATE AND THE GENETIC EQUILIBRIUM

The inadequacy of the experimental foundations of the theory of natural selection must be admitted, I believe, by its followers as well as by its opponents. The work to date has been concentrated mainly on proving the reality of natural selection as a process actually going on in wild populations, and a fair degree of success has been achieved in this field. The evidence presented in the foregoing paragraphs seems to demonstrate the effectiveness of selection in bringing about certain adaptational changes. The manner of action of selection has been dealt with only theoretically, by means of mathematical analysis. The results of this theoretical work (Haldane, Fisher, Wright) are however invaluable as a guide for any future experimental attack on the problem.

The main assumption inherent in the theory of natural selection is that some hereditary types of a species may have a certain advantage over others in survival and reproduction. Mathematically, the simplest case is the so-called genic selection, when the gene allelomorphs a and A tend to be reproduced in each generation in the ratios $(1-s) : 1$. The selection coefficient, s, is a measure of the advantage or the disadvantage. Thus, if on the average only 999 gametes carrying a function where 1,000 gametes carrying A are retained, the value of s equals $+0.001$. If the frequencies of the genes A and a in the initial population are q and $(1-q)$ respectively, in the next generation they will become $(1-s) (1-q)/1-s (1-q)$ for a, and $q/1-s (1-q)$ for A. The change in the frequency of the gene A per generation is therefore $\Delta q = sq (1-q)/1-s (1-q)$. If the selection coefficient s is small, the value $s (1-q)$ is close enough to zero to be disregarded, and the formula becomes $\Delta q = sq (1-q)$.

Zygotic selection, when the carriers of the three genotypes aa, Aa, and AA reproduce in the ratios $(1-s')$, $(1-hs')$, and 1 respectively, is probably most important in nature. Here we have a selection coefficient s' to measure the advantage or disadvantage of one of the homozygotes, and hs' for that of the heterozygotes Aa. The rate of change in the gene frequencies q and $(1-q)$ under the zygotic selection may be calculated according to the formula $\Delta q = s'q (1-q) [1-q + h (2q-1)]$ given by Wright (1929, 1931). A special case of zygotic selection is when A is completely dominant over a, and consequently the viability of the heterozygote Aa is equal to that of the homozygote AA; this case is important owing to the prevalence of the recessive mutant genes in wild populations (Chapter III). It is easy to see that the formula for the gene frequency change turns out to be $\Delta q = s'q (1-q)^2$.

Very interesting computations of the speed with which a change in the genetic constitution of a population may occur under the influence of selection have been made by Haldane (1932). He finds that in general the progress of selection is fairly rapid, that is to say, it takes relatively few generations to accomplish a given amount of change, only for intermediate gene frequencies. If, on the other hand, the initial gene frequency is very small or very large, approaching either zero or one hundred per cent of the population, the progress of selection is appallingly slow even with appreciable selection coefficients. For example, with a selection rate $s' = 0.001$ it takes 11,739 generations to raise the frequency of a dominant gene from $q = 0.000,001$ to $q = 0.000,002$. A similar amount of change for a recessive gene would take as many as 321,444 generations. For the further progress of a dominant the following numbers of generations are needed:

Frequency	0.001—1%	1—50%	50—99%	99—99.999%
Generations	6,920	4,819	11,664	309,780

It is possible to calculate that even with selection coefficients smaller than 0.001 the process will go on, that is to say, the gene frequencies may eventually increase from a very small value to nearly one hundred per cent, or vice versa. Selective advantages or disadvantages for a given genetic type in the population, however small and insignificant they may seem, will be ultimately instrumental in bringing forth an evolutionary change. The number of generations, and consequently the amount of time needed for the change, may, however, be so tremendous that the efficiency of selection alone as an evolutionary agent may be open to doubt, and this even if time on a geological scale is provided.

The above calculations are, of course, based on the assumption that mutation does not occur. Combined with mutation, the process of selection may be either enhanced in speed or, vice versa, slowed down still further. If mutation to an allelomorph favored by selection takes place more frequently than away from that allelomorph, the speed of the process is greatly accentuated in its initial stages. After an initial increase in frequency, the relative importance of the mutation declines, and the further increase has to proceed by selection alone (unless, of course, the mutation rate is very high). Thus, with an allelomorph A mutating to a at a rate $1:1,000,000$, the change of the gene frequency from $q = 0.000,001$ to $q = 0.000,002$ is accomplished in a single generation, while a similar change requires $321,444$ generations for a recessive type with a selective advantage of 0.001 (see above).

If the predominant mutation rate is away from the allelomorph favored by selection, the gene frequency will never reach zero or unity, and a genetic equilibrium will be established instead. The position of the equilibrium will evidently be determined by the relation between the conflicting mutation and selection rates. The calculations of the equilibrium values for various mutation rates and types of selection is possible with the aid of simple formulae given by Wright (1929, 1931a), Fisher (1930), and others. Suppose that the gene A mutates to a at a rate u per generation, and that this mutation is opposed by a genic selection of the intensity s, conferring an advantage on the allelomorph A; the rate of change of the gene frequency q of A will be $\Delta q = -uq + sq(1-q)$. The attainment of an equilibrium means that no further change takes place, that is, $\Delta q = 0$. Solving the equation, we find the equilibrium value $q = 1 - u/s$. For a mutation opposed by a zygotic selection, the survival values of the individuals carrying the genotoypes aa, Aa, and AA are $(1-s')$, $(1-hs')$, and 1, respectively. The equilibrium value proves to be q $= 1 - u/hs'$, unless h approaches zero. For an unfavorable dominant mutation, in which the viability of the heterozygote Aa equals that of the homozygote AA, the equilibrium value for the wild type gene is q $= 1 - u/s'$. Finally, for an unfavorable recessive mutation the equilibrium for the wild type gene is q $= 1 - \sqrt{u/s'}$. Taking the mutation rate u to be 0.000,01, and the selection coefficient $s = 0.001$, the numerical values of the equilibrium points will be q $= 0.99$ and q $= 0.90$ for the normal allelomorph of a dominant and a recessive mutant gene, respectively. In other words, the unfavorable recessive mutations will be allowed to accumulate in the populations to a much grater extent than the equally disadvantageous dominants. This is, of course, exactly what the analysis of wild populations reveals.

Evolutionary changes engendered in a population by the combined action of the mutation and selection pressures will continue to take place as long as some genes have not yet reached their equilibrium frequencies. The attainment of the equilibrium for every gene would mean cessation of evolution. In practice such a contingency is, however, very remote, since the equilibrium values themselves are constantly changing. The magnitudes as well as signs of the selection, and to a lesser extent also of the mutation rates, are determined by the environment. The environment does not remain constant, not only in terms of geological periods but even from one year to the next. Selection and mutation rates, and hence the genetic equilibria, are in a state of perpetual flux. The nature of the genetic mechanism is therefore such that the composition of the species population is probably never static. A species that would remain long quiescent in the evolutionary sense is likely to be doomed to extinction. Whether the combined forces of mutation and selection are sufficient for a sustained progressive evolution is, nevertheless, not immediately clear. The extreme slowness with which the new favorable mutations that might arise in the species from time to time can acquire a hold on the species population has been pointed out already. The fluctuations of the gene frequencies caused by the variations in the equilibrium values for various genes make the process essentially reversible, since the restoration of the original environment after a temporary change brings the equilibria back to their old positions.

Granted that a complete reversion is unlikely, the evolutionary effectiveness of the process is still rather low (Wright 1931a). These considerations make it especially important to study the relations between selection and the factor discussed in the preceding chapter, namely the restriction of the population size.

SELECTION IN POPULATIONS OF DIFFERENT SIZE

The mode of action of natural selection has been described so far without reference to the size of the population on which the selection operates. As a matter of fact, most discussions of the evolutionary rôle of selection that can be found in the biological literature omit this point entirely. And yet it is known that the effectiveness of selection is to a certain degree a function of the population size; selection pressures of equal intensity may be effective in large and impotent in small populations. In the preceding paragraphs formulae for the genetic equilibria have been discussed; it must be emphasized that these formulae are strictly applicable only to infinitely large populations, or at least to populations in which the products $4Ns$ and $4Nu$ are much greater than unity. N is here the population number in the sense defined in Chapter V.

The relations between selection and the population size have been investigated chiefly by Wright (1931a and later work); his analysis is mathematically rather abstruse, forcing us to refrain from reviewing it here in detail. The end result is however simple enough. We have already seen in Chapter V that in small isolated populations the gene frequency, q, does not remain constant indefinitely, but fluctuates over a wide range of values. In the absence of mutation and selection, a fraction of the unfixed genes equal to $1/2N$ is eliminated in every generation on account of some allelomorphs reaching fixation and others being lost. The process as a whole leads to a gradual but relentless depletion of the store of hereditary variability. In large populations this scattering of the variability is unimportant, since the fraction $1/2N$ is negligibly small. Selection is intrinsically opposed to random variations in the gene frequencies; on the contrary, it tends toward the fixation of definite allelomorphs (namely of those that possess the highest adaptive value) and to the loss of the less favorable ones. The two processes are consequently

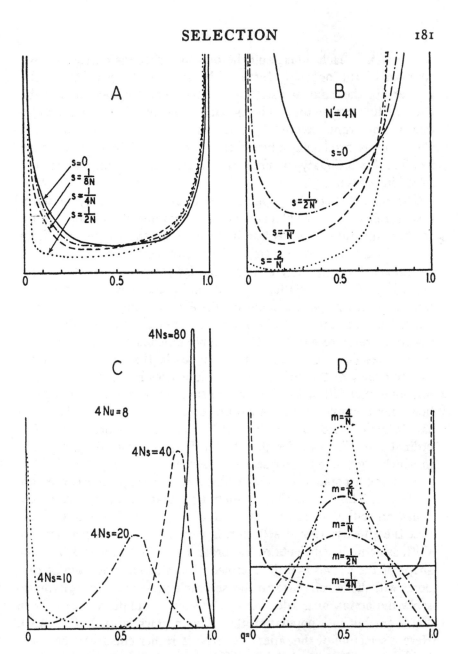

FIG. 15. Distribution of gene frequencies in populations of different size under different selection, mutation, and migration pressures. Further explanation in text. (From Wright.)

pitted against each other, and the outcome depends on the relative strength of the opposing forces. The smaller the population, the greater are the random variations of the gene frequencies and the less effective can be the small selection pressures. In a small population allelomorphs favored by selection may actually be lost, and allelomorphs producing adverse effects on the viability may reach fixation. Evolutionary changes may therefore proceed against the selection pressure.

The relation between population size and the efficiency of selection is represented graphically on the diagrams shown in Fig. 15. In these diagrams the abscissa indicates the gene frequencies from $q = 1$ (fixation) to $q = 0$ (loss). The ordinate may have two interpretations. First, we may consider the fate of the different variable genes in a single population, for natural populations may vary with respect to many genes. Some of these genes may reach fixation, others may be lost, and still others may remain unfixed, that is to say they are represented by two or more allelomorphs with different frequencies ($0 < q < 1$). The ordinates in the diagrams indicate the frequency of the different gene frequencies in a population. Second, one may follow the fate of the same gene in different populations, for example, in the subgroups of the same species, that is to say, in the isolated colonies into which the species is broken up. The ordinates in Fig. 15 refer then to the frequencies of the subgroups in which a given gene frequency is reached.

In a small population (Fig. 15A) the gene frequency curves are U-shaped. This means that in such a population a majority of the genes are either fixed or lost most of the time. The curves show that the effectiveness of selection is very low in small populations. With selection coefficients of the order $s = 1/8N$ to $s = 1/2N$, the shape of the curves is little modified. Genes are lost or fixed at random, with little reference to the selection pressure. Fig 15B represents the action of a selection of the same absolute intensity as in Fig. 15A, but in a population that is four times larger ($N' = 4N$). Here a selection of the order $s = 2/N'$ is rather effective; the curve is no longer U-shaped but has a definite maximum at the right, indicating that the gene allelomorphs favored by selection mostly reach fixation and supplant the less favored ones. The equation describing this family of curves is, according to Wright (1931), $y = Ce^{4Nsq} q^{-1}$ $(1-q)^{-1}$, where e is the basis of the natural logarithms, q the gene frequency, s the selection coefficient, and N the population number.

The interaction of the mutation pressure with selection and the population size factor presents naturally an even more complex problem. In certain conditions, namely, if the selection pressure is of the same order of magnitude as the mutation rate, and both are very small, the random variations of the gene frequencies in small populations become most important. The curve of the gene frequencies is U-shaped, indicating that gene allelomorphs are likely to reach fixation or loss largely irrespective of the mutation and selection pressures. Mutation and selection rates that may be regarded as small are such as make the products $4Nu$ and $4Ns$ smaller than unity. The greater the mutation and selection pressures, the greater is their effectiveness. An example borrowed from Wright (1931a) is reproduced in Fig 15C. It refers to a population of an intermediate size with a moderate mutation rate ($4Nu = 8$) opposed by a selection of varying intensity. With a selection of the order $4Ns = 10$, the mutation pressure gets the upper hand. The allelomorph favored by selection is mostly lost. A doubling of the selection intensity ($4Ns = 20$) changes the situation very appreciably. The gene frequencies fluctuate over a great range of values. In a species segregated into numerous isolated subgroups this means a great differentiation of the species population into local races, some of which will possess characteristics favored by selection and others the relatively unfavorable ones. With a selection becoming more stringent ($4Ns = 40$ and $4Ns = 80$), the amplitude of variation is gradually restricted, the gene frequencies being kept within rather narrow limits centered upon the equilibrium values. The equation describing the curves is $y = Ce^{4Nsq} q^{-1} (1-q)^{4Nu-1}$, where u is the mutation pressure.

The biologically highly significant corollary of the above analysis is that a species, broken up into isolated colonies, may differentiate mainly as a result of the restriction of the population size in these colonies. Even if the environment is homogeneous for all the colonies, and selection and mutation rates are the same and the initial composition of the populations identical, a sufficient lapse of time will bring about a differentiation. It must be stressed that the differentiation need not, theoretically, be of an adaptive kind. The latter point, however, must be much qualified by the consideration that in practice the separate colonies are exposed to a variety of environments. Although in the abstract the differentiation may be pictured as taking place under the influence of the restriction of the population size alone, or under the influence of the selection alone, in nature the process is going on because of both these factors, one or the other gaining the upper hand probably only temporarily. The genetic equilibrium in a living species population seems to be a delicately balanced system which can be modified by a number of agents.

The problem of the subdivision of the species population into colonies has already been discussed in Chapter V, and the meager observational evidence bearing on it has been reviewed there. It remains now only to point out that the genetic differentiation of the populations in the colonies, by whichever factors it may be caused, is counteracted by migration, that is by the interchange of individuals between the colonies. A complete isolation of colonies of one species from each other is probably a rather exceptional condition in nature. Whether the colonies are isolated geographically (which is probably the commonest method) or in any other way (for instance by the difference in the breeding seasons), a certain number of individuals from one colony is likely to pass to the adjacent one, and to join the breeding population of the latter. If the gene frequencies in the colonies are different, this migration pressure will evidently tend to decrease and eventually to level the difference. The racial differentiation in the semi-isolated colonies can therefore proceed only so long as the factors provoking it are more effective than the opposing migration pressure.

How large the migration pressure may be without annulment of the genetic differentiation in semi-isolated colonies has been studied mathematically by Wright (1931a). Let the frequency of a certain gene in the species population as a whole be 50 per cent ($q = 0.5$); the species is broken up into semi-isolated colonies, each exchanging the proportion m of its population with a random sample of the population of the species as a whole. With $m = 0$—that is, isolation being complete—the gene frequency in the separate colonies will eventually become different, q reaching zero in some and unity in other colonies. A similar result is accomplished with m = 1/4N, as shown by the U-shaped curve in Fig. 15D. With m = 1/2N, a critical value for the migration coefficient is reached, since above this the random fixation or loss of the gene allelomorphs in the separate colonies is unlikely. The value of 1/2N is a very low one, for it means an exchange of a single breeding individual every other generation. Wright points out however that an exchange between a colony and a random sample of the species population as a whole is only a rather theoretical situation, and that in practice migration occurs almost entirely between adjacent colonies. In a widely distributed species, subdivided in very many local populations, migration coefficients even greater than 1/2N will permit a considerable drifting apart of the genetic composition in the subgroups.

CONCLUSIONS

In the two preceding chapters an attempt has been made to analyze the different agents that are known or believed to influence the fate of the genetic diversity in natural populations. Gene mutations and chromosomal changes are constantly being produced and thrust into the genotype of the species. The great variety of the mutational and chromosomal changes, and the apparent identity of at least some of them with the genetic elements into which the differences between the actually existing races and species can be resolved, lead to the conclusion that the mutation process, in the wide sense of that term, is adequate to supply the building blocks of evolution. The processes that are enacted in the living populations are, however, in part transcendental to the mutation process proper, for they belong to the realm of the physiology of populations.

The scattering of the hereditary variability, which is due to the intrinsic properties of the mechanism of reproduction, is perhaps one of the most elemental processes of the latter kind. Its effect is twofold. In an isolated self-sufficient population it leads toward a genetic uniformity, a loss of variance, and consequently to a restriction of the adaptive potencies. In a species subdivided into numerous semi-isolated colonies, the same process leads toward a greater differentiation of the species population as a whole, which may mean an increase instead of a decrease of the potentialities for adaptation. The process of migration, which means in this case an exchange of individuals between the semi-isolated local colonies, counteracts the differentiating effects of isolation, and prevents the approach toward a genetic uniformity in the separate subgroups. Finally, natural selection combats the excessive accumulation of the gene variants that are unfavorable under a given set of environmental conditions, and detects and engenders the spread and the increase in frequency of the rare mutations that may increase the adaptive level of the species under the static conditions of the milieu. Natural selection is probably most important when the environment undergoes changes, for it is the sole known mechanism capable of producing a reconstruction of the genetic make-up of the species population from the existing elements. Such a reconstruction may be necessary in order that the species remain attuned to the demands of the environment and escape extinction.

Since evolution as a biogenic process obviously involves an interaction of all of the above agents, the problem of the relative importance of the different agents unavoidably presents itself. For years this problem has been the subject of discussion. The results of this discussion so far are notoriously inconclusive; the "theories of evolution" arrived at by different investigators seem to depend upon the personal predilections of the theorist. One of the possible sources of this situation may be that a theory which would fit the entire living world is in general unattainable, since the evolution of the different groups may be guided by different agents. To a certain extent this possibility is undoubtedly correct. In species that occur in great abundance in a fairly continuous area of distribution, the population size factor is bound to be less important than in species that are subdivided into numerous local colonies each having a small effective breeding population. In organisms whose environment is in the throes of a cataclysmic change, natural selection and mutation pressure are more important than in organisms living in a relatively constant environment. Nevertheless, one can hardly eschew trying to sketch some sort of a general picture of evolution. A very interesting attempt in this direction has been made by Wright (1931a, 1932), whose lead we may partly follow.

In an organism possessing only 1,000 genes each capable of producing ten allelomorphs, the number of the possible gene combinations that may be formed is 10^{1000}. Some, probably a great majority, of these combinations are discordant and have no survival value, but still very numerous ones may be supposed to be harmonious in the different ecological niches of the same environment, as well as in different environments. If the entire ideal field of possible gene combinations is graded with respect to the adaptive value, we may find numerous "adaptive peaks" separated by "valleys." The "peaks" are the groups of related gene combinations that make their carriers fit for survival in a given environment; the "valleys" are the more or less unfavorable combinations. Each living species or race may be thought of as occupying one of the available peaks in the field of gene combinations. The evolutionary possibilities are twofold. First, a change in the environment may make the old genotypes less fit than they were before. Symbolically we may say that the "field" has changed, some of the old peaks have been leveled off, and some of the old valleys or pits have risen to become peaks. The species may either become extinct, or it may reconstruct its genotype to arrive at the gene combinations that represent the new "peaks." The second type of evolution is for a species to find its way from one of the adaptive peaks to the others in the available field, which may be conceived as remaining relatively constant in its general relief. Here one must emphasize a very important point: the field of all the possible gene combinations is by no means immediately available to the species. Some of the combinations have undoubtedly never been found and tried (this is guaranteed by the stupendous number of them which is far greater than the number of individuals in any species). Hence, a shift from one adaptive peak to another, which may be even higher than the first but separated from it by an adaptive "valley," must involve a trial and error mechanism on a grand scale which will enable the species to "explore" the region around its own peak in order to finally encounter a gradient leading toward the other peak or peaks.

We may consider first the evolutionary possibilities of a very large undivided species (Fig. 16, Fig. 17A,B,C), where the effective breeding population is such that the products 4Nu and 4Ns are much larger than unity, or where the exchange of individuals between the colonies is so rapid that no isolation obtains. In such a species, each gene ultimately reaches its equilibrium frequency determined by the interaction of the mutation and the selection pressures. An increase of mutation or a relaxation of selection increases the variance in a non-

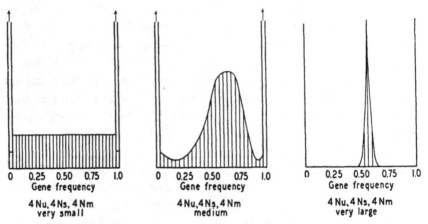

FIG. 16. Influence of population size on the distribution of gene frequencies. Further explanation in text. (After Wright, modified.)

adaptive fashion. The field occupied by the species (Fig 17A) spreads along the slopes of its adaptive peak. If the spread is sufficiently great, the species may find its way to a neighboring adaptive peak and occupy it as well. The new favorable mutations that may occur will be used to increase the height of the existing adaptive peak. A decrease of the mutation rate, or an increase in the stringency of selection (Fig. 17B), will force the species to withdraw to the highest level of its adaptive peak. The variability is reduced, the average adaptive level of the individuals is however increased. The possibility of a progressive evolution is curtailed, except through mutations that may be favorable at the very start. Since such mutations are probably very rare, this condition leads to an extreme specialization which may prove fatal if the environment changes.

A change in the environment provokes an alteration of the relief of the adaptive field. The relative elevations of the peaks and valleys may change, or the location of the peaks may shift in any direction (Fig. 17C). In extreme cases the adaptive peak may, figuratively speaking, escape from under the species, which may be left in a valley instead. An evolutionary change may become essential in order to escape extinction. A species with a large variable population will probably undergo a reconstruction consonant with the new demands of the environment, the genes will gradually reach new equilib-

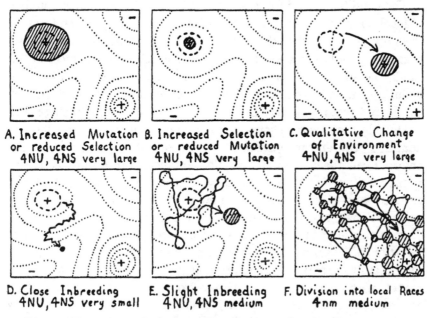

A. Increased Mutation or reduced Selection 4NU, 4NS very large B. Increased Selection or reduced Mutation 4NU, 4NS very large C. Qualitative Change of Environment 4NU, 4NS very large

D. Close Inbreeding 4NU, 4NS very small E. Slight Inbreeding 4NU, 4NS medium F. Division into local Races 4nm medium

FIG. 17. Fate of populations of different size and under different conditions in the "field" of gene combinations. Further explanation in text. (From Wright.)

rium values, and the new adaptive peak thus arrived at may prove to be higher than the old one.

A restriction of the population size which renders the breeding population very small (4Nu and 4Ns less than unity, Fig. 17D) leads to a depletion of the supply of the hereditary variability, and to fixation of genes more or less irrespective of their adaptive values. The part of the field occupied by the species is much reduced and the adaptive peak may be abandoned. This is a way to extinction. With an intermediate population size (Fig. 16 and 17E), the effect is about the same as with a relaxation of selection or increase of mutation in a large population; the species begins to wander on the slopes of the adaptive peak, and may find a gradient leading to a new one. The important difference is that a large population continues to occupy its old peak as well, while a smaller one moves in a body. The chance of extinction is obviously greater in the latter case, which consequently is unfavorable for evolution.

The conditions found in a species subdivided into a large number of local populations are very interesting (Fig. 17F). If the migration from one colony to the other is not very frequent, and if the effective breeding population sizes in some of the colonies are medium or small, the species has created numerous trial parties to explore the vicinity of its adaptive peak. The fate of each of these trial parties considered separately may be similar to that shown in Figs. 16 and 17D,E; in other words, some of them may wander too far from the adaptive peak and face extinction. With or without a change in the environment the species as a whole, however, is in a most advantageous position for an evolutionary change. For a colony that has reached a gradient leading to a new peak may climb it rapidly, increase in size, and either supplant the old species, or, more likely, form a new one that owes its allegiance to the new peak. Natural selection will deal here not only with individuals of the same population (intra-group selection) but also, and perhaps to a greater extent, with colonies as units (inter-group selection). In case of an environmental change, the colony that has first reached the vicinity of the new adaptive peak will obviously have the best chance of being the victorious one. Here, then, conditions are given both for a differentiation of a single species into derived ones, and for a movement of the species as a whole to a new status.

Wright (1931, 1932) argues very convincingly that the last discussed state of affairs in a species, that is, a differentiation into numerous semi-isolated colonies, is the most favorable one for a progressive evolution. Indeed, a number of considerations speak in its favor. The present writer is impressed by the fact that this scheme is best able to explain the old and familiar observation that races and species frequently differ in characteristics to which it is very hard to ascribe any adaptive value. Since in a semi-isolated colony of a species the fixation or loss of genes is to a certain degree independent of their adaptive values (owing to the restriction of the population size), a colony may become different from others simultaneously in several characters. One or a few of the latter may be adaptive, and may enable the population to conquer new territories or ecological stations. The rest of the characters may be neutral, with respect to adaptation, and yet they may spread concomitantly with the adaptive ones. For example, the chromosome structures that are so variable in *Drosophila pseudoobscura* can hardly be regarded as anything other than neutral characters, although some of them have become racial characteristics in subdivisions of the species population. This can be explained on the assumption that large parts of the territory now occupied by the species are inhabited by populations derived from relatively small colonies which were once living perhaps on the edge of the ancient species area. In such colonies certain chromosomal structures were established by chance, but now they have become the property of a widely distributed race.

Wright (1932) formulates his conclusions in the following sentences. "The most general conclusion is that evolution depends on a certain balance among its factors. There must be gene mutation, but an excessive rate gives an array of freaks, not evolution; there must be selection, but too severe a process destroys the field of variability, and thus the basis for further advance; prevalence of local inbreeding within a species has extremely important evolutionary consequences, but too close inbreeding leads merely to extinction. A certain amount of crossbreeding is favorable, but not too much. In this dependence on the balance the species is like a living organism. At all levels of organization life depends on the maintenance of a certain balance among its factors."

VII: POLYPLOIDY

I T HAS BEEN pointed out in Chapter III that the sudden origin of a new species by gene mutation is an impossibility in practice. The argument employed to prove this thesis is simple enough. Races of a species, and to a still greater extent species of a genus, differ from each other in many genes, and usually also in the chromosome structure. A mutation that would catapult a new species into being must, therefore, involve simultaneous changes in many gene loci, and in addition some chromosomal reconstructions. With the known mutation rates the probability of such an event is negligible. The process of species formation is apparently a slow and gradual one, consuming time on at least a quasi-geological scale.

It is highly remarkable, therefore, that alongside this slow method of species formation there should exist in nature a quite distinct mechanism causing a rapid, sudden, cataclysmic emergence of new species. It is remarkable, furthermore, that while the slow method seems to be encountered throughout the living world and may in this sense be called the general one, the cataclysmic origin of species is confined to some, though large, groups, mostly (so far as known) in the plant kingdom. The latter method of species formation is connected with a multiplication of the chromosome complement, called polyploidy. If the term "mutation" is used in its widest sense (Chapter II), the origin of polyploidy is, of course, a mutation by definition. The difference between a "mutation" causing polyploidy and a gene mutation is clear enough however, and the distinction between the two methods of species formation is therefore not likely to cause difficulties in practice. Moreover, one may note that species formation through gene mutation involves a gradual reconstruction of the genotype of the ancestral species to give rise to the genotypes of the derived ones. Species formation through polyploidy occurs mostly as a result of hybridization of two previously existing species. The chromosome complement is doubled in a hybrid, and the polyploid possesses all the genes which were present in both of its ancestors and no new genes. We are dealing here with a fusion of two old species to form a single derived species. Since however the ancestral species may continue to exist side by side with the polyploid, the total diver-

TABLE 17

HAPLOID CHROMOSOME NUMBERS IN FLOWERING PLANTS (*after Hernandes from Darlington, 1932*)

CHROMOSOME NUMBER	NUMBER OF SPECIES	CHROMOSOME NUMBER	NUMBER OF SPECIES
12	391	15	27
8	332	22	25
7	236	32	25
9	170	28	24
16	153	19	22
6	134	26	20
10	126	36	19
14	125	30	11
24	80	23	8
11	70	45	8
21	64	42	6
18	58	3	5
17	48	38	5
20	47	40	5
4	42	Others	39
27	31	Totals:	
13	30	Below 12	1,242
5	27	12 and above	1,171

sity in a given genus is augmented in the process. Both methods of species formation increase the organic diversity.

How widespread polyploidy is among plants is shown by an examination of the chromosome numbers in various genera. The fact that in many genera the species have chromosome numbers that are simple multiples of a minimum or basic number was noticed very early in the history of plant cytology. Classical examples of this are wheats (Triticum), where species with the diploid numbers 14, 28, and 42 (the basic haploid number is 7) are known, Chrysanthemum with 18, 36, 54, 72, and 90 (basic number 9), Solanum with 24, 36, 48, 60, 72, 96, $\pm$ 108, 120, and $\pm$ 144 (basic number 12), and

Papaver with 14, 28, 42, 70, 22, and 44 (basic numbers 7 and 11). The general statistics of the chromosome numbers among the flowering plants also reveal some interesting regularities (Table 17).

In a large sample containing 2,413 species, some numbers occur considerably more frequently than would be expected in a chance distribution. In a frequency distribution of numbers between 3 and 100, not a single peak is found at a prime number; and the most frequent prove to be the numbers with the lowest factors (8, 9, 12, 16, 18, 24, 27, 32, and 36). Moreover, the relative frequencies of certain low numbers (8, 7, 9, 10, 11, and 13) are repeated by those of their multiples (16, 14, 18, 20, 22, 26).

Winge (1917) was the first to emphasize the above regularities in the distribution of the chromosome numbers in plants. He concluded that the simplest way to account for them is to suppose that the reduplication of the chromosome complements, polyploidy, has taken place frequently in the phylogeny of some genera. Moreover, Winge put forward a suggestion that proved to be a prophetic one, namely, that the origin of a polyploid species is most likely to take place in the offspring of a hybrid between two diploid species, and more specifically where the chromosome pairing in the hybrid is decreased or lacking. The argument used by Winge in making this deduction is a teleological one: the "need" for a partner with which a chromosome could pair might stimulate the doubling of the whole complement which would provide such a partner for every chromosome. Nevertheless, the deduction has been borne out by the discovery of the fertile allopolyploids eight years later (Clausen and Goodspeed 1925). More recently, Müntzing (1936) has adduced some fairly convincing arguments to show that the doubling of the chromosomes in non-hybrid organisms (autopolyploidy) also must be regarded as an evolutionary factor of some consequence. The relatively greater rôle of the allopolyploids is however taken as established by most geneticists and cytologists.

Although species formation through polyploidy is restricted to certain groups of organisms, its interest can hardly be overestimated. In the particular groups where it occurs it seems to play a major rôle in phylogeny. Moreover, species formation through polyploidy has been studied experimentally in more detail, and is better understood than the more widespread method of the origin of species through the accumulation of mutational changes.

The origin within a species of individuals possessing twice as many chromosomes as normal (autotetraploidy) is one of the oldest known chromosomal aberrations. The gigas "mutation" repeatedly observed by De Vries in his classical experiments on *Oenothera Lamarckiana* proved to have 28 instead of the normal 14 chromosomes, and hence was a tetraploid. It is interesting that this first example of the chromosome reduplication may be classed either as a case of auto- or of allopolyploidy, since *O. Lamarckiana* is now known to be a permanent structural hybrid. This serves to emphasize the fact that the distinction between auto- and allopolyploidy is a matter of degree.

Tetraploids have been obtained experimentally in a number of organisms. As early as 1914, Gregory described tetraploid strains in *Primula sinensis*. Winkler (1916) induced polyploidy in Solanum by grafting together the species *S. lycopersicum* (tomato) and *S. nigrum*. The adventitious shoots that arose at the grafting point were sometimes tetraploid. Jorgensen and Crane (1927) and Jorgensen (1928) have shown that grafting is not essential in this process. Young plants of Solanum are decapitated, the formation of the axillary shoots is prevented, but the adventitious shoots that develop under the callus are permitted to grow. Some of the latter, about 6 per cent, are tetraploid. Evidently the cell divisions that take place in the vicinity of the scar tissue sometimes lead to a chromosome splitting without a corresponding cell fission, thus giving rise to cells with double the normal number of chromosomes. This method of obtaining polyploids seems applicable to a variety of plants besides Solanum (Karpenchenko 1935). Blakeslee and Belling (1924) have observed a similar process of the doubling of the chromosome number in the somatic tissues of the Jimson weed, *Datura stramonium*. Tetraploid branches arise in this plant both spontaneously and under the influence of cold treatments. Numerous tetraploids and higher polyploids were induced in various species and genera of mosses by Marshall and by Wettstein (1924, 1928). Polyploids have also appeared in species of Drosophila. Here they are due to the occasional production of gametes containing the somatic complement of chromosomes. Tetraploid spermatogonia and oögonia are not very rare in the gonads, and they presumably give rise to diploid sperms and egg cell. Fertilization of such diploid eggs by the normal haploid sperms results in triploid individuals.

Polyploid forms arising from diploid relatives through the doubling of the chromosome number occur in nature in some plants. The whole subject has been reviewed in detail by Müntzing (1936), who lists fifty-eight examples of this phenomenon; we may confine ourselves to the consideration of only two of them. *Biscutella laevigata*, a species of Cruciferae, is distributed over Central Europe and Italy. Among the races into which this species is subdivided, some have 18 chromosomes and are probably diploid, while others have 36 chromosomes and are tetraploid (Manton 1934). The distribution area of the tetraploid races is a continuous one, including the Alps, Carpathians, and the mountains of Italy and of the northern part of the Balkan peninsula. The area of the diploid races is much smaller in extent than that of the tetraploids, and in addition is sharply discontinuous. The diploids are confined to the valleys of Rhine, Elbe, Oder, upper Danube, and some of their tributaries. Although the origin of tetraploids in Biscutella has not been observed in experiments, it is virtually certain that they have arisen from diploid ancestors, since the change from a tetraploid to a diploid is much more rare than the opposite, and less likely to produce a viable type. The diploids may be regarded as relics of an ancestral population of *Biscutella laevigata*, and the tetraploids as their successors. Such an interpretation agrees with the known facts of the recent geological history of Europe. The diploids are confined to regions that were not covered by the ice sheet during the glacial period and which were consequently open to habitation by plants for a long time. On the other hand, the tetraploid races occur almost exclusively in the parts of the country that were under the ice till a geologically more recent time and must therefore be regarded as immigrants from elsewhere. Manton concludes that the diploid races represent interglacial, if not preglacial, relics.

Similar relationships are displayed by several American species of Tradescantia (Anderson and Woodson 1935, Anderson and Sax

1936). In each of these species, races are encountered that have twice as many chromosomes as others. Thus, *T. occidentalis* is distributed from the Rocky Mountains, over the prairie states, east to the Mississippi. Most of this area is occupied by tetraploids, but a restricted region in central and eastern Texas is inhabited by diploids. The range of *T. canaliculata* lies mostly East of that of *T. occidentalis*, but in a fairly broad strip of land west of the Mississippi the ranges of the two species overlap. The diploid form of *T. canaliculata* is, characteristically enough, confined to almost the same territory in which the diploid *T. occidentalis* is found. Anderson regards this fact as being due to more than a coincidence. The region that harbors the diploid races of both species mentioned (and of some others in addition) is geologically a very ancient territory which has been continuously open for plant habitation. It seems probable that the diploids are here also relics, and that the territory where they occur is a center from which the spread of the more successful tetraploid types has taken place; *T. occidentalis* has spread from there mostly north and northwest, while *T. canaliculata* has taken the direction east and northeast.

The doubling of the chromosome complement increases the number of genes, but the kind of genes and their quantitative relations, the genic balance, remain unchanged. Nevertheless, polyploids are as a rule distinguishable from the diploids from which they sprang in morphological as well as in physiological characters. This is probably due to the increase of the nuclear volume, which in turn produces various alterations in the cellular physiology. Most commonly the tetraploids possess the so-called gigas complex of characters, which was first observed in the gigas mutant of *Oenothera Lamarckiana*. According to Müntzing (1936), this is expressed in the tetraploid by a thicker stem, a greater height, larger, thicker and relatively shorter and broader leaves, a darker green pigmentation, larger flowers and seeds than in the diploid. Some of these characteristics are usually present both in the tetraploid races obtained in experiments and in the natural tetraploids, but exceptions are encountered. Some of the tetraploids are, for example, not taller than the diploids. Moreover, while some polyploid races are striking enough to have been noticed, described, and named by systematists, others differ from the diploid relatively only slightly, and have not been recognized as transgressing the limits of the ordinary individual variability. The lack of a perfect correlation between the chromosome multiplication and the external traits has been especially emphasized by Wettstein (1924, 1928). For example, cell size in some species of mosses is positively correlated with the chromosome number over the whole range of variations studied, from the haploid to octoploid and higher. In others a maximum cell size is reached, and a further chromosome multiplication produces no new increase, or may even result in a reduction of the cell size and a dwarfism of the plants.

Among the purely physiological consequences of polyploidy a decrease of the growth rate seems to be the most frequent one. Tetraploids develop more slowly, and therefore take more time to reach maturity and the flowering stage than the diploids. A decreased rate of cell division is the most probable explanation of this difference. Sansome and Zilva (1933) have compared the vitamin C concentration in diploid and tetraploid tomatoes and found that the latter contain about twice as much vitamin as the former. Furthermore, tetraploid tomatoes contain relatively more water, less cellulose and ash, and more nitrogen and protein than the diploids (Kostoff and Axamitnaja 1935). No such differences have been detected, however, in a comparison of diploid and tetraploid petunias made by the same authors. A greater hardiness of the polyploids under various environmental conditions is pointed out by several observers.

It seems clear enough that the doubling of the chromosome complement produces a change in the norm of reaction of the organism, and its effect is in this respect analogous to that of a gene mutation. As any other type of change in the reaction norm, that induced by polyploidy may be favorable for the organism under some and unfavorable under other environmental conditions. It is therefore not unexpected that the geographical distributions of some polyploids are different from those of the corresponding diploids. This is clear in the already discussed examples of Biscutella and Tradescentia, where the tetraploids are to all appearances more successful ecological types than the diploids, and have extended their geographical areas in geologically recent times, leaving the diploids behind as relics. According to Müntzing (1936), the geographical distribution of the diploids

and tetraploids is known or suspected to be different in thirty out of
the thirty-eight cases in which the appropriate data are available.
Where no geographical differences are present, the polyploids manifest some ecological peculiarities. Thus, the diploid *Allium schoeno-
prasum* is a subalpine plant in the Altai mountains, while the tetraploid is abundant at lower elevations. In other species the tetraploids
are, on the contrary, characterized by a more northern or alpine distribution than the diploids. In Eragrostis the diploid is annual and
adapted to moist habitats, while the tetraploid is perennial and stands
more dryness; finally, the octoploid is also perennial and grows in
sand dunes (Hagerup 1931).

An important property of the autopolyploids concerns the behavior of their chromosomes at meiosis (for a detailed discussion of
this topic see Darlington 1932). In a normal diploid organism every
chromosome has as a rule one and only one complete homologue. A
number of bivalents equal to the haploid chromosome number are
therefore formed, the disjunction at both meiotic divisions proceeds
normally, and all the gametes produced contain a complete haploid
set of chromosomes. In the autopolyploids every chromosome has
more than one homologue—two in a triploid, three in a tetraploid, five
in a hexaploid, etc. An opportunity for the formation of trivalents,
quadrivalents, and higher associations instead of bivalents presents
itself. A tetraploid may possess a haploid number of quadrivalents,
or pairing may be variable, with bivalents, trivalents, quadrivalents,
and univalents formed in varying proportions. The resulting disjunction is frequently abnormal. Instead of each pole's of the first division spindle receiving an equal number of chromosomes of each kind
(normal disjunction), the disjunction may be either genetically or
numerically unequal. In the former case the gametes will contain
some chromosomes in excess, and will be deficient for other chromosomes; in the latter, both the number and kind of chromosomes in
the division products will be unlike. Thus, a trivalent gives two
chromosomes to one pole and one to the other; a quadrivalent may
disjoin normally with two chromosomes going to each pole, but a
three-one distribution is also possible; the univalents are distributed
to the poles at random, but they may also undergo two equational
divisions. Now, for a tetraploid to breed true, it must produce only

gametes containing a complete diploid set of chromosomes. Any deviation from this condition results in an instability of the chromosome numbers in the progeny. A polyploid with an odd number of sets of chromosomes (triploid, pentaploid, etc.) in general can not breed true in this way.

The observations on the autopolyploids obtained experimentally show various abnormalities in their meiosis. In the tetraploid *Datura stramonium* mostly quadrivalents are formed (Blakeslee, Belling, and Farnham 1923). Nevertheless, among the pollen mother cells examined only 70 per cent had the normal 24-24 chromosome distribution to the poles. Among the rest, a majority of the cells had a 23-25 distribution, and a few cells a still more abnormal 22-26 and 21-27 distribution. In the tetraploid tomato, Jorgensen (1928) found mostly bivalents and a very regular distribution of chromosomes at meiosis, but Lesley and Lesley (1930) obtained quite different results in their strain of the same plant, namely a varying number of bivalents and quadrivalents and the formation of gametophytes with from 22 to 26 chromosomes. It is possible that the behavior of a tetraploid is influenced by the genetic composition of the ancestral diploid strain. In the *Primula sinensis* tetraploid, both quadrivalents and bivalents are formed, but the disjunction seems fairly regular (Darlington 1931). A part of the offspring of an autopolyploid is therefore likely to contain certain chromosomes in excess, and to be deficient for certain other chromosomes. Since such chromosome variations frequently reduce the viability of their carriers, the fertility of a polyploid may be reduced compared to that of the diploid.

There seems to be no doubt that the disturbances in the chromosomal mechanism inherent in autopolyploidy reduce in general its usefulness as an evolutionary method. Nevertheless, the fact that the autopolyploids in some species have been even more successful than their diploid ancestors shows that the disadvantages resulting from the chromosomal variations may be counterbalanced by the physiological superiority which some polyploids appear to possess. That the tetraploids found in nature have an impaired stability of the chromosomal mechanism similar to that encountered in experimental autopolyploids is shown by several observers. Manton (1934) finds that the tetraploid *Biscutella laevigata* is somewhat variable in chromosome number; among the 67 plants examined, only 54 had the normal number 36, while 2 had 34, 6 had 35, and 5 showed 37 chromosomes. In the diploids the number (18) seems to be constant. In this connection it is important, however, that the tetraploid is capable of reproducing itself not only by seed but also asexually, through the formation of adventitious buds on the roots. Such bud formation has not been found in the diploids. Asexual reproduction is, however, a method resorted to by many organisms whose chromosomal apparatus for some reason ceases to function normally (cf. Chapter X). Müntzing (1936) finds that the fertility of the natural autopolyploids is diminished on the average less than in the experimental ones, and concludes that the autopolyploids may become established only in those species where they happen not to reduce fertility beyond a minimum permitted by natural selection.

EXPERIMENTAL ALLOPOLYPLOIDS

The intergeneric hybrids between radish (*Raphanus sativus*) and cabbage (*Brassica oleracea*) may serve as an illustration of the results obtained when the chromosome complement is reduplicated in crosses of taxonomically rather remote forms (Karpechenko 1927a, b, 1928). Both parents have the same chromosome number, viz. eighteen diploid. The cross succeeds fairly easily; from 202 cross-pollinated flowers 123 hybrids were obtained. The F_1 hybrids have 18 chromosomes, 9 from the radish and 9 from the cabbage parent. No chromosome pairing takes place, 18 univalents are present at the metaphase of the first division, and are distributed at random to the poles. At the second division the univalents split, giving rise to cells with a varying number of chromosomes, mostly from 6 to 12. In some of the pollen mother cells, however, the first division is abortive, and nuclei are formed that include all of the 18 univalents. The second division then gives rise to two spores, and subsequently to pollen grains containing the full haploid complement of the radish as well as of the cabbage chromosomes.

These F_1 hybrids are nearly sterile; most plants produce no seeds at all, but some do produce a few (821 seeds from 90 plants). A cytological examination shows that most of the F_2 hybrids (213 out of 229) have 36 chromosomes in their somatic cells. The origin of such plants is in all probability due to the union of the exceptional gametes possessing the full chromosome complement of the F_1 hybrid. The F_2 plants contain, therefore, 18 radish and 18 cabbage chromosomes, in other words a diploid complement of the chromosomes of each parental species. Such F_2 hybrids are tetraploid. The meiotic divisions are very regular, in striking contrast with the abnormalities observed at meiosis in the F_1 hybrids. In these tetraploids, 18 bivalents are formed, disjunction is normal, and the resulting cells, with few exceptions, contain 18 chromosomes each. It is practically certain that the 18 bivalents that appear at meiosis are due to the pairing of 9 radish chromosomes with their 9 radish homologues, and of 9 cabbage chromosomes with their 9 cabbage homologues. The pairing is consequently between similar chromosomes (autosyndesis), rather than between the chromosomes of different species (allosyndesis). With a normal disjunction, the gametes produced by a tetraploid plant are identical with the exceptional gametes of the F_1 hybrid, and carry 9 cabbage and 9 radish chromosomes.

The tetraploid plants are fully fertile. They produce numerous flowers and fruits, and from a single fruit between 900 and 1,000 seeds can be obtained. A fruit of cabbage has between 1,200 and 1,300 seeds, and that of radish about 600. Moreover, the offspring of a tetraploid, fertilized by another tetraploid, are like their parents in appearance as well as in the chromosome number. No segregation that might be expected in the offspring of a hybrid is observed. The tetraploids are hence both a fertile and a true breeding type. It is important to understand the causation of this phenomenon. The sterility of the F_1 hybrid is apparently due to the irregularity of meiosis. The random distribution of the chromosomes at the first meiotic division gives rise to gametophytes having unbalanced chromosome complements, i.e., neither a full haploid set of radish nor a full set of cabbage chromosomes. Such gametophytes are inviable or non-functional. Only the exceptional gametes of the F_1 hybrid that carry the full sets of the chromosomes of both species are viable. In the tetraploid, all the gametes produced are like each other, and like the exceptional gametes of the F_1. They are chromosomally balanced, and hence are viable and functional. The tetraploid is fertile because its meiosis is regular and only autosyndetic pairing of chromosomes

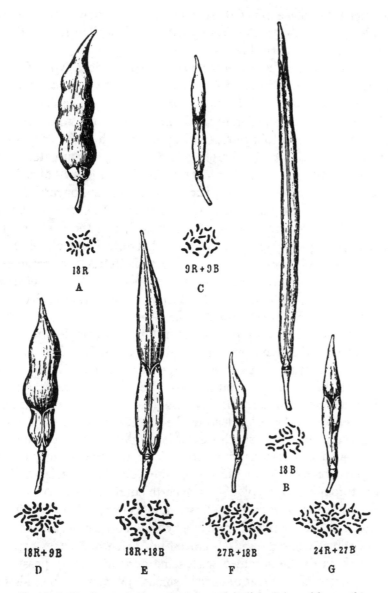

FIG. 18. Fruits and chromosomes of radish (a), cabbage (b), and of hybrids between them. (c) Diploid F₁ hybrid; (e) the tetraploid Raphanobrassica; (d) triploid; (f) pentaploid. (From Karpechenko.)

mosomes takes place. Since there is no recombination of the chromo-
of radish and cabbage, the prerequisite of Mendelian segregation is
lacking, and the hybrid breeds true.

The tetraploid hybrids represent a morphological type which is
distinct both from radish and from cabbage. Some characters are intermediate, in others the influence of one of the parents predominates,
and still others are peculiar to the hybrid. As chance would have it,
the tetraploid has a foliage resembling the radish and a root resembling the cabbage, and is, therefore, worthless as a cultivated plant.
The fruit structure (Fig. 18) is very interesting. The fruit of radish is
spindle shaped, non-dehiscent, that of cabbage is elongate and de-
hisces in two valves. The base of the radish fruit is, however, homologous to the two-valved part of the cabbage, and the cabbage fruit
has an apical part resembling the radish fruit. The fruit of the F_1
hybrid (Fig. 18C) is clearly a compromise structure, resembling
cabbage in the lower, and radish in the upper, part. The tetraploid
resembles the F_1 diploid, but in the former (Fig. 18E) the size of the
fruit is appreciably greater than in the latter.

The tetraploid hybrids possess also the gigas complex of characters discussed above in connection with the autopolyploidy. They are
very large luxuriant plants with large cells. Their viability is not
inferior to that of the parental species. The uniformity and constancy
of their characters, and the obvious differences between the tetraploid
and either parental species, lead to the conclusion that we are dealing
with a full-fledged new species experimentally created, to which the
name Raphanobrassica is given to indicate the manner of its origin.

Raphanobrassica is by no means the only new species that has
arisen through allopolyploidy in experimental cultures. A synthetic
species, *Nicotiana digluta* (Clausen and Goodspeed 1925, Clausen
1928a), has been obtained from the hybrids between *N. tabacum* (24
chromosomes haploid) and *N. glutinosa* (12 chromosomes haploid).
The F_1 hybrids are obtained with some difficulty. They possess the
expected somatic number of chromosomes, 36, which at meiosis form
some bivalents, but are mostly left as univalents. A majority of the
F_1 plants were completely sterile, but a single plant proved to be an
exception and was fertile. The offspring of this exceptional plant was
externally like the F_1 but showed some gigas characteristics. The

number of the chromosomes was found to be 72, which is equal to the sum of the diploid numbers of the parental species (48 + 24 = 72). At meiosis 36 bivalents and no univalents are formed, and the gametes produced contain 36 chromosomes, which is the sum of the haploid numbers of *N. tabacum* and *N. glutinosa*. The F_2 plants are fertile and breed true; they and their offspring are the new species, *N. digluta*. The history of *N. digluta* is, then, on the whole parallel to that of Raphanobrassica, with one important difference. The origin of Raphanobrassica was due to the production by the F_1 hybrid of exceptional gametes containing the entire somatic complement of chromosomes. The union of such gametes gave rise to an allotetraploid carrying diploid sets of the chromosomes of both parents. *N. digluta*, however, arose from a single F_1 plant which stood out among its sibs because of its fertility. The exceptional plant had its chromosomes doubled apparently in the somatic as well as in the germinal tissues, and hence was already an allopolyploid. These two methods of the origin of polyploidy, through the production of unreduced gametes and through a somatic doubling of the chromosome number, are encountered both among the auto- and among the allopolyploids.

Darlington (1932) lists thirty-seven examples of experimental allopolyploids that had been obtained up to that time. Among these, one of the oldest and best known is *Primula Kewensis* (18 chromosomes haploid, 36 diploid), which arose from the hybrid *P. floribunda* (n = 9) × P. verticillata (n = 9) described by Digby (1912) and Newton and Pellew (1929). The diploid hybrid shows a virtually complete pairing (9 bivalents at meiosis) but is nevertheless highly sterile. The tetraploid forms mostly 18 bivalents, but sometimes there are 4 quadrivalents, and the hybrid is fertile. *P. Kewensis* is not a strictly true breeding type, since it occasionally throws some aberrant individuals whose origin is apparently due to quadrivalent formation at meiosis. The cross *Digatalis purpurea* (n = 28) × D. *ambigua* (n = 28) gives a semi-sterile F_1 generation having from 5 to 12 bivalents and from 32 to 46 univalents at meiosis. The allotetraploid *D. mertonensis* was obtained by a doubling of the chromosome complement of the F_1; it is much more fertile than its diploid progenitor, and regularly forms 56 bivalents and no quadrivalents (Buxton and Newton 1928).

Uncommonly interesting results have been obtained by crossing various species of Crepis (cf. Babcock and Navashin 1930). The cross *C. biennis* (n = 20) × *C. setosa* (n = 4) produces a hybrid with 24 somatic chromosomes which is semi-fertile. In its meiotic division 10 bivalents and 4 univalents appear; the former disjoin normally, giving 10 chromosomes to each pole, while the latter are distributed at random. This condition must be interpreted as indicating that *C. biennis* is itself a polyploid, and contains in its haploid set two groups of 10 chromosomes, which are at least partly homologous to each other and form bivalents in the hybrid. The 4 univalents are then the chromosomes of *C. setosa*. In F_2 and further generations a variety of types appear, most of which are sterile because of disharmonious chromosome complements. However, one of these segregants proved fertile and true breeding, and received the name *Crepis artificialis* (Collins, Hollingshead and Avery 1929). A cytological examination of *C. artificialis* showed 24 somatic chromosomes that formed 12 bivalents at meiosis. The origin of this chromosome complement is interpreted as due to the presence of 10 pairs of *biennis* and of 2 pairs of *setosa* chromosomes. The four *setosa* chromosomes were identified cytologically in the chromosome group of *artificialis;* the identification was possible owing to the fact that the chromosomes of *setosa* differ in size and other characteristics from those of *biennis* and are therefore recognizable in the hybrids. *C. artificialis* is not an allopolyploid in the sense that its chromosome group is not a sum total of those of the parental species; it is rather a segregant which happens to possess a viable combination of chromosomes that gives rise to nothing but bivalents at meiosis.

The hybrids between *Crepis capillaris* (n = 3) and *C. tectorum* (n = 4) described by Hollingshead (1930b), and those between *C. rubra* (n = 5) and *C. foetida* (n = 5) studied by Poole (1931, 1932) are characterized by the formation of bivalents at meiosis in the F_1 generation. The species crossed are here evidently more closely related to each other than those in the examples discussed previously, and their chromosomes still possess the pairing "affinity" for each other. Allotetraploids derived from them showed various abnormalities at meiosis. The *rubra* × *foetida* tetraploid has for example, from 0 to 5 quadrivalents and from 0 to 10 bivalents. While the F_1 hybrids were semi-fertile, the tetraploids not only showed no improvement in fertility compared to the F_1, but in fact manifested a deterioration.

There seems to exist a curious relation between the behavior of the diploid hybrid and that of an allopolyploid derived from it. Where the diploid hybrid forms mostly or only univalents and is nearly sterile, the doubling of the chromosome complement results in a restoration of the fertility and a regular formation of bivalents (e.g., Raphanobrassica). Where, on the other hand, the pairing in the diploid is retained, the tetraploid may show quadrivalents and other variations at meiosis, and may be largely sterile. This rule (which is not quite without exceptions, e.g., *Primula Kewensis*) was first formulated by Darlington (1928, 1932), and corroborated by newer data. We shall discuss its significance in Chapter IX. For the time being it is sufficient to point out that the allopolyploids derived from the hybrids between closely related forms constitute a transition to the autopolyploids, and share with the latter the disturbances in the chromosome mechanism produced by the presence of more than two homologous chromosomes in a single nucleus.

ALLOPOLYPLOIDS IN NATURE

The experimental allopolyploids represent an impressive array of novel types created artificially by combining various previously existing species. Some of these synthetic types can be properly regarded as new species. Indeed, they possess complexes of morphological and physiological traits not present in any known species, and, in addition, are fertile and true breeding forms similar in this respect to any "good species" found in the natural state. Last but not least, the allopolyploids are, as we shall see below, isolated from their progenitors by the barriers of incompatibility and of sterility of the hybrids. It is no exaggeration to say that the production of allopolyploids is the most powerful tool yet available to a geneticist for molding living matter into new shapes. We may turn now to a consideration of evidence that shows that the same tool has been resorted to on a grand scale for the production of new species in nature. At least in one case, that of *Galeopsis Tetrahit*, an existing species has been experimentally resynthesized from its putative ancestors. In at least two further cases the presumptive evidence is fairly complete to make it very probable that a species found in nature has arisen from two known species. Finally, in numerous cases the origin of species through allopolyploidy may be regarded as certain, although the ancestral species can be only conjectured.

In his monographic work on the genus Galeopsis, *Müntzing* (1930, 1932) shows that six out of the eight species investigated have the haploid number of chromosomes, 8, and the two remaining ones have n = 16. Among the former are the species G. *pubescens* and G. *speciosa,* and among the latter is G. *Tetrahit.* The cross *pubescens* × *speciosa* succeeds easily when *pubescens* is used as the female parent. F_1 hybrid is highly sterile; its anthers contain only 8.9 per cent to 22.3 per cent of visibly good pollen, and few good ovules are produced. At meiosis varying numbers of bivalents and univalents are formed. In the F_2 generation a single plant was found that proved to be triploid (3n = 24). Its origin is probably due to the union of a gamete containing the somatic complement of the F_1 hybrid (i.e., 8 chromosomes of *pubescens* and a like number from *speciosa*) with a gamete carrying 8 chromosomes, the proportions of the chromosomes of the two parental species in the latter gamete not being clear. This triploid plant was backcrossed to a pure *pubescens.* A single seed resulted from the backcross. It gave rise to a plant which proved to be a tetraploid (4n = 32). This tetraploid plant was fertile, and became the progenitor of a strain which, for reasons stated below, has been named "artificial Tetrahit." The origin of the tetraploid was due to a union of an unreduced gamete of the triploid with a normal one of G. *pubescens.*

The triploid hybrid between G. *pubescens* and G. *speciosa,* as well as its tetraploid derivatives, exhibit a striking resemblance to G. *Tetrahit.* The latter species has however taken no part in the production of the hybrids. The resemblance, which according to Müntzing reaches in some individuals an apparent identity, suggests that the true G. *Tetrahit* arose as an allotetraploid from the cross G. *pubescens* × G. *speciosa* or types similar to these species. A series of tests were imposed to prove the validity of this hypothesis.

The artificial *Tetrahit* is like the real *Galeopsis Tetrahit* in possessing 32 chromosomes in somatic cells and 16 bivalents at meiosis. The meiotic divisions are, with few exceptions, normal. The disturbance of the meiosis encountered in the F_1 hybrids between *pubescens* and *speciosa* is abated in the artificial *Tetrahit*. A cross between the artificial and the natural *Tetrahit* gives normally developed offspring which are externally similar to either parent. The fertility is complete in some individuals, while others are partially sterile; it must be taken into consideration that a partial sterility has been observed by Müntzing (1930) in some lines of the pure *G. Tetrahit* as well. The meiotic divisions are normal; 16 bivalents are formed which undergo a regular disjunction. In short, the artificial and the natural *G. Tetrahit* are similar both morphologically and in their genetic and cytological behavior.

Although the origin of the natural *Galeopsis Tetrahit* from a cross between *pubescens* and *speciosa* is very probable, it remains unknown when and where the event took place. The data of Huskins (1931) pertaining to the grass species *Spartina Townsendii* are in this respect more complete. This species was discovered in 1870 in a single locality in southern England. Soon thereafter a rapid spread of *S. Townsendii* was recorded. By 1902 its known distribution area had expanded to cover many thousands of acres along the coast of England, and in 1906 it appeared on the coast of France. A further expansion ensued, and owing to its desirable agricultural properties, *S. Townsendii* has been introduced artificially into other parts of the world.

The origin of *Spartina Townsendii* has been a subject of discussion for some time. A comparative morphological study of various species of the genus Spartina has led systematists to believe that *S. Townsendii* is a hybrid between two other species, namely *S. stricta* and *S. alterniflora*. Among these, *S. stricta* is an indigenous European species known for about 300 years; *S. alterniflora* is native in America, but it has been introduced into England, and has spread somewhat, becoming common in some localities. It is suggestive in this connection that in the process of its expansion *S. Townsendii* has overrun some of the territory previously occupied by *S. stricta* and has supplanted the latter. *S. alterniflora* has suffered much the same fate where it has come into contact with *S. Townsendii*.

The chromosome number of *Spartina stricta* has been found by Huskins to be 56 (diploid). That of *S. alterniflora* is $2n = 70$. On the supposition that *S. Townsendii* is an allotetraploid derivative from the two others, it should have a chromosome number equal to the sum of the diploid numbers of its putative ancestors, $56 + 70 = 126$. The number actually found is 126, with a possible variation of ± 2 chromosomes. Meiosis in *S. Townsendii* shows mostly bivalents, but also some multivalents. The presence of the latter may account for a rather wide variation of the external characteristics observed in this species. One can assert with some confidence that *S. Townsendii* arose in the nineteenth or the eighteenth centuries, probably on the southern coast of England.

A somewhat different complex of evidence is presented by Anderson (1936) to prove that *Iris versicolor* is an allopolyploid derivative of two other species, *I. virginica* and *I. setosa*. The area inhabited by *I. virginica* extends from Virginia along the coast of the Atlantic and the Gulf of Mexico; a race of the same species occupies the territory south of the Great Lakes, and the valleys of Mississippi and Ohio. The distribution region of *I. setosa* is broken up in two parts; one race inhabits the Pacific and the Bering Sea coasts of Alaska, and the other is found in Labrador, Newfoundland, and Nova Scotia. Both species are regarded as survivors from pre-glacial times. On the other hand, *I. versicolor* is likely to be a more youthful species, since it occurs in the northeastern states, and from Labrador, along the St. Lawrence and Great Lakes, to Wisconsin and Winnipeg. A major part of this area was glaciated, and the flora is conjectured to be of a more recent origin.

A cytological examination has fully corroborated the hypothesis of the origin of *Iris versicolor* from *setosa* and *virginica*. Indeed, the diploid chromosome numbers are 38 for *setosa*, 70 to 72 for *virginica*, and $38 + 70 = 108$ for *versicolor*. Both *virginica* and *versicolor* show some multivalent chromosome associations at meiosis. This fact, as well as the chromosome numbers, suggest that *virginica* is itself an allotetraploid derivative of two unknown species which had chromosome numbers of the order of magnitude now present in *setosa;* its origin is hidden in antiquity. *I. versicolor* must then be an allohexaploid which has received twice as many chromosomes from the *virginica*-like as from the *setosa*-like parent. The morphological characteristics of *versicolor* may therefore be expected to approach more closely those of *virginica* than those of *setosa*.

A careful morphological comparison of all three species has shown first of all that *Iris versicolor* is closer to the Alaskan than to the Labradorean race of *I. setosa*. In fact, *versicolor* is in most characters intermediate between *virginica* and the Alaskan *setosa*, closer to the former than to the latter. Some of the traits of *versicolor* could not be found however in either species. This might be accounted for either on the supposition that *versicolor* arose from an undiscovered or extinct variety of one of its putative parents, or that its peculiar traits are a concomitant of polyploidy. A search has shown that the former possibility is more probable, because a new race, *I. setosa* variety *interior*, has been found in the Yukon valley in Alaska, and this race possesses the requisite complex of characters. It is apparently significant that the new race occupies a territory which is adjacent to, but was not covered by, the continental ice sheet.

An artificial hybridization of *Iris virginica* and *I. setosa* has been attempted, but the results are inconclusive. Owing to the difference in the flowering seasons, the crossing of the two species is difficult; some pollinations resulted in some hybrid seeds of which none germinated. This result does not, of course, preclude the possibility that sometimes hybridization may be successful.

Similar, though perhaps less complete, evidence is available on the origin of a few more existing species. *Rosa Wilsonii* is an allohexaploid derived from *R. pimpinellifolia* and *R. tomentosa* (Blackburn and Harrison, 1924); *Aesculus carnea* is a hybrid between *A. hippocastanum* and *A. Pavia* (Skovsted 1929); and the hexaploid *Phleum pratense* is a product of crossing a diploid *P. pratense* with a tetraploid *P. alpinum* (Gregor and Sansome 1930). The origin of *Penstemon neotericus* from *P. laetus* and *P. azureus* has been made virtually certain by Clausen (1933).

HYBRIDS BETWEEN POLYPLOID SPECIES

An allopolyploid possesses sets of chromosomes that are present separately in two or more species that went into its makeup. Thus, an individual of Raphanobrassica carries nine pairs of radish and nine pairs of cabbage chromosomes. It is reasonable to suppose that in the allopolyploids that have arisen in a remote past the chromosomes have had time to undergo changes both of the nature of gene mutations and genic rearrangements, and are no longer wholly identical with the chromosomes of the ancestral species. Nonetheless, a certain similarity between the chromosomes of a polyploid and those of its diploid relatives may persist for a long time, and may manifest itself in pairing and bivalent formation in the hybrids.

As early as 1909, Rosenberg described the chromosome behavior in a cross between an apparently diploid species *Drosera longifolia* (n = 10) and a tetraploid *D. rotundifolia* (n = 20). The hybrid has 30 somatic chromosomes and is therefore a triploid. At meiosis, 10 bivalents and 10 univalents are produced; the former disjoin regularly, while the latter are distributed at random. In this particular case it is impossible to determine which chromosomes are giving rise to the bivalents. One may suppose that the 10 *D. longifolia* chromosomes unite with 10 *D. rotundifolia* ones, and the remaining 10 *D. rotundifolia* chromosomes are left over as univalents; this assumes that the pairing is between the chromosomes of different species (allosyndisis). Or else, the 20 *D. rotundifolia* chromosomes may unite in 10 pairs, and leave the *D. longifolia* chromosomes unpaired; this would be a pairing between chromosomes of the same species, autosyndesis. If only similar chromosomes pair, autosyndesis would mean that the two sets of chromosomes of *D. rotundifolia* are more similar to each other than either of them is to *D. longifolia* chromosomes. Allosyndesis would indicate that the *D. longifolia* chromosomes resemble one set of the *D. rotundifolia* chromosomes more than they do the other, and are more alike than either of these two sets; this, in turn, might suggest the inference that *D. longifolia* is one of the ancestors of the polyploid *D. rotundifolia*.

Such an inference appears justified by analogy with the behavior of the experimental allopolyploids when crossed to their known ancestral species. The synthetic *Nicotiana digluta* (see above) has 72 chromosomes in its somatic cells, 48 of which (24 pairs) are derived from *N. tabacum,* and 24 (12 pairs) from the *N. glutinosa* ancestors. The number 12 being basic in the genus Nicotiana, *N. digluta* must be regarded as an allohexaploid. A backcross of *N. digluta* to *N. tabacum* results in a hybrid carrying 60 chromosomes, i.e., a pentaploid (Clausen 1928a). The origin of these 60 chromosomes is as follows: 36 of them have been contributed by the *N. digluta* gamete, which in turn is known to contain 24 chromosomes of *N. tabacum* and 12 of *N. glutinosa;* the remaining 24 have been introduced by the *N. tabacum* gamete. The hybrid shows 24 bivalents and 12 univalents at meiosis. The bivalents are here the *N. tabacum* chromosomes, and the univalents are the remaining *N. glutinosa* ones. The cross *N. digluta* × *N. glutinosa* gives rise to a hybrid with 48 chromosomes, which forms 12 bivalents and 24 univalents at meiosis. The univalents are here the *N. tabacum* and the bivalents the *N. glutinosa* chromosomes (Clausen 1928a, 1928b).

The study of the chromosome pairing in the hybrids between polyploid species becomes a method for tracing phylogenetic relationships. The limitations and the pitfalls of this method are however rather serious, and should be clearly realized. The difficulty of distinguishing between auto- and allosyndesis has already been mentioned. In fact, both of these processes are known to occur, sometimes separately and at other times in combinations. In any given case the evidence must be carefully scrutinized to distinguish between the two alternatives. More serious still, the assumption is implicit in this method that the chromosomes that pair in a hybrid are "similar." The similarity of chromosomes is however a concept which is ambiguous unless an attempt is made to define it precisely. We know that chromosomes may become differentiated through the occurrence of gene mutations, or through changes in the gene arrangement (translocations, inversions, etc.), or both. We know, furthermore, that these changes do not necessarily go hand in hand. Apparently equally remote species may be rather similar or may be very different in the gene arrangement (Chapter IV). Which of these changes determine the occurrence or the failure of the meiotic pairing? If chromosome pairing is a function of the similarity in the gene arrangement (cf. Chapter IX), the formation of bivalents does not exclude the possibility that the partners differ greatly in the allelomorphic state of their genes.

The fact that in the hybrids between *Nicotiana digluta* and *N. tabacum* and *N. glutinosa* the bivalents are formed through a union of like chromosomes is, of course, not surprising, for it is indeed clear that some of the chromosomes of *N. digluta* are in every respect similar to those of *N. tabacum,* and others to those of *N. glutinosa.* But if the assumption is made that the chromosome pairing in the *Drosera rotundifolia* × *D. longifolia* hybrid is allosyndetic, it still would not necessarily follow that the latter species is an ancestor of the former. The real ancestor of *D. rotundifolia* might have been a quite distinct species which possessed a gene arrangement somewhat similar to that present in the now living *D. longifolia.* At best a phylogeny based on studies on the bivalents formation in the hybrids between the extant species is capable of giving only a rough approximation to the actual situation. Nevertheless, data of this kind do furnish conclusive evidence of the compound nature of polyploid species, and this alone makes them valuable.

The relationships between species of Nicotiana have been studied by Clausen and Goodspeed (1925), Goodspeed and Clausen (1928), East (1921), Clausen (1928a, b), Brieger (1928), Lammerts (1931), Webber (1930), Rybin (1927, 1929), and others. In this genus there exist a group of species with the haploid number of chromosomes equal to 12 (*N. sylvestris, tomentosa, Rusbyi, paniculata, glutinosa*), and a group having n = 24 (*N. tabacum, rustica,* and others). Other chromosome numbers also occur. The synthesis of a new species *N. digluta* with n = 36 from *N. tabacum* and *N. glutinosa* has been discussed above. Data are present, however, to show that the twenty-four chromosome species are in turn allopolyploids derived from a natural crossing of twelve chromosome species. The cross *N. tabacum* × *N.* sylvestris gives a triploid hybrid (2n = 36) that forms 12 bivalents and 12 univalents at meiosis. Since the haploid *N. tabacum* (n = 24) is known to have no bivalents, the formation of the bivalents in the *N. tabacum* × *N. sylvestris* hybrid is allosyndetic. The conclusion follows that the chromosome complement of *N. tabacum* contains a set of 12 chromosomes which are structurally similar to those of *N. sylvestris,* and another set of twelve that are structurally different. The crosses *N. tabacum* × *N. tomentosa* and *N. tabacum* × *N. Rusbyi* gave also 12 bivalents and 12 univalents at meiosis in the F_1 hybrids. Since the cross *N. tomentosa* × *N. Rusbyi* gives 12 bivalents, these two species have structurally similar chromosomes.

Hence, the complement of *N. tabacum* contains 12 chromosomes similar to either *tomentosa* or *Rusbyi*. The question now is this: are the chromosomes of *N. sylvestris* similar to those of *N. tomentosa* and *N. Rusbyi?* A negative answer to this question follows from the fact that the *N. tomentosa* × *N. sylvestris* and *N. Rusbyi* × *N. sylvestris* hybrids form 24 univalents and no bivalents. Hence, the haploid set of *N. tabacum* (n = 24) is composed of two dissimilar groups of chromosomes, twelve in each, which are related to the chromosomes of *N. sylvestris* and *N. tomentosa*—*N. Rusbyi* respectively. The inference that these species, or some other species with similar chromosomes, have taken part in the production of *N. tabacum* is a probable one.

The relationships between *Nicotiana rustica* (n = 24) and *N. paniculata* (n = 12) are also interesting. The hybrid between these species produces 12 bivalents and 12 univalents. Some of the gametes that are formed by this hybrid contain the somatic chromosome number (36), and their union with normal gametes of *N. paniculata* gives rise to a tetraploid hybrid (2n = 48). This tetraploid is, however, highly unstable and forms numerous multivalent associations at meiosis, some of the chromosomes remaining as univalents. It seems, then, that the complement of *N. rustica* contains a set of 12 chromosomes similar to *N. paniculata,* and also a set of chromosomes the origin of which is unknown. Segregates may be obtained from the hybrids that have 2n = 24, the number characteristic for *N. paniculata,* and yet are clearly distinct from the latter species. The origin of these segregates, some of which breed true, is apparently due to exchanges between the two different chromosome sets of *N. rustica.*

Numerous cyto-genetic investigations on species of wheat (Triticum), on the related genus Aegilops, and on hybrids between them furnish abundant material from which inferences may be drawn on the phylogeny of this group of forms. The literature of this subject is voluminous. Here belongs the work of Kihara and his school (Kihara 1919, 1924, Kihara and Nishiyama 1930, Kihara and Lilienfeld 1932, 1935, Lilienfeld and Kihara 1934) and also of Sax (1922), Sax and Sax (1924), Sapehins (1928), Watkins (1930, 1932), Bleier (1928, 1933), and many others. The described species of wheat fall into three groups differing in the number of chromosomes:

THE EINKORN GROUP	THE EMMER GROUP	VULGARE GROUP
(n = 7, 2n = 14)	(n = 14, 2n = 28)	(n = 21, 2n = 42)
T. aegilopoides	T. dicoccoides	T. spelta
T. thaoudar	T. dicoccum	T. vulgare
T. monococcum	T. durum	T. compactum
	T. turgidum	T. sphaerococcum
	T. pyramidale	
	T. polonicum	
	T. persicum	
	T. Timopheevi	

In Aegilops, species with 2n = 14 and 2n = 28 are known. The basic chromosome number is evidently n = 7; the einkorn group is diploid, the emmers tetraploid, and the vulgare group (soft wheats) hexaploid. With few exceptions, the hybrids between species possessing the same number of chromosomes show only bivalents at meiosis, and are fully fertile. The hybrids between the representatives of the vulgare and emmer groups are pentaploid (21 + 14 = 35), and show at meiosis 14 bivalents and 7 univalents. The triploid hybrids (emmer × einkorn, 14 + 7 = 21) have from 4 to 7 bivalents and from 7 to 13 univalents. The vulgare × einkorn cross (21 + 7 = 28) produces from none to as many as 10 bivalents, 7 being the usual number at least in certain crosses.

These relationships have been interpreted to mean that the einkorn, emmer, and vulgare groups have, respectively, one, two, and three sets of seven chromosomes which are different from each other. These sets are denoted as A, B, and D. The hybrids AAB and AABD have as a rule seven, and AABBD fourteen bivalents. The emmer group arose as an allotetraploid derivative from the hybrids between an einkorn species furnishing the genom A and some other plant that gave the genom B. Species of the vulgare group are allohexaploids, and their origin is due to a cross of an emmer (AB) with something supplying the genom D. This something is supposed to be a species of Aegilops which may possess D.

This hypothesis has on the whole withstood the tests imposed on it, but its schematic simplicity is largely a thing of the past. Variations in the bivalent number in some crosses have been mentioned already. In addition, Kihara and Nishiyama (1930) have described the formation of trivalents at meiosis in the hybrids *Triticum aegilopoides* × *T. dicoccum, T. aegilopoides* × *T. spelta,* and *T. durum* × *T. vulgare.* The presence of the trivalents may indicate that some of the chromosomes of the supposedly different genoms preserve enough similarity so that they can pair; it may also be due to some of the *T. aegilopoides* chromosomes (genom A) consisting of parts that lie in different chromosomes of the same set in *T. dicoccum* and *T. spelta.* The genoms denoted by the same letter in different species, may be distinct, due to changes by translocations, inversions, etc.*

The divergent evolution of the once identical chromosome sets is nicely illustrated by the results of Lilienfeld and Kihara (1934) on *Triticum Timopheevi,* a species of the emmer group. As such, *T. Timopheevi* is expected to carry the chromosome sets A and B encountered in all emmers. The crosses *T. Timopheevi* × *T. pyramidale, T. Timopheevi* × *T. dicoccum, T. Timopheevi* × *T. persicum* and *T. Timopheevi* × *T. durum* give, nevertheless, semi-sterile F_1 hybrids which have quadrivalents, trivalents, bivalents, and univalents in varying proportions at meiosis. In Lilienfeld and Kihara's figures the number of univalents varies from 1 to 8, and that of bivalents from 5 to 12. The cross *T. Timopheevi* × *T. aegilopoides* gives however a result similar to that obtained in the crosses between *T. aegilopoides* (an einkorn) and any other species of the emmer group. The chromosome set A is common to the einkorns and to the emmers, including *T. Timopheevi.* The difference between *T. Timopheevi* and the other emmers lies evidently in the set B. Lilienfeld and Kihara cut the Gordian Knot by saying that *T. Timopheevi* possesses a special chromosome set (genom) G, not present in any other known wheat. The "formula" of *T. Timopheevi* is AG, while all other emmers are AB. From their own data it follows, however, that G is more like B than anything else; denoting this chromosome set by a sepa-

---

*The terms "genom" and "genom analysis" unfortunately seem to have taken root in the field of wheat cyto-genetics, in spite of the fact that they are highly misleading. These terms imply that definite chromosome sets contain similar (or, respectively, different) gene complexes. Yet, the "genom analysis" has nothing to do with gene qualities. What is being studied is the chromosome pairing, which is in all likelihood determined by the gene arrangement, the gross structure of the chromosomes. To assume that the genic differentiation is proportional to the structural diversity would be gratuitous. A non-committal expression like "set of chromosomes" would be more suitable.

rate letter does not seem to be a particularly helpful expedient. There is no assurance that *T. Timopheevi* is distinct in its origin from other emmers; the differentiation of one of its chromosome sets may have taken place before or after it has become a tetraploid species. The differentiation of the chromosome sets is a gradual process, as Kihara himself has pointed out.

A comparison of Triticum species with those of Aegilops shows especially clearly that the "similarity" and "dissimilarity" of the chromosome sets is a matter of degree (Kihara and Lilienfeld 1932, and others). In a hybrid between two races of *Ae. ventricosa,* all chromosomes pair, but two circles of four chromosomes are formed. Obviously, translocations have taken place in the phylogeny. *Ae. speltoides* × *Triticum monococcum,* a hybrid between two species each possessing 2n = 14, gives from 0 to 7 bivalents. The chromosome sets present in the two species are denoted as S and A respectively; the occasional formation of 7 bivalents shows that S and A are related, and yet they are by no means identical. The cross *Ae. Aucheri* (a species related to *Ae. speltoides*) × *T. durum* leads Kihara and Lilienfeld to believe that some of the chromosomes of the set S of Aegilops are more closely related to the chromosomes of the set A and others to those of the set B of *Tr. durum.* The tetraploid species *Ae. cylindrica* has two sets denoted C and D; the set D is somewhat related to the chromosomes that differentiate the vulgare group from the emmer group of wheats (see above). And yet, the hybrid *Ae. cylindrica* × *Tr. aegilopoides,* which contains the three different sets A, C, and D, has from 4 to 8 bivalents. Some of the chromosomes of these sets are obviously related.

Kihara and Lilienfeld introduce subscripts to distinguish between the supposedly similar sets of chromosomes in different species. Thus, A_{eink}, A_{em}, and A_D are the chromosomes of the A set from einkorn, emmer, and vulgare, wheats respectively. A_{eink} is clearly different from the two other, but all of them are related to the S set of the Sitopsis section of the genus Aegilops. There are four semi-homologous C sets in different Aegilops species, and they are also somewhat related to the sets S, T, E, F, and D found in other Aegilops, and to A of the Triticum species. Kihara and Lilienfeld have sketched a phylogenetic tree of the chromosome sets of Aegilops and Triticum. This phylogeny is merely a graphic representation of the frequencies of pairing between the chromosomes of the different sets, and it may or may not have anything to do with the phylogeny of the species that are the carriers of these chromosomes.

The evolutionary changes among wheats are of two kinds. The species formation through allopolyploidy, that is, through the emergence of new combinations of the chromosome sets, dovetails with the processes of the differentiation of the chromosome sets themselves by gene mutation and by changes in the gene arrangement. The latter class of changes is perhaps less spectacular than the results of the polyploidy, but its existence can be deduced from the available data. The occurrence of allopolyploids implies as an antecedent a differentiation of the chromosomes in the ancestral species by translocation, inversion, and other means.

POLYPLOIDY IN ANIMALS

The prevalence of the polyploid series of chromosome numbers in plants and their relative scarcity among animals constitutes the greatest known difference between the evolutionary patterns in the two kingdoms. Muller (1925) has pointed out that this difference is probably caused by the preponderance of hermaphrodites among the higher plants, and of forms with separate sexes among the genetically and cytologically studied animals. Wherever the sex determination is due to a heterochromosome mechanism, polyploidy may result in the production of sexually abnormal or sterile types.

The well-known data of Bridges and others demonstrate that sex in Drosophila is determined by the ratio or balance between the X-chromosomes which carry the femaleness and the autosomes which tend toward maleness. A zygote possessing equal numbers of X-chromosomes (X) and of sets of autosomes (A) develops in a female. Diploid (2X, 2A), triploid (3X, 3A), tetraploid (4X, 4A), and probably also haploid (1X, 1A) females are known. A ratio $X : A = 1$ is therefore female determining. A ratio $X : A = 0.5$, however, determines a male; a normal diploid male is 1X, 2A, and the unpublished data of Sturtevant show that a tetraploid 2X, 4A is also a male. Ratios intermediate between 1 and 0.5 give rise to intersexes (2X, 3A, or 3X, 4A) and zygotes of the constitution 3X, 2A and 1X, 3A develop in the so-called superfemales and supermales, respectively. The Y-chromosome of Drosophila has nothing to do with sex determination, although a male devoid of it is sterile.

A reduplication of the chromosome complement in an organism like Drosophila may give rise to tetraploid females (4X, 4A) and tetraploid males (2X, 4A). The reduction division in tetraploid males has not been studied, but on theoretical grounds it is believed likely to give rise mostly to 1X, 2A spermatozoa. Such spermatozoa uniting with the eggs of a normal diploid female (1X, 1A) would produce intersexes (2X, 3A). A tetraploid female produces diploid 2X, 2A eggs, which on fertilization by the sperm of a normal diploid male (1X, 1A and 1Y, 1A) give triploid females (3X, 3A) and intersexes (2X, 3A). A triploid female crossed to a diploid male produces triploid and diploid females, diploid males, intersexes, superfemales, and supermales. Even if tetraploid females and males should appear at once in such quantities that they are likely to mate with each other rather than with the normal diploids, the tetraploid race can not become established, for the offspring produced would consist of tetraploid females (4X, 4A) and intersexes (3X, 4A). In short, tetraploidy in an organism with separate sexes is likely to lead to complications with the sex determining machinery.

An obvious corollary to the above arguments is that the formation of polyploid races and species is not likely to occur in dioecious plants, and, contrariwise, may be expected in animals that are hermaphroditic or that propagate parthenogenetically or asexually. The available data, meager as they are, agree fairly well with these deductions. The willows, Salix, are typical dioecious plants. The cross *Salix caprea* × *S. viminalis* has produced an approximately tetraploid plant which was morphologically similar to a garden form known as *S. laurina*. Among *S. laurina* only female trees occur, and they are propagated vegetatively instead of by seed (Hakansson 1929).

A very interesting situation has been described by Vandel (1928, 1934) in a European species of sow bugs, *Trichoniscus elisabethae* (*T. provisorius*). Two races are present in this form, one of which has both males and females in normal proportions (1:1), while the other consists almost exclusively of females. The eggs of the bisexual race are incapable of developing without fertilization, while the unisexual race is parthenogenetic. The bisexual race inhabits chiefly the Mediterranean countries, while the parthenogenetic one occurs mostly in northern Europe; a fairly broad stretch of land has both races. Vandel shows that the bisexual race is diploid ($2n = 16$), and its spermatogenesis and oögenesis follow the customary course. The parthenogenetic race has 24 chromosomes, and is therefore triploid. The oögenesis in the triploid females is peculiar in that no chromosome pairing takes place at meiosis, the oöcytes undergo a single maturation division in which the chromosomes split equationally, and the resulting eggs with their 24 chromosomes require no fertilization to begin development.

Crossing parthenogenetic females to males of the bisexual race rarely succeeds because the males refuse to copulate with the triploid females. Where copulation does take place, the eggs apparently fail to be fertilized and develop parthenogenetically. The parthenogenetic race occasionally produces a few males (the sex ratio is about 100 ♀ ♀ to 1.6 ♂ ♂) which carry 24 chromosomes like their mothers. The spermatogenesis in the exceptional triploid males involves no chromosome pairing at meiosis; two equational divisions are observed, and the spermatozoa that develop contain the whole complement of 24 chromosomes. The triploid males copulate with the triploid females, and the seminal receptacles of the latter may be filled with sperm; the sperm, however, is non-functional and the fertilized females produce parthenogenetic eggs as usual. The reason why some triploid individuals are females and other males remains quite obscure. In any event, the formation of a polyploid race in Trichoniscus is accompanied by a modification of the oögenesis that enables the organism to reproduce parthenogenetically.

A similar situation has been found in the shrimp *Artemia salina*, which inhabits salt water lakes and pools all over the world and forms numerous phenotypically distinguishable local races, the differences between which are due in part to a direct modification by the environment. In a series of papers Artom (see Artom 1931 for further references, also Gross 1932) shows that these local races fall into three groups. The first group consists of diploid ($2n = 42$) bisexual animals with a normal gametogenesis. The eggs of this race fail to develop without fertilization. The second group is likewise

diploid, but males are very rare and the females produce parthenogenetic eggs. Finally, the third group is tetraploid ($2n = 84$) and
the females are parthenogenetic. In the tetraploid female, no chromosome pairing is observed at meiosis, and the single maturation division is equational, as in Trichoniscus. A few tetraploid males are
also known; their spermatogenesis is, according to Artom, normal.
No copulation between the males and the parthenogenetic females
seems to take place. The origin of the rare males in the parthenogenetic Artemia remains as much of a mystery as the similar process
in Trichoniscus.

Some unique mechanisms have developed in connection with the
polyploidy in the moth *Solenobia triquetrella* studied by Seiler
(1927). This species has a tetraploid ($2n = 120$) parthenogenetic
race which is rather widespread in central Europe, and also a diploid
($2n = 60$) bisexual one which is very rare and has been found only
in a certain narrowly circumscribed locality near Nürnberg, Germany. The caterpillar of Solenobia, like those of other representatives of the family Psychidae to which it belongs, builds a special
"house" which it carries about as it walks and feeds. The pupation
takes place in this house. The females are wingless, while the males
have wings and are good fliers. After hatching from the pupa, a
parthenogenetic female proceeds almost immediately to deposit her
eggs in the now empty house, and dies. A bisexual female lingers until a male arrives and copulation takes place, after which it also deposits its eggs in the house, and perishes.

The chromosome cycle in the diploid race is normal. The parthenogenetic race consists of females only; the oögenesis also proceeds at
first normally. The 120 chromosomes unite to form 60 bivalents. Two
maturation divisions follow the normal course, and the female pronucleus with 60 chromosomes emerges from the process. It migrates
to the periphery of the egg, and after some delay proceeds to cleave,
of course without fertilization. Two cleavage divisions take place,
resulting in the appearance of four nuclei, 60 chromosomes in each.
These four nuclei form two pairs, and fuse, giving rise to two nuclei
now possessing 120 chromosomes, which is the tetraploid complement. The cleavage is then resumed, and a parthenogenetic female
develops.

If a male is available when the parthenogenetic female has just emerged from the pupa, a cross between the two races may take place. Moreover, the normally parthenogenetic eggs admit the spermatozoa into their cytoplasm; the penetration of several spermatozoa into a single egg (polyspermy) is frequent. The subsequent processes are rather chaotic. The female pronucleus divides twice and forms four 60-chromosome nuclei. The spermatozoa have in the meanwhile transformed into male pronuclei containing 30 chromosomes each. All four female nuclei may fuse with the sperm nuclei, giving rise to products that contain $60 + 30 = 90$ chromosomes. Or else the female nuclei may fuse in pairs, as they do in unfertilized eggs, and form 120 chromosome derivatives. Two of the female nuclei may fuse, and the remaining two may combine with the sperm nuclei, so that the organism has two kinds of cells, some bearing 120 and others 90 chromosomes. Some of the sperm nuclei can take part in the development without union with anything else, resulting in cells with only 30 chromosomes. Finally, several sperm nuclei can fuse with a single female one, which raises the chromosome number to as high as 240.

It is not surprising that the hybrids between the two races of Solenobia develop into strange creatures. A whole series of individuals, beginning with male-like and ending with female-like ones, is obtained. A majority are intersexual, or show combinations of female and male parts with very frequent asymmetry. Seiler does not state whether any of them are fertile, and if so what kind of offspring they can produce. The interracial hybridization, even if it sometimes occurs in nature, is unable to give rise to a viable hybrid form that could compete with the parental races.

The examples just discussed are sufficient to give an idea about the rather extraordinary circumstances which must occur in order that nature may connive at the development of polyploidy in animals that normally reproduce bisexually. Otherwise, the polyploids in animals have the same general characteristics as the polyploid plants. The multiplication of the chromosomes results in an increase of the cell size, and frequently in a general gigantism. The polyploid Solenobia and Artemia are larger than the diploid ones. The triploid females of Drosophila display no noticeable increase of the body size, but they differ from the diploids in their habitus. How frequent is polyploidy in those groups of animals that are normally hermaphroditic or asexual is for the time being unknown. A very special kind of polyploidy encountered in Hymenoptera and some other insects (diploid females, haploid males) is connected with the peculiar, and as yet insufficiently understood, method of the sex determination in these forms, and can not be directly compared with the polyploidy known in plants and in other animals.

MUTATION IN POLYPLOIDS

The occurrence of allopolyploidy in a group of organisms results in changes in the evolutionary pattern of the group. Among these changes the most obvious one concerns the appearance of the phylogenetic tree. The phylogenetic development is customarily represented by treelike diagrams, wherein the basal trunks symbolize the ancestral form. The branching of the trunk represents the differentiation of the ancestral species, giving rise to the variety of existing and extinct species derived from it. A more adequate image of the evolutionary process may be drawn if the trunk and the branches of the phylogenetic tree are pictured as cables consisting of numerous strings, running on the whole parallel to each other, but occasionally branching or coming to an end (Anderson 1936). Such a representation has the advantage that it takes into consideration the fact that most species and races are composites made up of numerous semi-isolated colonies that are to a certain degree independent evolutionary units.

In groups where allopolyploidy is not encountered, the cables will preserve their independence from each other. They may go parallel, or diverge, or converge somewhat, or branch further, but only in rare cases can they coalesce or intertwine where a hybridization of the separate species takes place. The emergence of an allopolyploid signifies that one or several strings have been torn away from two cables representing the two ancestral species. These strings fuse, and presently become subdivided again into a number of new ones, forming a new cable of the polyploid species. In a phylogenetic scheme this event can be represented as a simultaneous bifurcation of two cables, followed by an anastomosis of two out of the resulting four branches. Where the species formation through allopolyploidy is frequent, the phylogenetic "tree" will tend to lose its treelike appearance on account of the fusion of its branches, and will come to resemble a reticulate or netlike structure. Some of the difficulties encountered by systematists in classifying certain plant genera may be due to this circumstance (cf. Anderson 1936).

A less evident, but no less important, consequence of polyploidy involves the gene mutation process. Whether a species is an auto- or an allopolyploid, it has at least a part of its genes represented more than twice in the somatic cells and more than once in the gametes. Thus, an einkorn wheat is A^1A^1, an emmer $A^1A^1A^2A^2$, and vulgare $A^1A^1A^2A^2A^3A^3$, where A^1, A^2, and A^3 are the allelomorphs of the same gene located in the different semi-homologous sets of chromosomes of these plants. A mutation from A to a recessive allelomorph a can easily manifest itself in the phenotype of the einkorn wheat, since a heterozygote A^1a^1 will produce on inbreeding one-quarter of the offspring homozygous for $a^1(a^1a^1)$. A similar mutation in an emmer wheat, if it occurs in one chromosome set only, will not be detectable. An individual $a^1a^1A^2A^2$, despite being homozygous for a^1, carries two dominant allelomorphs A^2 which suppress the effects of a^1 on the phenotype. The same argument is applicable *a fortiori* to a mutation in a hexaploid vulgare wheat, where the effects of the recessive a^1 are suppressed by those of the two pairs of dominants A^2 and A^3. The phenotypic manifestation of recessive mutant genes in the emmer and vulgare wheats may become possible only if similar mutations take place in the two or three chromosome sets, respectively. This is, however, a rather improbable event, and hence the frequency of the detectable recessive mutants may be expected to be lower in tetraploids than in diploids, and lower in hexaploids than in tetraploids.

Indirect evidence corroborating the above deduction is afforded by the classical experiments of Nilsson-Ehle (1911) and others on the identical (polymeric) genes in the polyploid species of cereals. For in the vulgare wheats some characteristics (for example the seed color) are determined by three pairs of identical factors. A cross between a recessive strain ($a^1a^1a^2a^2a^3a^3$) and a strain carrying three dominants $A^1A^1A^2A^2A^3A^3$) gives in the F_2 generation a segregation in the ratio 63 dominants : 1 recessive. Polymeric genes appear to be very common in polyploids, although they occur as an exception also in organisms that are regarded as diploids (e.g., Drosophila). Their presence in diploids may conceivably be due to the existence of the repeat regions in some chromosomes (Chapter IV), which make a generally diploid species polyploid for some chromosome sections.

The mutation frequencies in related species having different chromosome numbers have been compared by Stadler (1929, 1932). The diploid oats *Avena brevis* and *A. strigosa* (n = 7) were given similar amounts of X-ray treatments as the tetraploid *A. byzantina* and *A. sativa* (n = 14). Fourteen mutations were obtained in the former and none in the latter in approximately equally large samples. The experiments with wheat species gave comparable results. Stadler expresses the mutation frequency in terms of the number of mutations obtained per unit of the X-ray treatment (r-unit). The resulting figures are 10.4×10^{-6} for the diploid *Triticum monococcum*, 2.0×10^{-6} and 1.9×10^{-6} for the tetraploids *T. dicoccum* and *T. durum*, respectively. No mutations were obtained in the hexaploid *T. vulgare*. Although the experimental errors are large enough to make the exact numerical values for the mutation frequencies somewhat uncertain, the differences between the diploid, tetraploid, and hexaploid wheats are large enough to be significant. Stadler concludes that the data "support the hypothesis that the reduced frequency of mutation observed in polyploid species is low because of gene reduplication."

The suppression of the effects of a recessive mutant gene appearing in one of the chromosome sets of a polyploid by normal allelomorphs lying in other chromosome sets precludes the phenotypic manifestation of the mutant, and prevents its detection. This mechanism need not, however, decrease the mutation rates of the genes as such. It is reasonable to suppose, in the absence of any evidence to the contrary, that the genes in the chromosomes of a polyploid mutate just as frequently as they do in the diploid. A majority of the mutations are, however, deleterious in nature. In a diploid, deleterious mutations are eliminated, or at least are kept down in frequency, by natural selection. In a polyploid, an unfavorable recessive mutation, even a lethal or a gene deficiency, arising in one of the chromosome sets is "sheltered" from natural selection by the presence of the normal allelomorphs in the other sets. Only the dominant mutations, or such recessives as have appeared in all chromosome sets, can be kept under control by selection.

The sheltering of the destructive mutations produces a dislocation in the usual relationships between the mutation and the selection proc- esses. It would seem that polyploidy must inexorably lead to a progressive deterioration of the germ plasm. And yet polyploid species not only exist in nature, but are sometimes more successful than their diploid relatives. Evidently polyploids are not exempt from the control by selection. A suggestion put forward to resolve this apparent paradox is that the initially homologous genes lying in the different chromosome sets will mutate in different directions, and will gradually become so distinct as to be no longer allelomorphic. Such a differentiation of the chromosome sets might eventually transform a polyploid into a species that has most genes represented only once in the gametes and twice in the zygotes, and thus would in effect restore the diploid status despite the altered chromosome num- ber. The presence of the duplicate, triplicate, etc., genes may be re- garded as a temporary condition which is overcome by mutation. Polyploidy is thus a mechanism whereby both the gene number and the gene variety are increased.

Without wishing to cast doubts on the germanity of the above sug- gestion, we still believe that it does not remove the difficulty entirely. To begin with, nothing is known about the kind of mutation that may make the once allelomorphic genes non-allelomorphic (one may note that the definition of allelomorphism is one of the weak spots of theoretical genetics). Moreover, it would seem that the uncontrolled mutation process would lead sooner to a destruction of the gene than to the emergence of a new gene having an adaptive value and funda- mentally different from the ancestral one. It is hard to avoid the belief that the destructive mutations are somehow controlled by selection even if they appear in a single chromosome set and are re- cessive. Perhaps the destruction of a part of the genes in a chromo- some set without a simultaneous disappearance of the others in the same set would not be permitted because such a condition would leave the organism with some genes represented more times than others, and would thus cause an unbalance. More experimental data on the genetic behavior of polyploids is needed in order to solve the outstanding problems of this kind.

VIII: ISOLATING MECHANISMS

THE FUNDAMENTAL importance of isolation in the evolutionary process has been recognized for a long time. Lamarck and Darwin pointed out that interbreeding of groups of individuals that are hereditarily distinct results in dissolution and swamping of the differences by crossing. The only way to preserve the differences between organisms is to prevent their interbreeding, to introduce isolation. Among Darwin's immediate followers the rôle of isolation was stressed especially by M. Wagner, in whose view it has assumed the position of keystone of the whole theory of evolution. Romanes originated the oft-quoted maxim, "without isolation or the prevention of interbreeding, organic evolution is in no case possible" which if taken too literally overshoots the mark.

From the viewpoint of present knowledge it appears that these early ideas about the rôle of isolation confused two entirely different problems. First the differences between individuals and groups may be due to a single gene or a single chromosome change. Such differences can never be swamped by crossing, since in the offspring of a hybrid segregation takes place, and the ancestral traits reappear unmodified. No isolation is needed to preserve the variation due to changes in single genes, and if one consents to dignify gene mutation by applying the name evolution, the latter is independent of isolation. The bearing of the particulate, as opposed to the blending, theory of inheritance on the problem of the retention of hereditary variation has been discussed above (Chapter V). The second class of differences between individuals and groups is genetically more complex, owing to the coöperation of two or more genes. Races and species usually differ from each other in many genes and chromosomal alterations. Species are distinct because they carry different constellations of genes. Interbreeding of races and species results in a breakdown of these systems, although the gene differences as such are fully preserved. Hence, the maintenance of species as discrete units demands their isolation. Species formation without isolation is impossible.

On the lowest level of the evolutionary process, which is concerned with the origin of hereditary variability, with changes in the basic units such as genes and chromosomes, the rôle of isolation is naught. But on the next higher level the molding of the above elements into integrated systems takes place. The interactions of mutation pressure, selection, restriction of population size, and migration create not new genes but new genotypes, which, in the symbolic language of Wright (1932), occupy only infinitesimal fractions of the potential "field" of gene combinations. Moreover, and this is important, the part of a field occupied by a species is due not to chance alone, but corresponds to the location of one of the "adaptive peaks." Related species occupy each a separate peak, and numerous peaks in the same field may remain unoccupied, since some gene constellations have never been formed and tried out. The adaptive valleys intervening between the peaks are mostly uninhabited, and some of them are so low as to be uninhabitable.

The symbolic picture of a rugged field of gene combinations strewn with peaks and valleys helps to visualize the fact that the genotype of each species represents at least a tolerably harmonious system of genes and chromosome structures. Interbreeding of species results in the breakdown of the existing systems, and emergence of a mass of recombinations. Among the recombinations some might be as harmonious as the old gene patterns; some might be in fact better than the old ones, that is to say, new and higher adaptive peaks may be discovered. But a majority, and probably a vast majority, of the new patterns are discordant, and fall in the adaptive valleys.

We are confronted with an apparent antinomy. Isolation prevents the breakdown of the existing gene systems, and hence precludes the formation of many worthless gene combinations that are doomed to destruction. Its rôle is therefore positive. But on the other hand, isolation debars the organism from exploring greater and greater portions of the field of gene combinations, and hence decreases the chance of the discovery of new and higher adaptive peaks. Isolation is a conservative factor that slows down the evolutionary process.

The antinomy is removed if one realizes that an agent that is useful at one stage of the evolutionary process may be harmful at another stage. Gene combinations whose adaptive value has been tested by natural selection must be preserved and protected from disintegration if life is to endure. Without isolation the ravages of natural selection might be too great. But too early an isolation of the favorable gene combinations formed in the process of race differentiation would mean too extreme a specialization of the organism to the environmental conditions that may be only temporary. The end result may be extinction. Favorable conditions for a progressive evolution are created when a certain balance is struck: isolation is necessary but it must not come too early.

The mechanisms that prevent the interbreeding of groups of individuals, and consequently engender isolation, are remarkably diversified. It is an empirical fact that in different organisms, frequently even in fairly closely related ones, the isolation of species is accomplished by quite dissimilar means. Nor is it necessary that the interbreeding of a given pair of species be prevented by a single mechanism; on the contrary, one may observe that in many cases several mechanisms combine to make the isolation of two species more or less complete. It is important, however, that any agent that hinders the interbreeding of groups of individuals produces the same genetic effect, namely, it diminishes or reduces to zero the frequency of the exchange of genes between the groups. I have proposed (Dobzhansky, 1937a) the expression "isolating mechanisms" as a generic name for all such agents.

The isolating mechanisms may be divided into two large categories, the geographical and the physiological. Groups of individuals may be debarred from interbreeding by the mere fact that they live in different geographical regions, and hence never meet. Geographical isolation is believed by some investigators to be of paramount importance in the process of racial differentiation. The genetic nature of geographical races has been discussed above (Chapters III and IV). The probable rôle of the subdivision of the population of a species into semi-isolated local colonies has also been considered (Chapters V and VI). Here we may add that geographical isolation alone is in general only a temporary measure and need not lead to a permanent segregation of the groups so isolated. Any species has a tendency to expand the area of its distribution; the forms now living in separate regions may eventually come together and meet. If no intrinsic, physiological, isolating mechanisms have developed, interbreeding will begin, and the originally separate groups will fuse together, at least in the area where the distribution regions overlap. Many examples of such a situation have been recorded, especially in plants (see Chapter X).

Geographical isolation is therefore on a different plane from any kind of physiological one. This consideration has to be qualified, because the occupation of separate areas by two species may be due not only to the fact that they have developed there, but also to the presence of physiological characteristics that make each species attached to the environment (climate, etc.) available in one but not in the other region. In this case we are dealing however with a kind of physiological isolation which is expressed in geographical terms. If two or more groups of forms are known to inhabit non-overlapping regions, no conclusions can be drawn as to the presence or absence of physiological isolation between them. Such groups, when brought together artificially or in the natural course of events, may interbreed freely or may continue to keep apart because of physiological isolation. In any concrete case only an experiment can decide what will take place. The physiological isolating mechanisms may be subdivided as follows:

I. Mechanisms that prevent the production of the hybrid zygotes, or engender such disturbances in the development that no hybrids reach the reproductive stage. "Incongruity of the parental forms" may be used as a general term for such mechanisms.*
   A. The parental forms do not meet.
      a. Ecological isolation—the potential parents are confined to different habitats (ecological stations) in the same general region, and therefore seldom, or never, come together, at least during the reproductive age or season.
      b. Seasonal or temporal isolation—the representatives of two or more

---

* The writer has sometimes applied the word "incompatibility" instead of "incongruity," but unfortunately the former is used in a different sense in botanical literature.

species reach the adult stage each at a different season, or the breeding periods fall at different times of the year.

B. The parental forms occur together, but hybridization is excluded, or the development of the hybrids is arrested.

    a. Sexual or psychological isolation—copulation does not occur because of the lack of mutual attraction between the individuals of different species. This lack of attraction may in turn be due to differences in scents, courtship behavior, sexual recognition signs, etc.

    b. Mechanical isolation—copulation or crossing is difficult or impossible on account of the physical incompatibilities of the reproductive organs.

    c. The spermatozoa fail to reach the eggs or to penetrate into the eggs; in higher plants the pollen tube growth may be arrested if foreign pollen is placed on the stigma of the flower.

    d. Inviability of the hybrids—fertilization does take place, but the hybrid zygote dies at some stage of development before it becomes a sexually mature organism.

II. Hybrid sterility prevents the reproduction of hybrids that have reached the developmental stage at which the parents normally breed. Sterile hybrids produce either no functional gametes, or gametes that give rise to inviable zygotes. The classification of the phenomena of hybrid sterility will be discussed in Chapter IX.

A wealth of data on the occurrence of various isolating mechanisms in different subdivisions of the animal and plant kingdoms is scattered through biological literature. The genetic analysis of isolating mechanisms, with the possible exception of hybrid sterility, has however been left in abeyance. It is a fair presumption that the pessimistic attitude of some biologists (e.g., Goldschmidt, 1933c), who believe that genetics has learned a good deal about the origin of variations within a species, but next to nothing about that of the species themselves, is due to the dearth of information on the genetics of isolating mechanisms. The maintenance of the separation between species is due to the presence of physiological isolating mechanisms that hinder their free interbreeding; races of a species are as a rule not so isolated, or show only rudiments of isolation. So long as the genetics of the isolating mechanisms remains almost a terra incognita, an adequate understanding, not to say possible control, of the process of species formation is unattainable. In the following paragraphs we shall nevertheless try to assemble some facts and to out-line some suggestions that may throw light on these problems.

Data on the habitat of a species or race, as well as information on the time of year when breeding takes place, are customarily given in the systematic and ecological literature on any one group. A perusal of such literature usually reveals some examples of related species that differ in these respects. It seems clear enough that such differences may decrease the frequency, or preclude entirely, the inter-breeding of the populations concerned. Investigations that are especially directed towards ascertaining to what extent the ecological and seasonal isolations are actually responsible for the maintenance of separation between species are, however, very rare. A genetic analysis of this type of difference between species or races has, to the writer's knowledge, never been made. With these qualifications in mind, we may examine a few instances in which the effectiveness of ecological or seasonal isolation suggests itself.

The experiments of Dice (1933) show that species of the mouse Peromyscus are as a rule not crossable under laboratory conditions, while races (subspecies) of the same species can be crossed and produce offspring. In this connection it is especially interesting that some races occur in the same general geographical region, without, however, producing intermediates or losing their distinctness. Dice (1931) has made a study of two races of P. *maniculatus* whose distribution areas overlap in a part of the state of Michigan, and found that one of them lives almost exclusively in forests and the other on lake beaches. Two other races of the same species occur together in the region of Glacier National Park, Montana; but Murie (1933) reports that one of them is confined to forests and the other to prairie habitats, and this excludes crossing. According to Pictet (1926, 1928a, b), races of the moths *Lasiocampa quercus* and *Nemeophila plantaginis* occur in Switzerland at different altitudes. In the latter species the races encountered above 2,700 meters and below 1,700 meters above sea-level differ, according to Pictet, in a single gene. At 2,200 meters a hybrid population is encountered, composed exclusively of heterozygotes; when bred in the laboratory the off-spring include heterozygotes as well as both homozygotes, but in nature the homozygotes are always eliminated by natural selection. The evidence presented by Pictet is however incomplete, and his interpretations may be questioned.

The malarial mosquitoes united under the name *Anopheles maculipennis* are divided into a group of species or races that are isolated from each other, at least in part, by their habitats. A rather extensive literature is devoted to investigations of these mosquitoes. Roubaud (1920, 1932) finds in France two "biological races" of *maculipennis,* one of which preys chiefly on man and the other on domestic animals. In Holland two forms are found (de Buck and Swellengrebel 1931, de Buck, Torren, and Swellengrebel 1933), while in Italy at least four races are distinguished (Hackett, Martini, and Missiroli 1932; Missiroli, Hackett, and Martini 1933). Aside from minor differences in the morphology of the adults, the races differ in the larvae, and especially in the coloration of the eggs. All races seem to be potential carriers of the malarial Plasmodium, but only one or two of them have a preference for man's blood, or at least bite man and domestic animals indiscriminately. The geographical distribution of the latter races coincides, as might be expected, with that of the endemic malaria. The apparently well authenticated fact that the distribution of endemic malaria in Europe has contracted in historical times is supposed to be correlated with the increase in the number of domestic animals in the now malaria-free localities. Roubaud (1920) optimistically recommends, as a method of combating malaria still further, a "trophic education" of the population of Anopheles to train them to use animals instead of man.

The experiments of de Buck, Schoute, and Swellengrebel (1934) and others leave no doubt that the differences between the Anopheles races are hereditary. It is especially interesting that each race is restricted in nature to a fairly definite habitat, and does not occur elsewhere. Ecologically distinct races appear to be present also in the common mosquito, *Culex pipiens* (Weier 1935, de Buck 1935).

Seasonal isolation between two closely related species of the mollusk Sepia has been studied in detail by Cuénot (1933). One species breeds in the spring in the littoral zone of the Atlantic and the Mediterranean, while the other breeds in the same localities in winter and at greater depths. Dr. Edgar Anderson kindly informs me that one of the main factors isolating certain species of Iris is a difference in the flowering seasons; the same is true for *Hamamelis virginiana* and *H. vernalis*. Mr. C. N. Rudkin permits me to quote his observations on the time of appearance of the adults in certain related species of butterflies in southern California, which are specifically known to occur together in the same localities (although their general distribution ranges do not coincide).

{*Euphydryas chalcedona* (Doubleday & Hewitson)—April to June
}*Euphydryas editha wrightii* (Gunder)—March
{*Melitaea neumoegeni* (Skinner)—late March to early April
}*Melitaea wrightii* (Edwards)—late April to early June
{*Argynnis macaria* (Edwards)—late April to early June
}*Argynnis adiaste atossa* (Edwards)—late May to August
{*Philotes sonorensis* (Felder)—February to mid-April
}*Philotes battoides bernardino* (Barnes & McDunnough)—May

The exact flying times for each species are variable depending upon seasonal weather conditions and altitude, but in the localities where they occur together little or no overlapping is observed.

SEXUAL ISOLATION

An obvious prerequisite for a sexual union between individuals of the same or of different species is that the sexes meet and perform the series of acts that precede and enable fertilization to occur. In some forms this series of acts is relatively short and simple. In the oyster the chemical substance or substances that are released in water, together with the eggs and spermatozoa, stimulate other individuals within a certain range to spawn and to eject further masses of sex cells (Galtsoff 1930). In other animals the procedure is more complex. Sexual recognition marks of various kinds (specific scents, colorings, sounds, and various behavior patterns grouped under the name courtship) enable the individuals of either sex to discern potential mates. Any incongruity between the mating reactions of two groups of individuals may engender sexual isolation. The physiological basis of sexual isolation may, however, be as unlike in different instances as are the mating reactions themselves.

Specific scents play an important role in the sex life of moths, and probably of most other insects as well. The experiments of Standfuss, Fabre, and many others have shown that if a female moth is exposed even in very artificial surroundings, males of the same species appear and try to reach the female despite the obstacles that may be placed in their way. It is established that the males sense the presence of the females at a rather great distance; the acuteness of smell so demonstrated is remarkable. Only rarely are males of species other than that to which the female belongs attracted (Federley 1932). In an experiment of the writer (unpublished), cages with females of *Dicranura vinula* and *D. erminea* were exposed on opposite sides of a house. Although in the locality where the experiment was carried out *D. vinula* was much more abundant than *D. erminea,* males of each species presently appeared near the cages that contained females of their own species. It may be noted that *D. vinula* and *D. erminea* may be crossed in captivity and produce sterile offspring (Federley 1915b).

When brought together artificially, males of a given species of moth as a rule pay no attention to females of species other than their own. Interspecific crosses may, however, be accomplished by special techniques. Both Standfuss (1896) and Federley (1929b) recommend placing a cage containing females of species A and males of B side by side with a cage containing females of B and males of A. The scent, and perhaps the sight, of females of their own species makes the males so excited that they copulate with females that they would not approach otherwise. In extreme cases copulation between distant species may be attempted even in the presence of partners of their own species.

The extent to which the mating reactions of species can differ without making the production of hybrids impossible may be seen from the work of Leiner (1934) on the fish *Gasterosteus aculeatus* and *G. pungitius.* These fish build special nests into which the females are goaded by the males; after the eggs are deposited and fertilized the female is driven away and the male stays to take care of the young. The behavior patterns of the two species are as follows:

G. pungitius	*G. aculeatus*
The nest is built hanging on some water plants.	The nest is built on the bottom, in a furrow dug by the fish.
The nest is composed of soft materials.	Hard materials are used in the construction of the nest.
The nest has an entrance and an exit.	The nest has a single entrance.
No preference for light or dark building materials.	On a light bottom dark building materials are preferred.
The nest is not changed after the eggs are deposited.	After egg deposition the nest is somewhat altered.
The male swims toward the nest in zigzags, attracting the female to follow him.	The male makes some zigzags in front of the female, and then swims straight to the nest followed by her.
The process of leading the female to the nest and the mating play coincide.	A special mating play is enacted.
The female enters the nest with little prodding by the male.	The male forces the female into the nest.

Leiner has obtained several hybrids from the cross *G. aculeatus* ♀ × *G. pungitius* ♂ with the aid of artificial insemination, and he gives some, though inconclusive, data that suggest that this cross takes place sometimes also in nature.

In some animals, especially among the predacious forms, hybridization is difficult because one or both of the prospective parents evince a belligerent attitude toward each other, instead of a sexual response. Wild species of sheep are very likely to kill the domestic sheep and goats offered them for mates. There exist some authentic data (Bristowe and Locket, 1929, and others) to prove that the involved courtship antics practiced by male spiders as a preliminary to copulation tend to delay the assumption by the female of an aggressive attitude. A male of a different species is simply put to death before copulation can take place. It is interesting however that in higher animals where the mating reactions involve complex systems of unconditional and conditional reflexes, large deviations from the normal behavior can be induced in experiments. Much work in this direction has been done by the Russian school of animal husbandry (see review by Serebrovsky 1935). Stallions can be trained to mount willingly a stuffed effigy of a mare and even that of a cow, and the same is true of bulls, boars, and male sheep. Male turkeys were induced to attempt copulation with fowls. The practical application of these results is principally in the development of techniques for the collection of semen for artificial insemination. They are also interesting because they show that the bar to hybridization formed by sexual isolation can be surmounted experimentally. In general, hybrids that probably never occur in nature can be frequently obtained in properly conducted experiments.

Some experimental data on sexual isolation have accumulated in Drosophila literature. Sturtevant (1915, 1921) studied the process of courtship in several species of that genus, and found it to be frequently different. Sturtevant (1920-21) has also obtained indirect evidence to show that in mixed cultures D. melanogaster and D. simulans exhibit a preference for mating with partners of their own species. Similar observations have been published by Lancefield (1929) for A and B races of D. pseudoobscura. A quantitative study of sexual isolation has been made by Boche (unpublished). In his experiments equal numbers of freshly hatched virgin females of race A and race B of D. pseudoobscura were placed in the same vials with half the total number of freshly hatched males of one of the races. The males therefore were able to "choose" their mates. After about ninety-six hours all females were dissected, and the presence or absence of sperm in their seminal receptacles was determined by a microscopic examination. In every experiment males of race A fertilized more race A than race B females, and the opposite was the case for race B males. In one experiment the frequency of fertilization was determined after different time intervals: the result suggests that the males copulate first with females of their own race, but later also interracial matings occur, and finally practically all of the females are fertilized. No indication of even slight isolation between strains of the same race coming from different geographical regions was obtained.

Somewhat more extensive data were secured by the writer (unpublished) for Drosophila pseudoobscura and D. miranda. In some experiments the technique used was similar to Boche's: ten D. pseudoobscura and ten D. miranda females were confined with ten males of one species for about ninety-six hours. The results are summarized in Table 18.

Sexual isolation between *Drosophila miranda* and *D. pseudoobscura* is undoubtedly present. Moreover, the isolation between these species is relatively stronger than that between race A and race B of *D. pseudoobscura* (see above). Indeed, a series of experiments

TABLE 18

FREQUENCY OF INTRASPECIFIC AND INTERSPECIFIC MATINGS IN MIXED CULTURES OF *Drosophila pseudoobscura* AND *D. miranda*.

MALES	D. pseudoobscura FEMALES FROM STRAIN	INTRASPECIFIC		INTERSPECIFIC	
		Fertilized	Unfertilized	Fertilized	Unfertilized
D. pseudo-obscura	Seattle-6 (race B)	58	5	3	60
	Seattle-4 (race B)	54	1	1	54
	La Grande-2 (race A)	67	9	9	71
	Texas (race A)	57	2	8	52
	Oaxaca-5 (race A)	41	4	7	35
D. miranda	Seattle-6 (race B)	18	17	3	34
	Seattle-4 (race B)	12	31	—	43
	La Grande-2 (race A)	14	20	4	31
	Texas (race A)	11	22	—	34
	Oaxaca-5 (race A)	22	18	4	37

in which the frequency of intraspecific and interspecific matings has been determined after varying time intervals has shown that after five days practically all females of the same species as the male are impregnated. The number of interspecific matings also increases with time, but even after twenty-one days no more than 25 per cent of the females are fertilized. In a third series of experiments the technique was so modified that no choice of mates was available. *D. miranda* females were confined with an equal number of *D. pseudoobscura* males for nine days, after which the proportion of the females fertilized was determined by dissection. Different strains of *D. pseudoobscura* were used. Table 19 gives a summary of the results.

If *D. pseudoobscura* females are confined for nine days with males of their own species, practically 100 per cent of them are fertilized. Therefore, the results shown in Table 19 give further proof of sexual isolation between the two species. In addition, these results reveal an important fact that the isolation varies in extent if different strains of *D. pseudoobscura* are used. It can be seen at a glance that males of race B display on the average a greater aversion to mating with *D. miranda* than do the males of race A.

TABLE 19

THE FREQUENCY OF FERTILIZATION (IN PER CENTS) OF *Drosophila miranda* FEMALES BY *D. pseudoobscura* MALES FROM DIFFERENT STRAINS

RACE	STRAIN	FREQUENCY	RACE	STRAIN	FREQUENCY
B	Cowichan-6	10.2±2.3	A	Pavilion-5	31.3±4.3
B	Quilcene-4	11.3±2.7	A	Lassen-1	32.7±4.6
B	Sequoia-4	17.8±2.9	A	Shuswap-3	36.4±3.8
B	Sequoia-8	29.4±3.4	A	Estes Park-1	36.9±4.2
A	Yale-7	18.3±3.0	A	Sequoia-15	40.8±3.7
A	Oaxaca-5	21.4±3.2	A	Cuernavaca-2	46.0±3.6
A	Olympic-2	23.0±2.7	A	Grand Canyon-3	50.4±4.7
A	La Grande-2	25.0±4.1	A	Julian	52.2±4.3

Within a race, especially in race A, wide differences are also observed. The degree of sexual isolation seems to stand in relation to the geographical origin of the given strain. It may be noted that the distribution area of *D. miranda* is relatively small, comprising only the territory around Puget Sound in the Pacific Northwest. This area is included in that of race B of *D. pseudoobscura,* which inhabits the Pacific Coast from southern California northward; the area of race A extends much further east and south than that of race B, but it barely comes in contact with that of *D. miranda* (on the Olympic peninsula). Strains of either race of *D. pseudoobscura* coming from localities in or near the distribution area of *D. miranda* show the greatest degree of isolation (Cowichan, Quilcene, Olympic, and Yale). Strains from somewhat more remote localities (Pavilion, Shuswap, Lassen, Estes Park) show less, and those from still more remote places (Sequoia, Julian, Grand Canyon, Cuernavaca) show least isolation. The only exception to the above geographical rule is the Oaxaca strain (from Mexico) which displays an unexpectedly high degree of isolation.

The existence of such inheritable differences between strains of a species is interesting, for the genetic mechanism determining the isolation between species may be visualized as having arisen through a summation of intraspecific variations of this kind. The observed geographical regularity becomes doubly significant from this point of view. *Drosophila pseudoobscura* and *D. miranda* can be crossed, but the offspring produced are sterile. The occurrence of hybridization is evidently disadvantageous to the species, since it impairs the biotic potentials of both participants. Genetic factors increasing the sexual or any other isolation may therefore be favored by natural selection. The strengthening of isolation is however more immediately important for the populations of *D. pseudoobscura* that inhabit the territory close to that where *D. miranda* occurs than for the populations from more remote localities. The exceptionally strong isolation found in the Oaxaca strain (see above) may conceivably be accounted for by the fact that in Oaxaca *D. pseudoobscura* shares the same territory with another related species, *D. azteca*. The value of this conjecture is dubious however, because *D. azteca* occurs throughout Mexico, and the Mexican strains except Oaxaca do not seem to show a very great isolation from *D. miranda*.

Systematic studies on the crossabilities of different species and subspecies in the mouse Peromyscus have been made by Dice (1933). Ten races of *P. maniculatus*, five of *P. leucopus*, four of *P. eremicus*, and two of each *P. truei* and *P. californicus* have been tested in many combinations. The general conclusion reached by Dice is that races of a species can be crossed and produce hybrids, while the separate species (with the exception of *P. maniculatus* × *P. polionotus*) do not cross. The cause of the non-production of hybrids is not exactly known, but a sexual isolation may be suspected. It does not follow, of course, that sexual isolation may occur only between separate species and not between subdivisions thereof. Spett (1931) obtained some data that suggest the existence of a rudimentary sexual isolation between mutants of *Drosophila melanogaster;* his observations are contradicted however by those of Sturtevant (1915) and of Nikoro, Gussev, Pavlov, and Griasnov (1935).

MECHANICAL ISOLATION

The elaborate structure of the external genitalia and their accessories in many animals, especially among insects, has for a long time attracted the attention of morphologists and systematists. The reason for this interest has been in part a pragmatic one: closely related species that are distinguishable with great difficulty by their external structures can sometimes be accurately classified by the structure of their genitalia. Ormancey (1849) seems to have been the first to apply this method for distinguishing the species of a family of beetles, and soon thereafter the method was introduced in other orders of insects, and also in spiders, mollusks, fish (the forms possessing gonopodia), mammals (especially bats and rodents), and other groups.

Although some conservative taxonomists have made vitriolic protests against the introduction of studies of genitalia as a part of the regular routine in describing species, the method has such obvious practical advantages in many genera and families that it has taken a firm root in modern taxonomy. That species are frequently easily distinguishable by their genitalia is indeed a plain observational fact; on this fact much theoretical superstructure has however been built. The great French entomologist Leon Dufour has propounded the so-called "lock-and-key" theory, according to which the female and the male genitalia of the same species (at least in insects) are so exactly fitted to each other that even slight deviations in the structure of either make copulation physically impossible. The genitalia of each species are "a lock that can be opened by one key only," hence the different species are isolated from each other simply and safely by the non-correspondence of their genitalia. In justice to Dufour it must be noted that the whole theory originated in pre-Darwinian days.

Dufour's theory has been much elaborated by K. Jordan (1905), who established an interesting contrast between geographical and non-geographical variations: variants of species living in the same locality show no correlated characteristics in the genitalia, while separate species, and sometimes geographical races of a species, do differ in the structure of these organs. Jordan adduces some further evidence in favor of the lock-and-key theory by showing that the male and female genitalia of Papilio species relate to each other as a positive and negative image, and proceeds to argue that geographical races become isolated from each other by variations in the genitalia, and thus become separate species. Jordan's theory is attractive in its simplicity if we disregard his Lamarckian notions, which do not seem to constitute an integral part of it. Unfortunately, there is more evidence against it than for it.

First of all, the lock-and-key relationship between the female and male genitalia of the same species is one the whole a rather exceptional condition. In many groups (e.g., in Drosophila) where some species differ quite sharply in the structure of the male genitalia, the female genitalia are far more similar, and in addition are not sculptured as a negative image of the male parts. This situation seems to be common in diverse groups. To be sure, female genitalia are often as rich in specific characters as those of the males, but the specific differences may reside in the parts of the apparatus that are not immediately concerned with copulation. For example, among ladybirds (Coccinellidae) and leaf beetles (Chrysomelidae) female genitalia of related species are often distinct in the shape of the chitinous spermatheca and that of the duct uniting the spermatheca with the bursa copulatrix. Yet during copulation, the penis of the male is inserted into the bursa copulatrix but certainly does not penetrate as far as the spermatheca or its duct. Curiously enough, there is a certain correlation between the shape of the penis and that of the spermathecal duct: in some genera of the ladybirds the former has a very long appendage (flagellum), and the latter may be longer than the body length. The writer has ascertained specifically in such forms that the flagellum does not enter the duct. The external female genitalia are rather uniform in species of the same genus in Coccinellidae, contrasting with the variability of the corresponding male structures. In some families and genera, distinct species have very nearly similar genitalia; mechanical isolation can in no case be regarded as a universal method of isolation, even among insects.

The experimental evidence in favor of mechanical isolation is scanty, and is confined mostly to a single order, namely Lepidoptera. Standfuss (1896) has described crosses between species of moths where copulation leads to injuries to the female organs that result in death. Federley (1932), who is inclined to ascribe more importance to mechanical than to sexual isolation, states that the *Chaerocampa elpenor* male may copulate with a female of *Metop-*

silus porcellus (moths of the family Sphingidae), but is sometimes unable to withdraw its penis, making egg-deposition impossible. The reciprocal cross succeeds easily. Sturtevant (1921) has observed apparently successful copulation between *Drosophila melanogaster* males and *D. pseudoobscura* females, which, however, does not lead to the production of hybrid larvae or adults. Whether fertilization of the eggs takes place is unknown. Some pairs are however unable to separate, and die in copula. What causes the different outcome of this copulation is likewise unknown. No copulation between *D. pseudoobscura* males and *D. melanogaster* females has been recorded.

Against the above facts which tend to prove the effectiveness of mechanical isolation, one may set an array of observations on crosses between species with differently built genitalia which seems to cause no injury to either participant. Copulation between rather remote species is very frequently recorded in the entomological literature, although it generally remains obscure whether any offspring is produced thereby. The production of offspring is however immaterial as far as the problem of mechanical isolation is concerned, since copulation does not necessarily insure the occurrence of fertilization and development (see below). It is significant, however, that variations in body size within a species of insects have not been shown to hinder copulation. In Drosophila mutants, increasing and decreasing body size are known, and they can be crossed with consequent production of normal offspring (e.g., mutations giant and dwarf in *D. melanogaster*). The variations in body size due to the abundance or scarcity of food during the larval stage are likewise no impediment for copulation.

Kerkis (1931) has made a statistical study of the variability of the external characteristics and of the genitalia in the bug *Eurygaster integriceps,* and finds the latter no less variable than the former—a conclusion contradictory to the opinions of some systematists who regard the limited variability of the genitalia as an explanation of their usefulness in classification. In fact, the explanation is to be looked for in a different direction: the complexity of the structures of the genitalia is sometimes so great that the genetic differences between the species are more likely to be manifested in these structures than in the relatively simple external ones. The conjecture is corroborated by observations that show that in those genera and families where the structure of the genitalia is simple they are less useful for classification than in groups with complicated genitalia or accessory organs. It is justifiable to conclude that, although some mechanical isolation may be effective as a bar to crossing in some organisms, its significance has been exaggerated. Some systematists (e.g., Kinsey 1936) have come to the same conclusion.

The differences in the flower structure in related species of plants may prevent cross-fertilization because the flowers are pollinated by different insects. How effective this form of mechanical isolation is in nature is however obscure. That different plant families are adapted for pollination by different insects is of course well known, although some insects (e.g., the honeybee) visit a surprisingly wide range of plants. Whether species of the same genus are debarred from crossing by the same method has never been adequately studied. A perusal of the Knuth-Ainsworth Davis monograph of flower pollination (1906-09) shows that the lists of insects known to visit the flowers of related plant species are in some instances different, but it remains unclear to what extent this may be accounted for by the occupation of dissimilar ecological stations by the plants involved. Perhaps only in families with very specialized flower structures (orchids, Papilionaceae, and some others) can mechanical isolation play an important rôle.

FERTILIZATION IN SPECIES CROSSES

Copulation in animals with internal fertilization, or the release of the sexual products into the medium in forms with external fertilization, or the placing of the pollen on the stigma of the flower in plants, are followed by chains of reactions that bring about the actual union of the gametes, or fertilization proper. These reactions may be out of balance in representatives of different species, with a consequent hindrance or a complete prevention of the formation of hybrid zygotes. In animals, the processes of hybrid fertilization have been studied, for obvious technical reasons, almost exclusively in marine forms where the fertilization can be easily observed in vitro. Moreover, a majority of the experiments concern crosses between forms so remote (e.g., different orders, classes, and even phyla), that the significance of the results from an evolutionary standpoint is limited.

Lillie (1921) has crossed two species of sea-urchins, *Strongylocentrotus purpuratus* and *S. franciscanus*. Both species inhabit the shore waters in the same locality, although *S. purpuratus* occurs between the tidemarks and slightly below the low-water mark, while *S. franciscanus* rarely lives above the low-water mark and goes to greater depths than the former. There exists consequently a partial ecological isolation between the two. Eggs of each species were placed in sea-water containing spermatozoa of the same or of the other species in different concentrations; the percentage of eggs that formed fertilization membranes and that cleaved was recorded. The concentrations of the *S. franciscanus* sperm that give from 73.3 per cent to 100 per cent of fertilization of the eggs of the same species produce from 0 to 1.5 per cent of fertilization in *S. purpuratus* eggs. With a concentration of the sperm of *S. franciscanus* that is forty times greater than is necessary to produce a 100 per cent fertilization of *S. franciscanus* eggs, only 25 per cent of *S. purpuratus* eggs are fertilized. A similar, though perhaps somewhat less pronounced, disability of *S. purpuratus* sperms to fertilize the eggs of *S. franciscanus* was also detected. Moenkhaus (1910) found in the cross between the fish *Fundulus heteroclitus* and *F. majalis* up to 50 per cent of polyspermic eggs which do not normally occur in intraspecific fertilizations. It may be noted that placing the eggs and spermatozoa in water of varying pH concentration sometimes permits the fertilization to take place where it would not do so otherwise.

The environment of the spermatozoa in the reproductive organs of the female of another species may be unsuitable for them and may cause their death, or at least a loss of fertilizing ability. Spermatozoa of higher animals are known to be highly sensitive to any variations in their environment, particularly to those in osmotic pressure. Serebrovsky (1935) gives following data for the spermatozoa of mammals as shown in Table 20.

The sperm can be preserved for artificial insemination for a long time if a proper environment is created, but the fertilizing ability is lost very quickly otherwise. The sperm of a duck, a goose, and a cock has been injected in the genital ducts of female ducks. After 22 to 25 hours the birds were dissected, and large numbers of spermatozoa were found in the upper portions of the oviducts. But while those of the drake were alive and motile, a majority of the spermatozoa of the goose and cock were already dead (Serebrovsky 1935). Mixing the sperm of different forms may also be fatal for their

TABLE 20

MOLECULAR CONCENTRATION AND THE OSMOTIC PRESSURE OF SPERM
(*after Serebrovsky*)

ANIMAL	MOLECULAR CONCENTRATION	OSMOTIC PRESSURE
Man	0.297	7.5
Horse	0.302	7.6
Dog	0.319	8.1
Pig	0.335	8.4
Bull	0.335	8.4
Sheep	0.357	9.0

viability (Godlewski 1926). As far as the writer is aware, no data of a similar kind exist for crosses between closely related species.

More extensive observations of the difficulties encountered in fertilization in hybrids are available for plants. Mangelsdorf and Jones (1926) and others found that in crosses between sugary and non-sugary maize (*Zea mays*) appreciable deviations from the normal segregation ratios are obtained, the number of sugary kernels being below the expectation. Sugary differs from non-sugary in a single gene, and the results are interpreted as indicating that if a mixture of sugary and non-sugary pollen is applied to the silks of a plant containing the normal allelomorph of sugary, a competition between the pollen grains ensues, the rate of growth of sugary pollen tubes being smaller than that of the normal pollen tubes. The growth rates of the two kinds of pollen tubes on sugary silks are, however, alike. Demerec (1929b) has described an even more extreme case of incompatibility between popcorn and other varieties of maize. If popcorn is used as a female parent in crosses where non-pop pollen is applied, almost no seeds are formed. Crosses in which pop is used as a male succeed without difficulty. If a popcorn plant is double pollinated (i.e., if a mixture of pop and non-pop pollen is applied), many selfed and very few hybrid seeds are obtained. When the silks of an ear of popcorn were divided in two parts and one part was pollinated with pop and the other with non-pop pollen, the resulting ears had a full complement of seeds on the selfed side and almost no seeds on the crossed side (for further examples see Brieger 1930).

An extensive series of experiments with crosses between different species of Datura has been described in short preliminary communications by Buchholz, Williams, and Blakeslee (1935). They found that the speed of the pollen tube growth in the style of the same species is frequently greater than in the style of a foreign species. Species of Datura may differ in the length of their style, there being some correlation between the speed of the pollen tube growth and the style length. The crosses in which the species with a short style is used as the female parent and that with a long style as the male parent are in general more likely to succeed than the reciprocal crosses. Moreover, the pollen tubes may burst in the style of a foreign species before they reach the ovary, the frequency of the bursting pollen tubes being characteristic for each cross. The crossability of different species is, therefore, a function of several variables: the speed of pollen tube growth, length of the style, and the frequency of bursting pollen tubes. To this must be added also the sensitivity of the process to the environmental conditions, and the viability of the embryos (see below). The failure of the pollen grains to germinate on a foreign stigma has also been observed in some crosses.

The success of crossing of species of wheat (Watkins 1932) and of herbage grasses (Jenkin 1933) depends on several factors, one of which is the chromosome number in the parental species to be crossed. According to Watkins, the pollen tubes grow best in the styles of plants with the same chromosome number as the male parent (that is, if the ratio of the chromosome numbers in the pollen and in the style is 1:2). In the style of a species having a higher chromosomal number, the pollen tube growth is normal or reduced, while in the style with a lower chromosomal number it is much reduced. The possible rôle of self-sterility in interspecific crosses has been discussed by Anderson (1924).

The incongruity of the allopolyploid hybrids with the parental species is an exceptionally interesting fact. It may be recalled that Raphanobrassica is a synthetic new species obtained by a doubling of the chromosome complement in the hybrid between radish (*Raphanus sativus*) and cabbage (*Brassica*). Karpechenko (1928) and Karpechenko and Shchavinskaia (1929) have made systematic attempts to cross Raphanobrassica with radish, cabbage and other species of cruciferous plants. The cross Raphanobrassica ♀ × Raphanus ♂ produced only eleven seeds from 382 artificially pollinated flowers, Raphanus ♀ × Raphanobrassica ♂ eleven seeds from 143 flowers, Raphanobrassica ♀ × Brassica ♂ two seeds from 551 flowers, and Brassica ♀ × Raphanobrassica ♂ no seeds from 411 flowers. No more successful were the attempts to secure offspring from such crosses by open pollinations; when the three species are planted side by side, each of them produces almost exclusively a pure progeny. Raphanobrassica produces however some seeds if crossed to *Raphanus raphanistrum*, a species related to radish but crossable only with difficulty to the latter. Karpechenko believes that the incompatibility in the above crosses is due to a slow growth of the Raphanobrassica pollen on the Raphanus and Brassica styles, and vice versa. The pollen tube growth has however not been studied specifically, and it remains possible that inviability of the zygotes is involved. The latter has been observed in the crosses between the allotetraploid derivative of *Nicotiana rustica* × *N. paniculata* and *N. rustica* (Singleton 1932). Whatever is the mechanism, it is clear that an incongruity between an allotetraploid and its parents would be very helpful for the establishment of the former as a separate species in nature.

VIABILITY OF HYBRID ZYGOTES

The occurrence of a union between the gametes of different species gives no assurance that the zygotes so formed will produce an adult hybrid organism. As a matter of fact, the life of a hybrid zygote may be cut short at any stage, beginning with the first cleavage of the egg and up to the late embryonic or post-embryonic development. The physiology of the developmental disturbances that prevent the hybrid from reaching maturity is almost entirely unknown. The theory that a lack of "affinity" or of "coöperation" between the ancestral germ plasms is involved gets us nowhere.

Hybridization between very remote forms (echinoderms × molluscs, echinoderms × annelids, etc.) frequently results in the sperm nucleus being simply eliminated from the first cleavage spindle, or else the paternal, and sometimes also some of the maternal, chromosomes are discarded in the cytoplasm and perish. Similar, although less extreme, disturbances are observed in crosses between different families and genera of sea urchins and between families and genera of amphibians (a review in Hertwig 1936). In hybrids between different fish (Moenkhaus 1910, Newman 1914, 1915, Pinney 1918, 1922, and others) all sorts of disturbances may occur, from chromosome elimination during cleavage, and arrest of gastrulation and of organ formation, to death of the advanced embryos. The above authors emphasize that the early or late death of the embryos is not necessarily correlated with the systematic remoteness or closeness of the forms crossed. In this respect the data of Zimmermann (1936) and Strasburger (1936) are very instructive. They have investigated the races of a ladybird beetle *Epilachna chrysomelina*, which inhabits southern Europe, Africa, and western Asia. This area is subdivided into several smaller regions, each inhabited by a separate race (subspecies). The crosses between most of the races that were available for experiments gave hybrids without much difficulty. But the cross between the South African form, *E. capensis,* and *E. chrysomelina* produced no larvae on account of the profound disturbances in the embryonic development. Morphologically, *E. capensis* is not much more different from *E. chrysomelina* than the races of the latter species are from each other. On the other hand, Pictet (1936) states that the viability of the hybrids between the moths *Lasiocampa quercus* from different localities is inversely proportional to the distance between the localities. Similar results have been obtained by Pictet in another moth, *Nemeophila plantaginis,* where the hybridization of local races may result in the production of unfertilized eggs (no cytological study has, however, been made).

The death of the hybrid zygotes has been observed also in plants. In some crosses between species of Datura, the pollen tubes reach the ovary, fertilization takes place, but nevertheless no seeds are obtained. The hybrid *D. stramonium* × *D. metel* develops up to the eight-cell stage of the embryo, but no further. The development of the endosperm in the same hybrid proceeds apparently normally up to the seventh day after fertilization, and then stops (Satina and Blakeslee 1935). A similar situation is encountered in crosses between some species of Nicotiana (McCray 1933). According to Watkins (1932), the non-production of seeds in wheat species crosses may be due to a disharmony between the development of the embryo and that of the endosperm. On account of the double fertilization process in the higher plants, the numbers of chromosomes in the embryonic and in the endosperm tissues are normally as 2:3. If a species with a high chromosome number is used as the pollen parent and that with a low number as the mother, the ratio of the chromosome numbers in the embryo and the endosperm is > 2:3, and the embryo dies. The reciprocal cross, giving rise to a ratio < 2:3, is less deleterious for the viability of the hybrid.

In some instances the constitutional weakness of the hybrid organism entails no great disturbances in the fundamental life processes, and the application of certain treatments enables the experimenter to bring to maturity hybrids that do not survive otherwise. A remarkable example of this phenomenon is afforded by the work of Laibach (1925) on hybrids between species of flax. In the cross *Linum perenne* × *L. alpinum,* the hybrid seeds are able to germinate with some difficulty. The seeds from the cross *L. perenne* ♀ × *L. austriacum* ♂ fail to germinate if left to their own devices. If, however, the embryos are artificially freed from the seed coat (the seed coat being here a purely maternal tissue), germination does take place, and the young seedlings may give rise to luxuriant hybrid plants that are fertile and produce normal seeds of the F_2 generation. Still greater is the suppression of the seed development in the cross *L. austriacum* ♀ × *L. perenne* ♂, and yet it can also be surmounted. The diminutive embryos are extracted from the seeds and placed in a nutrient solution containing from 10 per cent to 20 per cent sugar, where they continue to grow; after some days they are transferred to moist filter paper, and allowed to germinate. The seedlings are then planted in soil.

In crosses between species of certain moths only males appear among the adult hybrids (*Chaerocampa elpenor* ♀ × *Metopsilus porcellus* ♂ and *Deilephila euphorbiae* ♀ × *D. galii* ♂); females are present among the caterpillars but they die in the pupal stage. The reciprocal crosses give hybrids of both sexes (Federley 1929). Bytinski-Salz (1933) implanted the ovaries of the pupae that normally die into the pupae of the parental species. The implants developed in the new host far beyond the stage at which they would die in the body whence they came, thus proving that the inviability of a hybrid as a whole need not extend to all its tissues.

The appearance of unisexual progenies recorded in the crosses just discussed is a fairly common phenomenon in interspecific hybrids in animals; individuals of one sex die, while the viability of the other sex is affected little or not at all. Haldane (1922) has formulated a rule that, with some exceptions, holds rather well: "when in the F_1 offspring of two different animal races one sex is absent, rare, or sterile, that sex is the heterozygous sex." In mammals, Amphibia, and most insects, males are known to be heterozygous (XY) and females homozygous (XX) for sex, and accordingly male hybrids are defective more frequently than females. On the contrary, in birds, butterflies, and moths, females are heterozygous (XY) and males homozygous (XX); here female hybrids tend to be less viable than males.

A possible mechanism that may underlie Haldane's rule was suggested by Dobzhansky (1937b). It is known (Chapter IV) that *Drosophila pseudoobscura* and *D. miranda* differ in gene arrangement, and, what is especially important for us now, some genes that lie in one of these species in the X chromosome lie in the other in the autosomes, and vice versa. The cross *D. miranda* ♀ × *D. pseudoobscura* ♂ produces fairly viable female and abnormal male hybrids; the reciprocal cross gives rise to viable females, but the males die off. Suppose that *D. pseudoobscura* has in its X chromosome a certain group of genes *A* that lie in the autosomes of *D. miranda*, and that a group of genes *B* which in *D. miranda* lies in the X chromosome is located in the autosomes of *D. pseudoobscura;* with respect to these genes, the constitution of the females of both species and of the female hybrids is alike, namely *AABB*. Males of *D. pseudoobscura* and the male hybrids from the cross *D. miranda* ♀ × *D.*

pseudoobscura ♂ are *ABB; D. miranda* males and the male offspring from the cross *D. pseudoobscura* ♀ × *D. miranda* ♂ are *AAB*. The genotypes of the pure species are evidently so adjusted by countless generations of natural selection that the constitution *ABB* in *D. pseudoobscura* and *AAB* in *D. miranda* permits the development of the "normal" males of the respective species. But the genotype of a hybrid is on the whole a compromise, an intermediate, between those of the parental species; the constitution *AABB* is normal for females of either parent and for the hybrid females as well. The constitution *ABB* is however incompatible with the genotype of *D. miranda*, and *AAB* with that of *D. pseudoobscura*. The hybrid males suffer from a disturbance of the genic balance, and consequently have an impaired viability. An explanation of this type is applicable only to hybrids between species that differ in the distribution of genes among the X chromosome and the autosomes; this is known to be the case for *D. pseudoobscura* and *D. miranda*, but in other species crosses, critical data are lacking. The explanation can be made more general if one assumes that many species have a balance of genes in the X chromosome and the autosomes peculiar to themselves and different from other species. This balance may remain undisturbed in the homozygous sex in the hybrids, but it is likely to be upset in the heterozygous sex.

From a geneticist's point of view, it is especially important that the viability of the hybrids between the same two species may depend on the particular strains of the parental species used in the cross. Such facts may throw some light on the mechanism of the origin of isolation. The outcome of the cross *Crepis capillaris* × *C. tectorum* is variable; in some cultures all the hybrid seedlings die in the cotyledon stage, in others only half of the seedlings die, and in still others the hybrids are viable. Hollingshead (1930a) has shown that certain strains of *C. tectorum* carry a dominant gene which in the pure species produces no visible effects, and in particular has no apparent influence on the germination of the seedlings. If, however, a hybrid between *tectorum* and *capillaris* carries this gene, it does not develop beyond the cotyledon stage. The crosses in which the *tectorum* parent is homozygous for the gene in question produce accordingly no viable seedlings, while 50 per cent or

100 per cent of such seedlings occur in cultures in which the gene is heterozygous or absent respectively. Further experiments have shown that the same gene is lethal for the seedlings of the hybrids *C. tectorum* × *C. leontodontoides* and *C. tectorum* × *C. bursifolia,* but not in the crosses *C. tectorum* × *C. setosa* and *C. tectorum* × *C. taraxacifolia.* It may be noted that the wild populations of *C. tectorum* from some localities carry the gene, and those from other localities are free from it.

An analogous situation has been observed in crosses between the fish *Platypoecilus maculatus* and *Xiphophorus helleri* by Bellamy (1922), Kosswig (1929), and others. The dominant sex-linked gene *N* causes in *Platypoecilus* an increase of the black pigmentation compared to the recessive condition. If a Platypoecilus carrying N is crossed to Xiphophorus, F_1 hybrids are obtained that are heterozygous for N, but which show a greater extension of the black pigment than in the case of either heterozygous (*Nn*) or homozygous (*NN*) Platypoecilus. A backcross of the F_1 individuals (*Nn*) to Xiphophorus (*nn*) gives some heterozygotes (*Nn*) with a pathologically over-developed black pigmentation, which results in the appearance of melanotic tumors. The gene *N* is therefore innocuous for viability on the genetic background of Platypoecilus, but becomes virtually a lethal when introduced into the genotype of Xiphophorous. The exaggeration of the unfavorable effects of certain genes of *Drosophila pseudoobscura* in the hybrids between race A and race B of this species will be discussed below (Chapter IX). According to Kostoff (1936), the cross *Nicotiana rustica* var. *humilis* × *N. glauca* gives hybrids that die as early embryos, while in the *N. rustica texana* × *N. glauca* viable hybrids are obtained. According to Backhouse (1916) and Meister and Tjumjakoff (1928), success in the crosses between wheat and rye depends on the varieties of the parental species used.

THE ORIGIN OF ISOLATION

Despite the appallingly insufficient attention that the problem of isolation has received in genetics, there is every reason to believe that a great variety of isolating mechanisms are at work in nature, preventing the exchange of genes between populations of different species. The mode of origin of these mechanisms remains a puzzle, however, and some writers (e.g., Bonnier, 1924, 1927) are inclined to believe that the known genetic principles are insufficient to account for it. The scheme that is outlined in the following paragraphs is to be taken as a working hypothesis that may or may not prove useful in further work.

It is indeed difficult to conceive how isolation between two groups of individuals might arise through a single mutation. Mutations that change the sexual instincts, or the structure of the genitalia, or the physiology of the gametes, or some other properties of their carriers that are essential for reproduction may occur. Such mutations may prevent crossbreeding of the modified and the ancestral types, but this is not yet sufficient to produce a workable isolating mechanism. For isolation encountered in nature has always two aspects: the crossing of individuals of group A with those of group B is made difficult or impossible, but individuals of A as well as of B are fully able to breed *inter se*. A mutation that would produce isolation must therefore not only prevent crossbreeding between the mutant and the original type, but must simultaneously insure the normal crossability of the individuals carrying the mutation. In other words, it is essential not only that interbreeding between A and B be debarred, but also that a new and harmonious system of physiological reactions arise that would allow the propagation of the new type. Such a coincidence can hardly be imagined to be a common occurrence.

This difficulty does not apply to the origin of ecological or seasonal isolations, since these isolating mechanisms do not necessarily involve a reconstruction of the morphology or physiology of the reproductive system. However, unless the species concerned is capable of self-fertilization, even here the origin of isolation through a single mutational step is rather unlikely, since the mutant can hardly become established in nature. Let us suppose, for example, that the ancestral form and a mutant reach sexual maturity at different seasons, or exist in different ecological niches. With mutation rates that are as low as those observed for most genes in the laboratory, the number of mutants produced in each generation would be so small that they could hardly find mates. Only where a partial seasonal or ecological isolation obtains can their origin by a single mutation be envisaged.*

It is more probable that the formation of isolating mechanisms entails building up of systems of complementary genes. Let us assume that the ancestral population from which two new species are evolved has a genetic constitution *aabb,* where *a* and *b* are single genes or groups of genes. Assume further that this ancestral population is broken up into two parts that are temporarily isolated from each other by secular causes, such as a geographical isolation. In one part of the population, *a* mutates to *A* and a local race *AAbb* is formed; in the other part, *b* mutates to *B,* giving rise to a local race *aaBB.* The individuals of the constitutions *aabb, Aabb,* and *AAbb* are able to interbreed freely with each other, hence there is no difficulty in the gene *A* becoming established in the population; the same is true for *aabb, aabB,* and *aaBB* individuals. But the cross *AAbb* × *aaBB* is difficult or impossible, because the interaction of *A* and *B* produces one of the physiological isolating mechanisms discussed above. It follows that when carriers of the genotypes *AAbb* and *aaBB* come again in contact (because they have surmounted the geographical boundaries that have separated them, for example, or due to a change in their environment), they will be prevented from interbreeding by physiological causes.

The scheme just outlined may appear fanciful; it is useful, however, to examine it further to see whether the assumptions it involves are justified by factual data. One of the basic postulates is that the development of physiological isolating mechanisms is preceded by a geographical isolation of parts of the original population. The observational studies on variation in nature furnish a good deal of evidence to support this thesis. Since Darwin, and especially since Wagner, it is regarded as probable that the formation of geographical races is an antecedent of species formation; more recently this

---

* The origin of allopolyploids accompanied by isolation from the ancestral species may appear to vitiate the above arguments. This is not the case. Allopolyploids may be produced *en masse* in the localities where the distributions of the ancestral species overlap, and furthermore only those allopolyploids may be established in nature which happen to be isolated from their ancestors. The reduplication of the chromosome complement induces an isolation automatically, since a cross tetraploid × diploid gives rise to triploid offspring which is always unstable if propagated sexually. In general, the effects of polyploidy could hardly be likened in this respect to those of gene mutations.

dictrine has been strongly supported by many investigators, among whom we may name K. Jordan (1905), D. Jordan (1905), Semenov-Tian-Shansky (1910), Rensch (1929), and Kinsey (1936). Some systematists regard it as one of the greatest generalizations that has resulted from their work. The distribution regions of races of the same species as a rule do not overlap, while the areas of separate species frequently do. Now, coexistence of distinct groups of individuals in some locality without formation of intermediates and of recombinations of characters is indirect evidence that these groups are isolated physiologically (provided, of course, the differences between them are due to more than one gene). The assumption that geographical isolation is a *conditio sine qua non* of species formation is, nevertheless, not a necessary one. We have seen that ecological isolation may conceivably arise from a single mutation, and it may enable the groups of individuals to develop other physiological mechanisms. All that is necessary for the development of the latter is that some kind of isolation is present to start with. Once an isolating mechanism has appeared, the formation of additional mechanisms to strengthen the action of the first is made much easier. The interbreeding of two species is frequently prevented not by a single but by several mechanisms reinforcing each other's action (Dobzhansky 1937a). The question which of these mechanisms has developed first can be answered only by conjectures, however.

A geographical isolation of parts of a population may be followed by the appearance in the subgroups of inheritable changes that engender a permanent isolation between them. It follows that we may witness in nature isolating mechanisms *in statu nascendi*, when some individuals are already isolated and others not yet isolated from other species. The experiments of Hollingshead (1930a) discussed above furnish an admirable example of such a situation. A part of the population of *Crepis tectorum* is still able to produce viable offspring when crossed to *C. capillaris* and to certain other species, while other individuals produce only inviable hybrids. The gene responsible for the death of the hybrids has not yet permeated the entire population of *C. tectorum*. The variable intensity of sexual isolation between *Drosophila pseudoobscura* and *D. miranda* (see above) presents the same picture. Further experiments may be expected to uncover many new instances of this kind.

The spread within a population of genes that may eventually induce isolation between populations is probably due to their properties other than those concerned with isolation. What these properties are is a moot question, and here is the weakest point of the whole theory. What, for example, is the rôle played within *Crepis tectorum* by the gene that causes the death of the hybrids with *C. capillaris?* Hollingshead was unable to detect any effects of this gene, except that manifested in the hybrids. Isolation is in general a concomitant of the genetic differentiation of separate populations. It may be noted, however, that only those genetically distinct types that have developed isolation can subsequently coexist in the same region without a breakdown of the differences between them due to crossing. Therefore, isolation becomes advantageous for species whose distributions overlap, provided that each species represents a more harmonious genetic system than the hybrids between them. Under these conditions the genes that produce or strengthen isolation become advantageous on that ground alone, and may be favored by by natural selection. This may be at least a partial solution of the difficulty stressed above.

Lastly, one may inquire whether the genes that within a species are harmless or even useful can become deleterious in combination with other genes in a hybrid. The inviability (and, in part, the sterility) of hybrids appears to be due to such an action of genes, each of which taken separately produces no disturbance in its carrier. The observations of Hollingshead on Crepis and of Bellamy and Kosswig on Platypoecilus and Xiphophorus furnish incontrovertible evidence in favor of this assumption. The appearance of "novel" characters in hybrids is indeed one of the well known genetic phenomena. In this respect the recent observations of Irwin and Cole (1936) and Irwin, Cole, and Gordon (1936) on doves and pigeons are very suggestive. By means of immunogenetic reactions these investigators have detected in the blood certain species-specific substances. The hybrids have however not only the substances characteristic for their parents, but also certain "hybrid" substances that are absent in the parental species. The amount of hemoglobin in the blood of the hybrids between the yak and domestic cattle exceeds that present in either parent (Kosharin and Samochwalowa 1933).

IX: HYBRID STERILITY

INTRODUCTION

THE PROBLEM of hybrid sterility goes back at least to Aristotle, who in the "De generatione animalium" discussed at length the sterility of the mule. Aristotle's explanation of the sterility of mules has only an historical interest. In recent times much work has been done, and many valuable observations collected on sterile hybrids in various animals and plants. The phenomenology of hybrid sterility is now fairly well known, but its causal analysis is confronted with difficulties which have been only partly overcome. The time for a synthetic treatment of the subject is probably not yet at hand.

Sterile hybrids are frequently vigorous somatically, but their reproductive organs, more precisely the gonads, show derangements that prevent the development of functional sex-cells. This contrast is characteristic of sterile hybrids. The mule appears to have as harmoniously organized a system as either of its parents; in fact, under some conditions mules are superior in viability to the parental species. And yet, the testes and the ovaries of mules are manifestly abnormal, no spermatozoa or mature eggs being formed in them. The reduction of the viability and the sterility of hybrids are distinct phenomena. One might perhaps object to making such a distinction on the ground that the gonads are the place of least resistance, and their deterioration is a sign of some general weakness of the whole organism. This objection is invalid; a constitutional weakness in pure species is by no means always accompanied by sterility, and many hybrids with reduced viability are not sterile (e.g., the hybrids between the flax species described by Laibach, cf. Chapter VIII).

A dissociation between the processes taking place in the gonads of a hybrid and those in its soma has been observed in *Drosophila pseudoobscura* by Dobzhansky and Beadle (1936). The male hybrids between race A and race B of this species are sterile, their spermatogenesis being very abnormal. Testes of hybrid larvae were implanted in larvae of pure races, and vice versa. When the adult flies emerged, the testes of the host as well as the implants were examined cytologically. The development of the testes was found to proceed autonomously in all cases studied. If a testis of a pure race implanted into a hybrid becomes attached to the sexual duct of the latter, the hybrid is fertile, the functional sperm coming, of course, from the implanted testis. The spermatogenesis in the testis of a hybrid developed in the body of a normal male shows as great a disturbance as is customary in a hybrid. Testes of the host are in no way affected by the presence of implants of a different genetic constitution. Ephrussi and Beadle (1935) have obtained fertile eggs from the ovaries of *Drosophila simulans* implanted in *D. melanogaster,* although the hybrids between these species are completely sterile. All such experiments are probably open to the criticism that the transplantation is performed too late to change the course of the development of the implanted gonad. Nevertheless, it is most probable that, at least in the race A × race B hybrids in *D. pseudoobscura,* the sterility is due to processes taking place in the gonad itself, and not to interactions between the gonads and the somatic tissues.

In any case, the structure of the reproductive organs, and especially the chromosome behavior in the processes of gonogenesis,* have been the principal concern of most investigators working on the problem of hybrid sterility. The pioneer work in this field was that of Federley (1913, 1914, 1915, 1916), who described the chromosome behavior in hybrids between certain species of moths. He discovered that the chromosomes of the parental species frequently fail to pair and to form bivalents at meiosis in sterile and semi-sterile hybrids. The species *Pygaera anachoreta, P. curtula,* and *P. pigra* have 30, 29, and 23 chromosomes respectively (haploid); the hybrids have in their tissues the sum of the haploid chromosome numbers of the parental species. Little or no meiotic pairing is observed in the spematogenesis of the hybrids; most

---

* The terms "gones" and "gonogenesis" have been proposed by Lotsy to cover both the gametes of animals and the gametophytes in plants.

chromosomes remain univalent. In Pygaera hybrids, the univalents split at both meiotic divisions, and the spermatids tend to possess the somatic number of chromosomes. Various abnormalities (fused spindles, failures of cell division) were observed in the spermatocytes; the spermatids usually degenerate. The hybrids are semisterile.

Federley's observations have been confirmed and extended by many investigators working on interspecific hybrids in both plants and animals. The essential fact is that a failure of pairing between chromosomes of different species is observed at meiosis in most (although not in all) sterile hybrids. In detail the situation varies greatly. Some chromosomes may pair and form bivalents, others remain univalent. The proportion of bivalents and univalents varies not only in different hybrids but also in individuals and cells of the same hybrid. The bivalents disjoin normally, that is, at each division half of the bivalent passes to each pole of the spindle. The univalents split as a rule either at the first or at the second meiotic division; at the division at which no splitting takes place the univalents are distributed at random to the poles of the spindle, so that the daughter cells may come to have unequal numbers of chromosomes. The cell division mechanism itself may break down, and all sorts of degenerative phenomena appear in the cells. Few or no functional gones are produced.

The failure of the meiotic chromosome pairing is frequently the starting point of the abnormalities in the gonogenesis in sterile hybrids. This fact naturally leads to the supposition that a cause and effect relationship exists between these phenomena, and hence that the sterility of a hybrid is due ultimately to the failure of chromosome pairing in meiosis. As a general theory this supposition is vitiated, however, by the none too rare sterile hybrids that have an apparently normal chromosome pairing and disjunction at meiosis. To this class belong, for example, the hybrids *Ribes sanguineum × R. aurem* (Tischler 1906), *Digitalis lanata × D. micranta, D. purpurea × D. ambigua* (Haase-Bessell 1921), *Epilobium montanum × E. hirsutum* (Håkansson 1934), *Pisum humile × P. sativum* (Lutkow 1930), *Lolium perenne × Festuca pratensis* (Peto 1933). The hybrids between race A and race B of *Drosophila pseudoobscura* are very instructive in this respect. The F_1 hybrid males are always sterile, but the amount of meiotic chromosome pairing is highly variable (Dobzhansky 1934b). Within each race strains are encountered that produce hybrids in which no bivalents are formed at meiosis, and other strains that produce hybrids with bivalents only and no univalents. The meiotic divisions are, however, abnormal in either case: the first division spindle elongates enormously, bends into a ring, the cell body fails to divide, the second meiotic division is absent, and the giant binucleate spermatids which are formed degenerate. The failure of chromosome pairing is here an effect rather than the cause of the general disturbances in the process of spermatogenesis, and the presence or absence of unpaired chromosomes has no determining influence on the course of the meiotic divisions. The degenerative changes in the prospective sex cells begin however mainly after the meiotic divisions.

In other hybrids the cells of the gonads degenerate before the advent of the meiotic stages. In the hybrids between *Drosophila melanogaster* and *D. simulans,* the gonads are rudimentary, and, as shown by Kerkis (1933), spermatogenesis and oögenesis do not advance beyond spermatogonia and oögonia. Apparently similar conditions are encountered in some of the hybrids between species of gallinaceous birds (Poll 1910, 1920), and perhaps also in those between species of mammals (horse × zebra, yak × domestic cow). The malformations of flowers and anthers observed in some plant hybrids may belong to the same category (Lehmann and Schwemmle 1927, Müntzing 1930).

The degeneration of the germinal tissues leading to sterility of hybrids may set in before, during, or after the meiotic divisions, with chromosomes failing to pair, or with bivalents being formed. This great diversity of the phenomena of sterility is not in itself a decisive argument against the supposition that the failure of chromosome pairing may be a cause of the degeneration of the gametes, and consequently of the sterility. It merely suggests that more than one sterility mechanism is encountered in nature. Federley (1928) and Renner (1929) distinguish between gametic and zygotic sterility; the former consists in production of degenerate non-functional gametes (or gones), and the latter in production of gametes which form inviable zygotes. Müntzing (1930) prefers the terms "haplontic" and "diplontic" sterility; the former is due to the lethality of the haplophase, presumably caused by the genotypic constitution of the haplonts, and the latter to disturbances in the diploid part of the life cycle. These terms may be convenient for descriptive purposes, but such a classification hardly penetrates much below the surface of the phenomena. More basic seems to be the distinction between genic and chromosomal sterility (Dobzhansky 1933b) which is defined in the following paragraph.

GENIC AND CHROMOSOMAL STERILITY

Disturbances in chromosome pairing, which in turn lead to abnormalities in further stages of meiosis and ultimately to the degeneration of the gones, may be due to two classes of causes, which correspond to the two known methods of differentiation of the chromosomal materials. First, there is the mutation process which alters the structure of individual genes in the chromosomes. Second, we observe structural changes of a grosser nature, due to alterations in the relative positions of blocks of genes with respect to each other.

Meiosis, like any other physiological process, is ultimately controlled by the genotype of the organism. The normal course of meiosis involves a definite succession of events, which are so delicately adjusted that a failure of any one of them, or simply a change in the normal time relationships, may cause a derangement of the whole process. Gene mutations can attack this process at any stage. Consequently, chromosome conjugation may fail despite the presence of pairs of chromosomes having similar genes arranged in identical linear series. The resulting sterility will be due not to dissimilarities in the gross structure of the chromosomes, but to the genetic constitution of the organism. This is *genic sterility*.

On the other hand, a chromosome may fail to pair and to form a bivalent with another chromosome because it has no structurally similar partner. Two chromosomes may contain identical genes, but in one of them the arrangement may be *abcde*, and in the other *caebd;* or else, two chromosomes may have some genes in common, but each of them may contain also genes not present in the other (for instance, *abcde* and *abfgh*). Sterility due to such structural dissimilarities is termed *chromosomal sterility*.

The fact that the disturbances in the sex organs of hybrids may be initiated before meiotic chromosome pairing, as well as after the meiotic pairing and the divisions have been successfully completed, proves that genic sterility does not necessarily involve any interference with the chromosomal mechanism. On the other hand, where the abnormalities in the gonogenesis of a hybrid begin with a failure of chromosome pairing, the sterility may be either chromosomal or genic. The possibility that the sterility of some hybrids may be caused by a combination of the two causes must be also considered, since the interbreeding of many species is known to be prevented not by a single but by several isolating mechanisms which reinforce each other's action.

GENIC STERILITY WITHIN A SPECIES

Before proceeding with the discussion of the sterility in interspecific hybrids, it may be useful to consider briefly some similar phenomena encountered within a species, which may throw light on the more involved problems of species crosses. Decreased fertility and complete sterility are not rare in some species, and may be inherited as a property of a given strain. Gametogenesis may be modified in a manner resembling that in sterile interspecific hybrids, including failures of chromosome pairing and irregular disjunction of chromosomes at meiosis.

Sterility of the genic type has been described in several organisms. The "asynaptic" gene studied in *Zea mays* by Beadle (1930, 1933) may serve as an illustration of the kind of changes that may be observed. Maize plants heterozygous for this gene appear normal in every respect, but homozygous plants are distinguished by abortion of most of the pollen and of a majority of the ovules. The gene is, therefore, a recessive, and inbreeding of heterozygous plants produces segregations in the ratio 3 fertile : 1 sterile plants. Cytological examination of pollen mother cells in plants homozygous for asynaptic shows that the beginning of meiosis is normal, and the chromosomes pair to form ten double strands. However, no chiasmata are formed between the paired strands in most chromosomes, members of a pair fall apart, and univalents are present at the first meiotic division. The spindle elongates abnormally and aberrant chromosome complements are formed in the resulting daughter cells, most of which degenerate.

Beadle (1931, 1932a, b, c) has described also other genes causing sterility in maize. They are recessive, and the abnormalities engendered by them become manifest at different stages of meiosis. Thus, in individuals homozygous for the "sticky" gene meiosis is abnormal from the beginning, while another gene causes supernumerary cell divisions after the completion of meiosis. Some of these genes affect meiosis in both sexes, while others modify the meiotic processes in the male line only (i.e., in the microsporogenesis). In the third chromosome of *Drosophila melanogaster*, a recessive mutant gene is known that suppresses crossing over and causes abnormal disjunction of all chromosomes (Gowen 1931). Males homozygous for this gene seem to be quite normal, but a part of the eggs deposited by homozygous females are inviable because of being deficient for some chromosomes and having others in duplicate. Some of the eggs contain a full diploid chromosome complement; fertilized by normal spermatozoa, such eggs give rise to triploid females and intersexes. In *D. simulans* a gene in the third chromosome (not allelomorphic to the gene just described in *D. melanogaster*) disturbs the reduction division as well as chromosome disjunction during the cleavage of the eggs (Sturtevant 1929). This gene is also recessive, and affects meiosis in females but not in males. Genes modifying meiosis and causing partial sterility are known also in Crepis (Hollingshead 1930b, Richardson 1935), Datura (Bergner, Cartledge, and Blakeslee 1934), oats and wheat (Huskins and Hearne 1933), and in Sorghum (Huskins and Smith 1934).

Mutant genes that make their carriers sterile because of malformations in the flowers (e.g., in Antirrhinum, Baur 1924) or in the reproductive organs or the genitalia (the mutant "rotated abdomen" in *Drosophila melanogaster*), are not very rare. The recessive sexlinked mutant gene fused in *D. melanogaster* is especially interesting. Females homozygous for fused deposit eggs that are visibly abnormal in shape and that produce no viable larvae or adults when fertilized by the spermatozoa of fused males. The same eggs, are, however, viable if fertilized by spermatozoa not carrying fused. The eggs of females heterozygous for fused or free from it develop normally when fertilized by spermatozoa of fused males (Lynch 1919). Here, then, we are dealing with a gene whose action is somehow impressed on the cytoplasm of the egg during oögenesis (maternal effect). The fate of the egg is not sealed, however, until after fertilization, since the genes brought in by the spermatozoon can either mitigate the unfavorable maternal effects or make the resulting genetic constitution lethal. It will be shown below that similar phenomena are encountered also in interspecific hybrids.

We may conclude that the sterility genes within a species can affect any phase of the process of meiosis, as well as the development of the reproductive cells before and after meiosis. To that extent intraspecific sterility is analogous to the sterility of hybrids. The fundamental difference between the two phenomena must, however, be clearly realized. The term "hybrid sterility" applies where two forms each of which is fully fertile inter se (e.g., horse and ass) produce a hybrid that is partly or completely sterile (mule). The intraspecific sterility just discussed does not involve the presence of two fertile types: one type, the normal or the original one, is fertile, while the other is sterile. The heterozygote is usually fertile. Where (as in Crepis, Hollingshead 1930a) the intraspecific sterility is due to a dominant mutant gene that is active in heterozygotes, it is not proved that individuals homozygous for that gene would be ferile. Hybrid sterility could be produced by a single gene only provided that this gene would make both homozygotes (AA and aa) fertile and the heterozygotes (Aa) sterile. Such genes are theoretically possible, but they have never been observed, perhaps because they can not become established in nature. The only instances of genic hybrid sterility in which anything like a satisfactory genetic analysis has been made are those of the hybrids between races of the gypsy moth (*Lymantria dispar*), and of the hybrids between race A and race B of *Drosophila pseudoobscura*. In both these instances the sterility is due to the action of complementary genetic factors (see below), and it seems probable that hybrid sterility in general is due to at least two genes contributed by both parents entering the cross. Neither of these genes alone makes its carrier sterile, but they do so when combined.

Stern (1929, 1936) has made a very elegant attempt to devise a synthetic model of hybrid sterility within the species *Drosophila melanogaster*. Males devoid of a Y chromosome are sterile, and, moreover, Stern has found that at least two different sections of this chromosome are concerned with fertility, so that both sections must be present in order that the male may be fertile. Females carrying a whole Y chromosome or a part of it in addition to their X chromosomes are fertile. In a translocation the Y has been broken, and a section of it containing one of the "fertility genes" has become attached to the X. A race can be obtained (Fig. 19, upper right) in

TRANSLOCATION BETWEEN
X-AND Y- CHROMOSOME (Y WITH ♂-FERTILITY FACTOR)

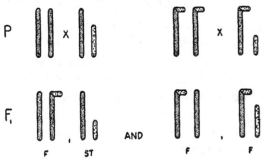

FIG. 19. Two strains of *Drosophila melanogaster* that produce sterile male hybrids when crossed. X-chromosome, stippled; Y-chromosome, dotted. Upper left, normal female; upper right, male from the translocation strain; below, the hybrids. F, fertile; ST, sterile. (From Stern.)

which females have two X chromosomes with sections of the Y attached to them, and males have a similar X and a fragment of the Y that is deficient for just the section that is attached to the X. A male of this race is fertile because it carries simultaneously both sections of the Y chromosome that are needed for fertility. If females and males of this artificial race are crossed to each other, no sterile offspring results, and the race breeds true. If normal females are crossed to the males of the artificial race, the resulting offspring are fertile females and sterile males (Fig. 19, lower left); a cross of normal males to females of the artificial race gives rise to a fertile F_1 generation (Fig. 19, lower right).

It must be granted that Stern's model has the essential features of hybrid sterility: two fertile and true breeding types produce some sterile offspring when crossed. The sterility of Drosophila males caused by the absence of the Y chromosome seems, however, to represent a very special case. According to Shen (1932), spermatogenesis in such males is normal, but the sperm loses its motility in the seminal vesicles, that is, outside of the gonad proper. No hybrids between natural species or races are known that would display this type of sterility; this may be due merely to an insufficiency of available data.

CHROMOSOMAL STERILITY WITHIN A SPECIES

The production of sex cells carrying abnormal gene-complements in individuals heterozygous for various chromosomal aberrations leads to a certain degree of sterility. Suppose that two normal chromosomes of a species carry the genes *ABCD* and *EFGHI*, respectively. A translocation involving an exchange of sections of these chromosomes takes place, as a result of which two "new" chromosomes, *ABFE* and *DCGHI*, are produced. Homozygous normals and the translocation homozygotes produce sex cells which carry every gene once and only once. But in a translocation heterozygote at least six classes of sex cells can be produced: (1) *ABFE, DCGHI*, (2) *ABCD, EFGHI*, (3) *ABFE, EFGHI*, (4) *ABCD, DCGHI*, (5) *ABFE, ABCD*, and (6) *DCGHI, EFGHI*. Classes 1 and 2 carry normal gene complements, but in classes 3 to 6 certain genes are deficient and other genes are present in duplicate; 1 and 2 are termed regular or orthoploid, and 3 to 6 exceptional or heteroploid gones or sex cells.

The fate of the exceptional gones is different in animals on one hand and in the higher plants on the other. In animals the meiotic divisions are followed by a rapid transformation of the resulting cells into eggs or spermatozoa respectively, and the animal gametes retain their functional ability even if they carry grossly unbalanced gene complements, that is, have deficiencies or duplications for large blocks of genes. The exceptional gametes give rise to zygotes. A proof of the correctness of the propositions just stated has been secured with the aid of simple experiments (Muller and Settles 1927, Dobzhansky 1930b, and others). The chromosomes involved in a translocation are marked by mutant genes, so that the course of every chromosome in the inheritance can be followed by simple inspection of the phenotype of the parents and offspring. Individuals heterozygous for the same translocation are intercrossed. Among the exceptional gametes, it may be noted, classes 3 and 4, and 5 and 6 are complementary to each other, since 3 and 5 are deficient for just the same genes that are carried in duplicate in 4 and 6 respectively, and 4 and 6 lack the genes that are present in excess in 3 and 5. The union of 3 and 5, or 4 and 6, gives rise to zygotes that carry every gene twice; such zygotes develop into viable individuals that can be identified. On the other hand, if translocation heterozygotes (Fig. 4D) are outcrossed to homozygous normals (A) or to translocation homozygotes (E), the exceptional gametes produce offspring carrying deficiencies for some genes and duplications for others. Such offspring are, as a rule, inviable or abnormal. The sterility of translocation heterozygotes in animals is due to the inviability of a part of their progeny.

In plants the meiotic divisions give rise to female and male gametophytes which undergo a process involving several cell divisions before the generative cells, the gametes proper, are produced. The exceptional gones are here aborted before fertilization can take place; only small duplications, and, rarely, deficiencies may pass through the gametophyte without causing abortion. Translocation heterozygotes are therefore characterized by abortion of a part of the pollen and ovules (Brink 1929, Brink and Burnham 1929, Burnham 1930 in maize, Belling 1927, Bergner, Satina, and Blakeslee 1933 in Datura, Håkansson 1929 in Pisum, and many others). In some plants the semi-sterility is used as a method for the detection of translocations. The translocation homozygotes are, in accordance with expectations, fertile.

The degree of the sterility of a translocation heterozygote is evidently dependent upon the relative frequency of the regular and exceptional gametes it produces. If the six classes of the gametes discussed above should be equally frequent, the regular ones would amount to 33.3 per cent of the total, and the translocation would be 66.7 per cent sterile. The origin of the exceptional gametes involves, however, the passing of certain sections of chromosome carrying homologous genes to the same pole of the spindle at the meiotic division; if all homologous sections should disjoin and pass to the opposite poles at the meiotic divisions, only regular gametes could be formed (cf. Figs. 4 and 5). Exceptional gametes are due to non-disjunction of chromosomes or chromosome sections at meiosis. Any factor that raises the frequency of non-disjunction increases sterility, and vice versa. The problem so stated is open to an experimental attack.

A correlation has been detected in translocation heterozygotes in Drosophila between the disjunction of the chromosome sections involved and the amount of crossing-over taking place in them (Dobzhansky and Sturtevant 1931, Dobzhansky 1933a, Glass 1935, and others). It is well known that the amount of crossing-over in a chromosome or a chromosome section is constant under normal conditions, that is, in the absence of chromosomal rearrangements and in a definite environment. In translocation heterozygotes, the frequency of crossing-over is reduced, especially in the vicinity of the points at which the chromosomes have been broken and reattached. If in a translocation a short section of one chromosome has been transposed to another chromosome, the crossing-over in that section is suppressed relatively more than in a long transposed section. The chromosomes and sections that suffer the greatest reduction of the frequency of crossing-over undergo non-disjunction at meiosis more frequently than sections in which crossing-over remains more nearly normal. The introduction of a factor that suppresses crossing-over (e.g., an inversion) in a given chromosome or section leads to a proportionate increase of the frequency of non-disjunction. These and other facts have led to a theory according to which the chromosomes and chromosome sections in structural heterozygotes compete with each other for pairing with their homologues at meiosis (Dobzhansky 1931, 1933a, 1934a).

In a normal, structurally homozygous, individual every chromosome has one and only one homologue having an identical gene arrangement. In translocation heterozygotes, some chromosomes consist of sections that are homologous to parts of two or more other chromosomes (Fig. 5). In inversion heterozygotes, certain chromosomes have homologues containing the same genes arranged in a different linear sequence. The meiotic pairing is due to a mutual attraction between homologous loci rather than between chromosomes as such (Belling 1927, 1928, 1931, and others). In structural heterozygotes parts of the same chromosome, in an attempt to pair with their respective homologues, may be pulled in different directions simultaneously. Pairing of some chromosome sections may be delayed or not attained at all. The lack of pairing engenders a reduction of the frequency of crossing-over, and, as known from cytological observations (cf. Darlington 1931), a failure of disjunction. Short chromosome sections are weaker than long ones in the competition for pairing. Hence, the crossing-over is strongly suppressed in short sections, and they frequently undergo non-disjunction, forming exceptional gametes. The more extensive the differences in gene arrangement between the chromosomes of the parents, the greater is the competition for pairing at meiosis, the more frequent are the failures of pairing and disjunction. Other things being equal, a structural heterozygote having chromosomes with more differences in the gene arrangement will produce more gametes with abnormal gene complements than a heterozygote with more nearly similar chromosomes. In different organisms the conditions of the meiotic pairings are, however, not necessarily alike; thus, the translocation heterozygotes in Oenothera produce relatively fewer exceptional gametes than the translocations in Drosophila.

Failures of the chromosome pairing at meiosis have been observed cytologically in some structural heterozygotes. McClintock (1932, 1933) and Burnham (1932) have found in maize that the chromosome sections adjacent to the loci of breakages in translocation heterozygotes frequently remain unpaired. Short chromosome sections transposed to other chromosomes fail to pair with their homologues more often than long sections. In Drosophila the meiotic prophases are technically difficult to study, but the chromosome pairing in the salivary gland nuclei is believed to represent a fairly accurate model of the meiotic pairing process. In translocation and inversion heterozygotes, the pairing is frequently disturbed in the vicinity of the breakage points, although cells with a complete pairing of all sections can sometimes be selected. In complex inversion heterozygotes in *Drosophila pseudoobscura* (Sturtevant and Dobzhansky 1936, also unpublished data) long sections of the chromosomes involved may fail to pair, and some short sections remain un-paired in all cells. Oliver and Van Atta (1932) observed a disturbance of the somatic pairing due to translocations and inversions. In hy-brids between races and species, failures of chromosome pairing are, on the whole, proportional to the structural differentiation of the chromosomes. Race A and race B of *Drosophila pseudoobscura* usu-ally differ in four inverted sections, and the chromosomes in the salivary gland nuclei pair except in the vicinity of the breakage points (Tan 1935, Koller 1936). *D. pseudoobscura* and *D. miranda* differ greatly in gene arrangement, and most nuclei in the hybrids show only a few chromosome sections paired (Dobzhansky and Tan 1936). *D. azteca* and *D. athabasca* have even more profoundly dif-ferent gene arrangements, and most of the salivary gland cells of the hybrid larvae show no pairing whatever, only exceptional cells having a few short sections paired (Bauer and Dobzhansky, unpub-lished).

Certain observations seemed, on the other hand, to contradict the theory of competitive pairing. Thus, Beadle (1932d) and other workers have shown that the frequency of crossing-over may be modified not only in translocation heterozygotes but also in trans-location homozygotes, although no competition for pairing can be supposed to take place in the latter. Dubinin, Sokolov, Tiniakov, and Sacharov (1935) and Dobzhansky (1936e) have pointed out, how-ever, that changes in the gene arrangement in the chromosomes may modify the frequency of crossing-over not only through the competi-tion for pairing, but also by altering the relative positions of the chromosomes with respect to each other in the nucleus, which is to a certain extent fixed (for example, in at least some organisms the spindle attachments of all chromosomes tend to be adjacent to one another in the resting nucleus as well as during the prophases and telophases). The modifications of the frequency of crossing-over in translocation homozygotes may be thus accounted for.

In making comparisons between the sterility of the structural heterozygotes within a species and the sterility of interspecific hy-brids, the following consideration should never be disregarded. In animals, the gametes bearing unbalanced gene complements have their fertilizing ability unimpaired, and the sterility of an individual producing such gametes is due to the elimination of a part of its off-spring. On the other hand, many sterile interspecific hybrids fail to produce anything resembling mature and functional eggs and sperma-tozoa. This distinction appears to be qualitative and not one of degree. It would, nevertheless, be unjustified to dismiss chromosomal sterility as a possible cause of the infertility of species hybrids in ani-

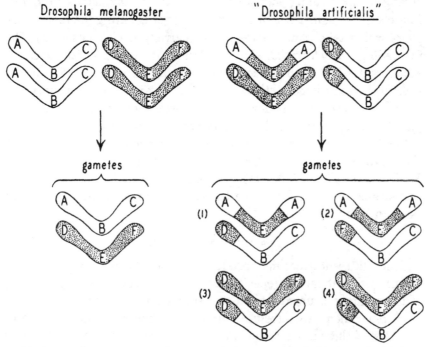

FIG. 20. Arrangement of parts of the second (white) and third (stippled) chromosomes in *Drosophila melanogaster* and *D. artificialis*. (After Kozhevnikov.)

mals. The presence of many unpaired chromosomes, univalents, at the meiotic divisions might, conceivably, disrupt the normal func-tioning of the cell division mechanism and thus interfere with the production of functional gametes. The proof that this possibility is ever realized is, however, lacking, and the claims of some investi-gators that the sterility of any hybrid in which chromosome pairing is lacking may be due to that fact are unjustified.

Interesting experiments on the production of isolation between

strains of *Drosophila melanogaster* by chromosomal changes have
been reported by Kozhevnikov (1936). By combining two different
translocations involving the second and the third chromosomes a
strain is obtained which Kozhevnikov designates "*Drosophila artificialis.*" This strain has one normal third chromosome and three
other chromosomes composed of sections of the second and third
arranged in a manner indicated in Fig. 20. *D. artificialis* produces
four classes of gametes shown in Fig. 20 (not counting the rare crossovers which take place between the rearranged chromosomes). On
inbreeding *D. artificialis,* sixteen classes of zygotes arise in equal
numbers, twelve of which die on account of deficiencies and duplications for chromosome sections, and only four survive. The surviving
ones are, however, identical in chromosome constitution with the
parents. The strain breeds true. A cross *D. melanogaster* ♀ × *D.
artificialis* ♂ produces no viable offspring whatever, since none of
the zygotes formed contain balanced chromosome complements.
Kozhevnikov's model depicts a situation in which a cross between
two strains produces no viable hybrids. It has obviously nothing to
do with hybrid sterility.

INTERSEXUALITY IN HYBRIDS

Since different genes may produce diverse physiological effects and
the processes of gametogenesis may be disturbed in a variety of ways,
it is perhaps not unexpected that several physiologically fairly distinct phenomena are subsumed under the general name genic sterility. One of these phenomena is the unbalance of the sex-determining
factors, making the hybrids intersexual. We have seen above that the
sex of an individual is decided by the balance of the female-determin-
ing and the male-determining factors it carries. In Drosophila, the
genes for femaleness are borne in the X chromosome and those for
maleness in the autosomes. In the gypsy moth, *Lymantria dispar,* the
X chromosome carries maleness, while the factor for femaleness is
transmitted, according to Goldschmidt, through the cytoplasm. Within a race the action of the sex-determining genes is so adjusted that
only females and males but no intersexes are produced. In Drosophila,
two X chromosomes combined with two sets of autosomes give a female, and one X chromosome a male. In Lymantria, where the female

sex is heterozygous, two X's give rise to a male and one X to a female. It can be readily seen, however, that in different races and
species the sex genes might vary in strength. In the hybrids, combinations of the female and male determiners may occur that would
produce neither females nor males but intersexes. Intersexes are frequently sterile. The appearance of sterile intersexes among hybrids is
a special case of genic sterility engendered by an unbalance of the
sex-determining factors.

A penetrating analysis of the intersexuality observed in crosses between geographical races of *Lymantria dispar* has been made by Goldschmidt (1932a, 1934b). The races from the northern island of Japan
(Hokkaido) and from Europe have "weak" sex-determining factors.
The sex determiners in the strains coming from Russian Turkestan,
Manchuria, and southwest Japan are "half weak" or "neutral." Those
in the strains from the middle and north Japan are "strong." According
to Goldschmidt, an extensive series of multiple allelomorphs of the
sex genes underlies the differences between the races, although some
minor modifying genes are also recorded.

A cross weak ♀ × strong ♂ produces in the F_1 generation normal
males and females that are intersexual. The reciprocal cross, strong
♀ × weak ♂, gives normal females and males in F_1 but in the F_2
half of the males are intersexual. Denoting the female determining
factor by F, the male determiner by M, the weak and strong allelomorphs by the subscripts w and s respectively, the above results can
be interpreted thus:

$F_w M_w$ (weak ♀) × $M_s M_s$ (strong ♂) = $F_w M_s$ (intersex) and
$F_w M_s M_w$ (♂)

$F_s M_s$ (strong ♀) × $M_w M_w$ (weak ♂) = $F_s M_w$ (♀) and $F_s M_w$
M_s (♂)

One F is normally sufficient to suppress the effects of a single M
and to produce a female; but an F_w is not strong enough to over-
power an M_s, hence an $F_w M_s$ individual is an intersex. Likewise, the
individuals of the constitution $F_s M_w M_w$ appearing in the F_2 from
the cross strong ♀ × weak ♂ are not males but intersexes. The degree
of intersexuality may be very different in different crosses. As shown
by Goldschmidt and his collaborators, intersexes are individuals that

begin their development as representatives of one sex but end it as representatives of the other. The greater the portion of the development which is dominated by either the female or the male determiners, the greater is the resemblance of the resulting individuals to normal females or males respectively. If, however, the turning point falls somewhere in the middle of development, the intersexes possess various mixtures of female and male parts. In some crosses where the "strength" of the sex determiners in the parental races differs only slightly, the intersexes are so much like normal individuals that they remain fertile. Where the difference between the parental races is greater, the intersexes have the gonads, the ducts of the reproductive system, the genitalia, the secondary sexual characters, and the sexual behavior pattern modified so extensively that they are sterile. The degenerative processes in the gonads may be quite extensive, but so far as is known the chromosome behavior at meiosis is unaffected (at least where the cells of the gonads reach the meiotic stages). Finally, the crosses between very strong and very weak races give intersexes that are transformed completely into individuals of the sex opposite to that which they should have had according to their chromosomal constitution (XX instead of XY females, and XY instead of XX males).

The geographical distribution of the Lymantria races with weak, neutral, and strong sex determiners has been indicated above. Goldschmidt has studied it with the utmost care; he concludes that races possessing sufficiently different sex genes to produce sterile hybrids when crossed never inhabit adjacent territories. The sole exceptions are north Japan, where very strong races occur, and the neighboring island of Hokkaido inhabited by very weak races; the islands are however separated by the Tsugaru Strait which is broad enough to prevent the mingling of races living on its opposite shores. The attempts of Goldschmidt to detect some correlation between the variations in the sex factors present in a given race and the climatic characteristics of the territory inhabited by it are thus far unavailing. The adaptational value of the different allelomorphs of the sex genes is unknown, and the mechanism that brings about the racial differentiation with respect to the sex genes defies analysis.

Sterile intersexual hybrids are known in several groups of animals, although in general they do not appear to be very frequent. In most sterile hybrids the disturbances are confined to the gonads, while in intersexes other parts of the reproductive system and the secondary sexual characters are affected as well. According to Goldschmidt's analysis (1931) of the data of Keilin and Nuttal, the crosses between the head and the body lice (*Pediculus capitis* × *P. vestimenti*) produce some intersexes in the F_2 and F_3 generations. A similar situation has been observed by Guyénot and Duszynska (1935) in F_2 to F_5 of the cross *Cavia apera* × *C. cobaya* (guinea pigs), and by Crew and Koller (1936) in the F_1 of the cross *Anas* × *Cairina* (ducks). In the latter cross spermatogenesis in the sterile males has been investigated; meiotic chromosome pairing appears to be normal, but various abnormalities in the second spermatocytes are recorded.

Years ago semi-sterile F_1 hybrids were obtained by Standfuss from the cross between the moth species *Saturnia pyri* ♀ × *S. pavonia* ♂. The hybrid males can be backcrossed to *S. pavonia* females; the progeny consists of males and what Standfuss has described as "gynandromorphs." The latter have been shown by Pariser (1927) and Goldschmidt (1931) to be intersexes. The origin of these intersexes is different from those in the Lymantria crosses. *S. pavonia* and *S. pyri* have 29 and 30 chromosomes, respectively (haploid). The F_1 hybrid has 59 chromosomes, most of which fail to form bivalents at meiosis (Pariser has counted from 45 to 51 bodies in the spermatocytes of the F_1, which suggests that from 14 to 8 bivalents and from 31 to 43 univalents are formed). Meiosis in the hybrids proceeds apparently according to the scheme described by Federley for Pygaera (see p. 260), and the backcross offspring are subtriploid (no complete triploids were observed, presumably because, due to the presence of some bivalents in the F_1 meiosis, no spermatozoa contain the full diploid chromosome complement). The Saturnia intersexes are, therefore, comparable to the triploid intersexes in Drosophila (cf. Chapter VII). Inasmuch as the sterility of some of the backcross individuals in Saturnia is due to their intersexuality, this case may be classed as belonging to the genic sterility type. The lack of chromosome pairing in the gametogenesis of the F_1 may, however, be due either to the dissimilarities in the gene arrangements in the chromosomes of the parental species (chromosomal sterility), or to the effects of complementary genetic factors (genic sterility).

GENIC STERILITY IN INTERRACIAL HYBRIDS OF DROSOPHILA PSEUDOOBSCURA

Species of Drosophila have few rivals as convenient objects for many types of genetic investigation. Until recently it has been however regarded as patent that other objects were preferable to Drosophila for studies concerning hybrid sterility. The only known example of interspecific hybridization in that genus was the cross *D. melanogaster* × *D. simulans,* which gives rise to completely sterile hybrids in the F_1 generation (Sturtevant 1920-1921, 1929). The impossibility of carrying the analysis beyond the F_1 has imposed a serious limitation on the progress of the work. A new state of affairs has been inaugurated by the discovery of the two "races" in the species *D. pseudoobscura,* which when crossed produce sterile males and fertile female hybrids. The latter can be backcrossed to males of either parental race; both sterile and fertile individuals are encountered in the further generations (Lancefield 1929). *D. pseudoobscura* combines the advantages of a first class laboratory animal with those of material in which a genetic analysis of sterility can embrace several generations of hybrids.

Because of the lack of visible external differences between the "races" they have been designated simply as race A and race B, respectively. The cross B ♀ × A ♂ gives in F_1 sterile males having small testes; the reciprocal cross, A ♀ × B ♂, produces sterile male hybrids that have testes of normal size (that is, similar in size to those of the males of the parental races). The backcrosses of the F_1 females to race A or race B males give progenies in which the males have testes of variable size, ranging from normal to very small. Males with small testes are always sterile, those with large ones are sometimes fertile (Lancefield 1929). The sterility is due to a profound modification of the process of spermatogenesis. The meiotic chromosome pairing is variable; no univalents, some univalents, or only univalents may be present at the first meiotic division. Irrespective of the numbers of the bivalents and univalents formed, only a single meiotic division takes place, with is characterized by an anomalous behavior of the spindle. The spermatids degenerate without being transformed into spermatozoa (Dobzhansky 1934b). The disturbances in spermatogenesis are in general greater the smaller the testes in a hybrid. Testis size is therefore a measure of the degree of departure from the normal course of the spermatogenesis that may be observed in a given individual.

As a basis for an attempt to analyze the causation of the sterility in the race A × race B hybrids, one must take cognizance of the following facts. The disturbances leading to the sterility of the hybrids are confined to their gonads, while the rest of the reproductive system (the sexual ducts, external genitalia) and the secondary sexual characters appear to be entirely normal (Dobzhansky and Boche 1933). The sterile males are therefore not intersexual. Some real intersexes have been found in *Drosophila pseudoobscura,* and they proved to be quite distinct in appearance from hybrid males. The gene arrangements in the chromosomes of race A and race B have been compared by Tan (1935) and Koller (1936), who have found in the hybrids four inverted sections: two in the X chromosome, and one each in the second and the third chromosomes. It is very doubtful, however, that the sterility of the hybrids is due to these differences in the gene arrangements in the chromosomes of the parental races. Indeed, individuals of *D. melanogaster* heterozygous for four and even for five inverted sections have been obtained, and proved to be fertile. Within each race of *D. pseudoobscura,* inversion heterozygotes are frequently encountered in nature, and they are fully fertile (Sturtevant and Dobzhansky 1936). The hybrids between certain strains of race A and race B show a complete chromosome pairing at meiosis, and yet the spermatogenesis in such hybrids remains abnormal, and they are sterile. One of the chromosomes (the fourth) is similar in gene arrangement in both races, but in the hybrids the fourth chromosomes sometimes also fail to pair (Dobzhansky 1934b).

The supposition that the sterility of the A × B hybrids may be of chromosomal type is further invalidated by the observations on the chromosome behavior in the tetraploid spermatocytes (Dobzhansky 1933b). Groups of such spermatocytes are frequently encountered in the testes of hybrid males. Their origin is apparently due to the failure of cell fission in one of the spermatogonial divisions, followed by the formation of binucleate cells, and the fusion of the telophasic groups of chromosomes at the next division. Tetraploid spermatocytes contain two full sets of chromosomes of race A as well as of race B. Each chromosome has, consequently, a partner exactly similar to it in gene arrangement. It is known that in some plant hybrids in which no chromosome pairing is observed in the diploid, the reduplication of the chromosome complement enables all chromosomes to pair and to form bivalents, and the hybrid to become fertile (Chapter VII). The proportion of the chromosomes that become paired in the tetraploid spermatocytes of *D. pseudoobscura* hybrids is however no greater than in the diploid ones. Moreover, after the meiotic division the cells derived from the tetraploid spermatocytes undergo the same series of degenerative changes as their diploid relatives.

It is virtually impossible to reconcile the facts just outlined with the supposition that the sterility of the interracial hybrids in *Drosophila pseudoobscura* is of the chromosomal type. The hypothesis of genic sterility offers an alternative explanation. This hypothesis (Dobzhansky 1933b, 1934b) involves the assumption that the hybrid inherits from each of its parents a gene, or a group of genes, the interaction of which causes sterility. Suppose that one of the parents carries the genes *SStt* and the other *ssTT;* the hybrid is *SsTt*. Individuals possessing *S* alone, or *T* alone, are fertile, but the simultaneous presence of *S* and *T* makes the carrier sterile.

Experiments were undertaken to test the validity of the above hypothesis. It is first of all necessary to determine which of the cellular elements are the carriers of the sterility genes, theoretically postulated. Lancefield (1929) and Koller (1932) have shown that the backcross males carrying an X chromosome similar in racial origin to their father have large testes, while males receiving from their mother the X chromosome of the other race have small testes and are sterile. Since all backcross males normally derive their Y chromosome from the father, the findings of Lancefield and Koller might have suggested that sterile males with small testes are those in which the X and Y chromosomes are of different race. Observations on hybrid males devoid of Y chromosome (XO males) have served to exclude this possibility. Males without a Y have testes of the same size as their sibs carrying a Y. The Y is not connected with hybrid sterility in *Drosophila pseudoobscura* (Dobzhansky 1933d).

In the progenies of F_1 hybrid females, backcrossed to males of either parental race, individuals appear that carry various combinations of chromosomes of race A and race B (Figs. 21 and 22). Some have all chromosomes of one race; others have an X chromosome of one race and autosomes of the other; still others carry various mixtures of the chromosomes of both races. Experiments can be so arranged that every chromosome (with the exception of the very small fifth) is marked with a mutant gene; the constellation of the chromosomes present in a given backcross individual is, then, recognizable from a simple inspection of its external appearance. It is a relatively simple matter to determine which combinations of chromosomes cause a male to have small testes and to be sterile, and which permit fertility. Moreover, if the original cross is made by using race A as a female (A ♀ × B ♂), the entire offspring has race A cytoplasm, while the cross B ♀ × A ♂ gives a progeny with race B cytoplasm. Several crosses in which the chromosomes of the parents were marked as indicated above have been studied by Dobzhansky (1936b). The testis size in the backcross males of different genetic constitutions was measured, and the fertility of some types determined. It became clear at once that fertility or sterility in the backcross males depends upon their chromosomal constitution, and not upon the ultimate source of their cytoplasm. Males whose grandmother was a race A female are similar to those descended from a race B grandmother, provided they have similar chromosome complements.

Backcross males with X chromosome and autosomes of the same race have large testes and are usually fertile (Figs. 21 and 22). The more dissimilar the X and the autosomes become in racial origin the smaller is the testis size. The smallest testes are observed in males with the X of one race and all autosomes of the other. With a single exception,* all chromosomes act alike, and their action is cumulative.

---

* The exception is that in backcrosses to race B males the sons having race A third chromosome and race B X chromosome have larger testes than their brothers homozygous for the race B third. This curious relation is due to a maternal effect (cf. Dobzhansky 1936b).

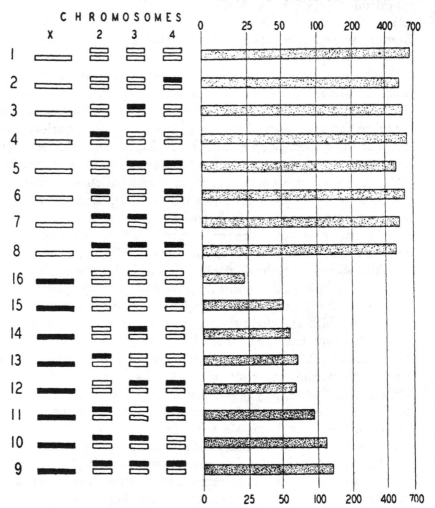

FIG. 21. Testis size in hybrids between race A and race B of *Droso-phila pseudoobscura*. Chromosomes of race A are represented white; those of race B, black. (From Dobzhansky.)

Thus, individuals carrying a race A X and race B autosomes have very small testes (class 16, Fig. 22). The introduction of one fourth or one third or one second chromosome of race A increases the testis size (classes 13-15, Fig. 22). The simultaneous introduction of the

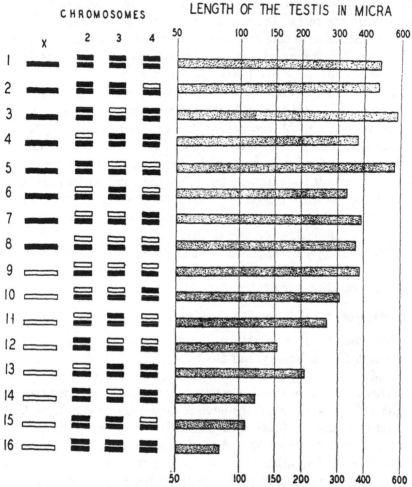

FIG. 22. Testis size in hybrids between race A and race B of *Drosophila pseudoobscura*. Chromosomes of race A are represented white; those of race B, black. (From Dobzhansky.)

fourth and third (class 12), or second and third (class 10), or second and fourth (class 11), or second, third, and fourth (class 9) chromosomes of race A increases the testis size more than each of these chromosomes does alone. Males having all chromosomes of the same race, or having one third or one fourth chromosome of the opposite race from the rest of the chromosomes, are fertile (classes 2 and 3).

But if both the third and fourth chromosomes disagree in racial origin with the rest of the complement, the male is sterile.

It is justifiable to conclude that, with the exception of the Y and of the small fifth chromosome, all chromosomes of race A and race B of *pseudoobscura* carry genes concerned with the sterility of the interracial hybrids. Moreover, Dobzhansky (1936b) found that the X, the second, and the third chromosome have each at least two such genes, and the unpublished data of the writer show that the fourth chromosome has also no less than two. The minimum number of genes concerned with the sterility is therefore eight, but the actual number is almost certainly greater. What the physiological reactions are through which these genes produce their effects culminating in the disturbance of spermatogenesis and in sterility is unknown. The only bit of information on this subject is that the fate of the testis is determined by its own genetic constitution and not by that of the body in which it develops (the transplantation experiments of Dobzhansky and Beadle, 1936, see above).

THE "STRONG" AND "WEAK" RACES OF DROSOPHILA PSEUDOOBSCURA

The cross race B ♀ × race A ♂ produces in F_1 sterile sons with small testes. Dobzhansky and Boche (1933) have shown that the exact size of the testes in such hybrids is a function of the strains of either race that are used. Certain strains of race B when crossed to the same race A strain, the environmental conditions being kept constant, produce hybrids with markedly larger testes than other strains. Likewise, certain race A strains produce much larger testes in the hybrids than others when crossed to the same race B strain. Strains that tend to give relatively large testes are designated as "weak" and those decreasing the testis size as "strong." The expressions "strength" and "weakness" as applied to *D. pseudoobscura* should not be confused with the similar terminology applied to Lymantria races (see above), for in the latter the "strength" of the sex determining factors is what is being referred to.

The strength of a given strain of *pseudoobscura* is an hereditary characteristic. The observed variation in strength is considerable: the linear dimensions of testes of a hybrid from a cross weak B ♀ × weak A ♂ may be three or more times greater than in those from the crosses of strong B ♀ × strong A ♂ . The significance of this variation is emphasized by the observation (see above) that the size of the testis in a hybrid is indicative of the degree of the disturbance in spermatogenesis (Dobzhansky 1934b). Although the F_1 hybrid males from crosses between race A and race B are always sterile, the hybrids between weak strains may display a complete chromosome pairing at meiosis, only bivalents being present, while in crosses between strong strains the meiotic pairing fails entirely, and not a single bivalent is formed. The disturbances in the processes following the meiotic division are also somewhat more pronounced in strong A × strong B than in weak A × weak B hybrids.

An analysis of the variations in strength has been undertaken by Dobzhansky (unpublished). The strong and weak races differ in their geographical distribution. In the region inhabited by race A strength increases rather gradually as one moves from northwest to southeast. British Columbia and Washington are populated chiefly by very weak or weak races, while in Mexico very strong variants are encountered. In race B the geographical regularity is less clear than in race A, although the strongest strains seem to come from the coastal regions of Oregon and Washington, and the weakest from the Sierra Nevada mountains. Strains coming from the same locality may differ in strength rather widely. Thus, both inter-group and intra-group variations are present. The genetic basis of the differences in strength is a rather complex one, genes modifying this character being found in all the chromosomes studied. Crosses between two strains of the same race often produce in the F_2 generation some segregants that are weaker or stronger than either parental strain. Wild populations are evidently quite variable with respect to the genes determining strength.

The importance of the above facts is revealed by observations which suggest that the genes determining strength are similar to, or identical with, the sterility genes. If race A × race B hybrid females are backcrossed to males of one of the parental races, one may obtain in the first or in the later backcross generations individuals that have most of the chromosomes of one race but carry one chromosome or a section of a chromosome of the other race. Parts of chromosomes or whole chromosomes of race A may be "transferred" by this method to race B, or vice versa. If the transferred chromosome or section is long, or carries powerful sterility genes, the male possessing it may be sterile; other chromosomes do not interfere with fertility to any appreciable extent. Sturtevant and the writer (unpublished) have accomplished a number of such "transfers" of chromosome sections to and from either race. The lines of race A and race B known to carry certain chromosome parts of the opposite race may then be tested for strength with the aid of the same method that is used for testing the wild strains coming from nature, i.e., the size of the testes is determined in the F_1 males from the crosses race B ♀ × race A ♂. Experiments of this type conducted by the writer have thus far given quite consistent results: the strength of a line of race A or of race B is decreased by the introduction of chromosomes or chromosome sections of the opposite race.

It has been shown that the sterility of the hybrids between race A and race B is due to the effects of a complex of sterility genes. Now it is known in addition that, by interracial crossing, this complex can be dismembered into its constituent elements, and, what is especially important, these elements prove to have the same action as the genes determining the differences in strength between the different lines of the same race. The gap between the interracial differences causing the sterility of hybrids and the intraracial variations is thus bridged. The implications are rather obvious. Within a race we find the genetic elements, the building stones, from which a mechanism causing hybrid sterility may, theoretically, be built. As the situation stands now the analysis is, of course, far from complete. We have not yet succeeded in synthesyzing two strains producing sterile hybrids from the genetic elements encountered within a race—a feat that is not beyond the range of possibility in theory. Moreover, the variations in "strength" encountered within a race may be interpreted either as the remains of the store of building materials from which the interracial difference had been built in the history of the species, or else as a sign of an occasional breakdown of the separation between the races. The latter interpretation would involve the assumption that race A and race B of *D. pseudoobscura* sometimes produce hybrids in nature, and that through such hybrids some of the genes of race A are transfused into race B, and vice versa. This possibility can not be entirely excluded at present, although the fact that the distinction between the races has never been found blurred, even in regions where their geographical distributions overlap, argues against it.

CHROMOSOMAL STERILITY IN SPECIES HYBRIDS

Chromosome pairing at meiosis in a hybrid may take place either between chromosomes of the different parents (allosyndesis), or among those of the same parent (autosyndesis). If all the chromosomes fail to find mates of either kind, only univalents are formed. In most hybrids the univalents are distributed at random to the poles of the spindle at the first or at the second meiotic division (see above). The resulting gones may contain either unequal or, due to chance, equal numbers of chromosomes. In either case they are likely to have mixtures of the chromosomes of both parents, and to be deficient for certain genes and carry duplications for others. Gones bearing an entire haploid set of chromosomes of one species and none of those of the other may sometimes also be produced, but the greater is the number of chromosomes in the parental species, the smaller the probability of such an event. Since deficiencies and duplications are frequently lethal to the zygotes (in animals) or to the haplonts (in higher plants), the hybrids having only univalents at meiosis are likely to be sterile.

The failure of pairing between the chromosomes of different species in a hybrid may, theoretically, be due to their structural dissimilarities, that is, to differences in gene arrangements. The question that immediately presents itself is whether species that produce hybrids with unpaired chromosomes differ in gene arrangement, and, if so, whether the failure of pairing is actually due to these differences? A definite answer to this question would be possible if we were in possession of accurate data on the gene arrangement in at least a fair sample of species producing sterile hybrids. Comparative studies on gene arrangement are, however, confronted with technical difficulties that are so formidable that to date even approximately satisfactory information has been secured only in a very few exceptionally favorable objects, for example, in some species of Drosophila in which the chromosomes can be studied with the aid of the salivary gland method. In race A and race B of *D. pseudoobscura*, the gene arrangements proved to be somewhat different, but the data presented above show conclusively that the sterility of the hybrids between these "races" is genic and not chromosomal. The gene arrangements in *D. pseudoobscura* and *D. miranda* are profoundly different (Dobzhansky and Tan 1936), but whether the sterility of the hybrids between them is chromosomal is as yet obscure.

One must admit, I think, that apart from the production of abortive eggs, ovules, and pollen by certain translocation heterozygotes of Drosophila, maize, peas, and other forms, the sterility of no hybrid has as yet been definitely proven to be of the chromosomal type. The supposition that the sterility of some interspecific hybrids is chromosomal rests on indirect evidence, which is however sometimes strong enough to permit making inferences, with a reasonable degree of assurance. The origin of fertile allopolyploids from sterile diploids (Chapter VII) is one of the principal sources of such evidence. For example, the diploid hybrids between radish and cabbage (Karpechenko 1927a, b, 1928) have 18 chromosomes, 9 from each parent. Only univalents are formed at meiosis, the distribution of the chromosomes to the poles of the spindle is random, a vast majority of the gones degenerate, and the hybrid is nearly sterile. But where the meiotic division is not completed, and all the chromosomes are included in the same nucleus, the gones containing full sets of radish and cabbage chromosomes are formed and become functional. It is therefore fairly certain that the degeneration of the majority of the gones is due to the presence in them of unbalanced chromosome complements.

The union of the exceptional gones containing full sets of the chromosomes of both parental species gives rise to a tetraploid hybrid, Raphanobrassica, which has nine pairs of radish and nine pairs of cabbage chromosomes. In contrast to the diploid hybrid, Raphanobrassica has only bivalents at meiosis, the disjunction is normal, practically all the gones formed contain full sets of radish and cabbage chromosomes, and the hybrid is fertile. The dependence of fertility or sterility on the chromosome pairing or the lack of it is clear enough. The failure of meiotic pairing between the chromosomes of raddish and cabbage may, a priori, be due either to dissimilarities in the gene arrangement in the chromosomes of these species, or to the formation in the hybrid of a combination of genes that suppresses pairing. No direct information on the gene arrangements in radish and cabbage is available; nevertheless, the case must not be dismissed without a consideration of the implications of the two hypotheses. If the pairing of chromosomes in the diploid hybrid is suppressed by a combination of genes, the perfect pairing observed in the tetraploid is very astonishing, for the genetic constitutions of the two are similar. The chromosomes have been merely reduplicated in the tetraploid, but the relative numbers of the chromosomes and genes of each kind have remained constant. One might assume that the change in the physiology of the organism due to polyploidy has been such that chromosome pairing became normal where it had been impossible heretofore. But even with such a supplementary hypothesis, a difficulty remains. Since in Raphanobrassica the radish chromosomes pair only with radish, and the cabbage with cabbage homologues, what is the agent preventing the pairing of the chromosomes of the two species with each other?

If, on the other hand, the lack of pairing in the diploid hybrid is due to a structural dissimilarity of the radish and cabbage chromosomes, the behavior of the tetraploid is what might be expected. In Raphanobrassica every chromosome has a mate which is structurally identical with it; a purely autosyndetic pairing consequently takes place; the chromosomes of radish and cabbage remain in the tetraploid as distinct from each other as they had been in the diploid ancestor, hence no allosyndetic pairing occurs in either. The conclusion arrived at may be formulated thus: where an allopolyploid derived from a sterile diploid is fertile, the sterility in the latter is of the chromosomal type. An obvious corollary to this conclusion is that in the hybrids whose sterility is genic, a reduplication of the chromosome complement will not entail a restoration of fertility. Indeed, the gene combination that suppresses chromosome pairing in the diploid persists unaltered in the tetraploid, and should, therefore, produce similar effects in the latter. The availability of an exact homologue for each chromosome in the tetraploid will not be sufficient to enable the pairing to take place. The hybrids between race A and race B of Drosophila pseudoobscura seem to provide a crucial test of the hypothesis. In these hybrids, whose sterility is genic, the tetraploid spermatocytes show the same amount of chromosomes pairing as do the diploid ones (Dobzhansky 1933b, 1934b).

Darlington (1928, 1932a) has established a rule, the substance of which seems paradoxical and inexplicable except in conjunction with the hypothesis of chromosomal sterility. Sterile diploid hybrids with little or no chromosome pairing at meiosis give rise to allopolyploids that are fertile and display mostly or only bivalents at the meiotic division. Contrariwise, the allopolyploids derived from the diploids that have many bivalents show an irregular chromosome pairing and disjunction. The fertility of an allopolyploid is in general inversely proportional to that of its diploid ancestor (cf. Chapter VII). The occurrence of pairing in a diploid hybrid indicates that the gene arrangement in the chromosomes of the parental species is similar enough for some or all chromosomes of one species to find approximate homologues among those of the other. The doubling of the chromosome complement gives rise to a situation in which each chromosome has one exact homologue and two potential mates that are more or less similar to it. In the competition for pairing that arises at meiosis, the pairing of the chromosomes of the same species is interfered with by the presence of the partial homologues among those of the other. Pairing is variable, bivalents, trivalents, quadrivalents, and univalents are formed in proportions that are inconstant from cell to cell, gones with unbalanced chromosome complements are produced, and the hybrid is more or less sterile. Where the chromosomes of the parental species fail to pair in the diploid on account of extensive dissimilarities in the gene arrangement, every chromosome in the allotetraploid has only one mate, with which it can pair with little or no interference from the chromosomes of the other species. Hence, only bivalents are produced, and meiosis is regular.

Where chromosome pairing in a diploid hybrid is suppressed by the genetic constitution rather than by dissimilarities in the gene arrangement, the same suppression should be encountered in the allotetraploid derived from it. Moreover, the greater the suppression in a diploid hybrid of such kind, the greater it will be in the tetraploid. In other words, Darlington's rule is expected not to apply to hybrids whose sterility is genic. It does not. In weak A × weak B hybrids of *Drosophila pseudoobscura,* only bivalents are present in the diploid spermatocytes, and in the tetraploid ones bivalents, quadrivalents, and some trivalents and univalents are found. Strong A × strong B crosses show little or no chromosome pairing either in diploid or in tetraploid spermatocytes.

The only hybrid whose behavior seems difficult to account for is *Primula verticillata* × *P. floribunda* (= *P. Kewensis*). According to Newton and Pellew (1929), the diploid has mostly bivalents, and yet it is nearly sterile; the tetraploid has again bivalents (with an occasional quadrivalent) and is fertile, although not quite true breeding. The hypothesis of chromosomal sterility furnishes still the most plausible explanation of this exceptional behavior. The gene arrangements in the chromosomes of *P. verticillata* and *P. floribunda* are probably different, but not different enough to prevent their pairing in the diploid hybrid. In the tetraploid hybrid each chromosome has a homologue with an exactly similar gene arrangement, and also two partial homologues. Due to competition for pairing, the complete homologues will unite to form bivalents more frequently than the partial ones. In a few cells, quadrivalents may however arise owing to the chance occurrence of pairing between parts of chromosomes of the two species. Now, the meiotic disjunction in the tetraploid will give rise to gones most of which will contain a full haploid set of *verticillata* and a full set of *floribunda* chromosomes. The gones formed in the diploid will have equal numbers of chromosomes, but a majority of them will be mixtures, including some of *verticillata* and some of *floribunda.* Gones with a pure complement of one species will sometimes also be produced, but the probability of such an event is rather small. Since each parental species has nine chromosomes (haploid), only one gone in 512 ($1:2^9$) may carry a full set of *verticillata* and 1 in 512 of *floribunda* chromosomes, and this without taking into consideration the crossing over that presumably takes place in the diploid hybrids. The inviability of the gones with mixed chromosomes may be due either to differences in the gene arrangement in the parental species, or to qualitative differences in their genes. Further data on the genetics of the diploid and tetraploid *Primula Kewensis* are necessary to test the validity of this explanation.

Whereas the extent of the differences in gene arrangement in the parental species that produce sterile diploid hybrids is elusive and difficult to evaluate and to describe accurately, the fact that such differences must exist can be fairly safely inferred if the reduplication of the chromosome complement in such a hybrid restores fertility. Manifestly, the hybrids whose sterility is genic must be guarded against in making such inferences, but the behavior of the allopolyploids seems to furnish the necessary evidence. Using the same method of reasoning, one may attempt to deduce the existence of structural differences between the chromosome sets in natural allopolyploids.

The species whose origin through allopolyploidy is known, or suspected, show as a rule only bivalents at meiosis; exceptions are known but they are on the whole rare. And yet in an allopolyploid most genes must be supposed to be represented four (tetraploid) or more times. The formation of multivalent chromosome associations is prevented by the competition for pairing: every chromosome is more likely to pair with its complete homologue than with a partial one.

Through the occurrence of parthenogenesis the egg of a polyploid may develop and give rise to an organism that is haploid with respect to its progenitor. But if the progenitor is itself a polyploid, the "haploid" derived from it has two or more sets of chromosomes that are in part homologous. Thus, Gaines and Aase (1926) have obtained a haploid from the wheat *Triticum compactum;* since *T. compactum* is a hexaploid, the "haploid" plant has in reality three sets of seven chromosomes each. None of the chromosomes in such a haploid has an exact homologue, but as many as three bivalents may be formed at meiosis, evidently due to the association of partly homologous chromosomes from different sets. From 5 to 12 bivalents have been observed by Jorgensen (1928) in the "haploid" derived from the hexaploid species *Solanum nigrum* (2n = 72, basic number 12). Similar observations have been made by Chipman and Goodspeed (1927) and Lammerts (1934) in the haploid *Nicotiana taba-*

cum (2n = 48, basic number 12), and by Buxton and Darlington (1932) in the haploid *Digitalis mertonensis* (2n = 112, basic number 7). Some chromosome pairing may be observed also in haploids derived from species that are not suspected of being polyploids (Emerson 1929, Catcheside 1932). Unless a non-homologous pairing like that discovered in maize by McClintock (1933) is here involved, these cases must be due to the presence of homologous segments in the different chromosomes of the same diploid set, similar to the "repeat" segments in the chromosomes of Drosophila (Bridges 1935, cf. Chapter IV).

In hybrids between polyploid and diploid species, or in those between polyploid ones, several chromosome sets of different origin are brought together. Such hybrids may show at meiosis only bivalents, or both bivalents and univalents, or only univalents. Pairing may take place between chromosomes of different parents (allosyndesis) or between those of the same parent (autosyndesis). As pointed out in Chapter VII, it is not always easy to determine the origin of the bivalents in a particular hybrid, but in some favorable objects the discrimination between the alternatives is possible. Autosyndetic pairing is what interests us at present, since its occurrence indicates a partial homology between chromosomes of the same species. The cross between two species of poppy, *Papaver nudicaule* (n = 7) and *P. striatocarpum* (n = 35), gives a fertile hybrid with 42 chromosomes that unite to form 21 bivalents at meiosis (Ljungdahl 1924). We are compelled to conclude that 7 chromosomes of *nudicaule* have paired with 7 of *striatocarpum*, and that the remaining 28 chromosomes of the latter species have formed 14 more bivalents. Since in the pure *P. striatocarpum* (2n = 70) 35 bivalents and no quadrivalents or higher associations are reported at meiosis, the pairing of the partially homologous chromosomes must be prevented by the competition of the complete homologues. In the hybrid, where complete homologues are absent, the partial homology asserts itself in the form of pairing.

Autosyndetic pairing is found also in the hybrid *Digitalis lutea* (n = 48) × *D. micrantha* (n = 24), which is described as absolutely sterile despite the presence of 36 bivalents and no univalents at meiosis (Haase-Bessel 1921). The sterility is here caused probably by the same mechanism as that in the diploid *Primula Kewensis*. A partial autosyndesis is observed in the hybrids between diploid and octoploid strawberries (Yarnell 1931), and also in those between species of wheats (Kihara and Nishiyama 1930, Kihara and Lilienfeld 1932, 1935, Liljefors 1936), *Crepis biennis* × *C. setosa* (Collins, Hollingshead and Avery 1929), species of violets (J. Clausen 1931), and species of cotton (Skovsted 1935).

It may be noted that all the hybrids whose sterility is supposed to be chromosomal belong to the plant kingdom. This suggests that hybrid sterility is usually chromosomal in plants and genic in animals, but so broad a generalization is decidedly premature at present. The apparent prevalence of chromosomal sterility among plants may be due simply to the fact that it has been inferred from observations on fertile allopolyploids. Allopolyploids are at least rare in animals, for reasons discussed in Chapter VII. In general, the discrimination between chromosomal and genic sterility is as yet possible in so few hybrids that any statistical treatment of the data would rest on too insecure a foundation.*

Aside from the *Drosophila pseudoobscura* hybrids, genic sterility is established also in those between *D. melanogaster* and *D. simulans*. The gene arrangements in these species are similar, except for an inversion in the third chromosome and perhaps a few minor alterations in the X and in other chromosomes (Sturtevant 1929, Pätau 1935, Kerkis 1936). Spermatogenesis and oögenesis in the hybrids are arrested before the advent of the meiotic stages (Kerkis 1933). A cross between triploid females of *D. melanogaster* and males of *D. simulans* gives rise to some triploid female hybrids that carry two chromosome sets from the *D. melanogaster* and one set from the *D. simulans* parent. If the sterility of the diploid hybrids were due to a lack of pairing partners among the chromosomes, one might expect that in the triploid the *D. melanogaster* chromosomes would pair and some functional eggs would be produced. Actually,

---

* P. Hertwig (1936) in her book on species hybrids in animals distinguishes between "Hemmungs-Sterilität" and "Konjugationssterilität," which correspond to our genic and chromosomal sterility. She places all hybrids in which a failure of chromosome pairing is observed (except the A × B hybrids in *Drosophila pseudoobscura*) in the latter category. The validity of this classification is doubtful, since disturbances of the meiotic pairing can be produced by gene action as well as by differences in gene arrangements.

the triploid hybrids are completely sterile (Schultz and Dobzhansky 1933, cf. Kozhevnikov 1933).

The sterility of hybrids in which gametogenesis is disrupted before meiosis is probably genic (*Drosophila melanogaster* $\times$ *D simulans*, some of the hybrids between species of birds, hybrids between the domestic cow and the European buffalo and yak, horse $\times$ zebra, and others). Where the meiotic divisions are successfully completed, and the gones nevertheless degenerate, the sterility may be either strictly genic, or due to the formation of lethal gene or chromosome combinations. Some of Federley's data on hybrids between species of Pygaera (Federley 1913, 1931) suggests a chromosomal sterility, since in the triploid hybrids having two chromosome sets of one species and a single set of the other the former give rise to bivalents and the latter remain as univalents. Chromosome pairing is therefore not suppressed by the genetic constitution of the hybrid; the degeneration of the spermatids after the meiotic divisions seems however to indicate that a genic sterility is involved as well.

MATERNAL EFFECTS IN HYBRIDS

Differences in the outcome of the reciprocal crosses between the same two species are not rare. For example, the cross *Drosophila melanogaster* ♀ $\times$ *D. simulans* ♂ gives females but not males, while the reciprocal one produces male, but few or no female, hybrids (Sturtevant 1920-21). A variety of mechanisms are responsible for such differences. The chromosomal constitutions of the hybrids of the heterozygous sex coming from reciprocal crosses are unlike. Federley (1929) found that the females from the cross *Chaerocampa elpenor* ♀ $\times$ *Metopsilus porcellus* ♂ have a combination of the X chromosome of the latter and the Y chromosome of the former species which acts as a lethal; the reciprocal cross gives rise to viable females that carry the X of *C. elpenor* and the Y of *M. porcellus*. A hybrid inherits its chromosomes from both parents, but its cytoplasm supposedly from the mother only. Interactions between the same chromosome complement and different cystoplasms may give dissimilar results. It must be kept in mind however that the characteristics of the cytoplasm of an egg from which a hybrid develops may be determined by its intrinsic properties, independent of the chromosomes it carries or has carried, or else by the properties of the chromosomes that have been present in the egg before the meiotic divisions and fertilization. The former mechanism is spoken of as cystoplasmic inheritance, while the latter, the predetermination of the cytoplasm by the chromosomes, is known as a maternal effect. It is amazing that even some modern authors fail to draw a distinction between two such fundamentally different phenomena; the resulting confusion in the genetic literature is indeed inexcusable. Here we are concerned with maternal effects inasmuch as they may contribute to the formation of isolating mechanisms, and of hybrid sterility in particular.

The difference between reciprocal crosses of race A and race B of *Drosophila pseudoobscura* proved to be amenable to a fairly exact analysis. The cross B ♀ × A ♂ produces F_1 hybrid males with small testes, while the F_1 hybrids coming from the cross A ♀ × B ♂ have large testes (Lancefield 1929). Males with small testes have consequently the cytoplasm and the X chromosome of race B and the Y chromosome of race A. Hybrid males with large testes have race A cytoplasm and X chromosome, and race B Y chromosome. Both kinds of males have hybrid autosomes, that is one set from race A and the other from race B. Dobzhansky (1933d, 1935c, 1936b) has studied the size of the testes in males coming from backcrosses of the F_1 hybrid females to males of both parental races, and also in males devoid of Y chromosome. The latter type arises from exceptional eggs that carry no X fertilized by X-bearing spermatozoa. The frequency of the production of such exceptional eggs is usually very low, but it is enhanced by treating females with X-rays. The data are summarized in Table 21.

The conclusions that may be deduced from the data presented in Table 21 are as follows: (1) among the backcross males, the testis size is determined exclusively by their chromosomal constitution and is independent of the source of their cytoplasm (compare classes 13 and 16, 14 and 17, 15 and 18); (2) the presence or absence of the Y does not influence the testis size (compare 1 and 5, 2 and 6, 4 and 7); (3) males of the same chromosomal constitution in the F_1 generation and in the backcrosses may differ in testis size (compare 3 and 15 and 18); (4) the backcross males have small testes whenever their X chromosome does not agree with the autosomes in racial origin, and large testes when the chromosomes are racially alike; (5) in F_1 males, small testes are found if hybrid autosomes are present with race B cytoplasm (classes 4, 7, 8), or if race B X chromosome is present together with race A cytoplasm (class 9); hybrid autosomes in race A cytoplasm give large testes in the pres-

TABLE 21

TESTIS SIZE IN *Drosophila pseudoobscura* MALES OF DIFFERENT
CONSTITUTION

NO.	RACE OF THE MOTHER	RACE OF THE FATHER	X CHROMOSOME	Y CHROMOSOME	AUTOSOMES	TESTIS SIZE
1	A	A	A	A	A	large
2	B	B	B	B	B	"
3	A	B	A	B	hybrid	"
4	B	A	B	A	"	small
5	A	A	A	n one	A	large
6	B	B	B	"	B	"
7	B	A	B	"	hybrid	small
8	B	A	A	"	"	"
9	A	B	B	"	"	"
10	F_1(A ♀ ×B♂) ♀	A	A	A	A	large
11	"	A	A	A	hybrid	small
12	"	A	B	A	"	
13	"	B	B	B	B	large
14	"	B	B	B	hybrid	small
15	"	B	A	B	"	0
16	F_1(B ♀ ×A♂) ♀	B	B	B	B	large
17	"	B	B	B	hybrid	small
18	"	B	A	B	"	"
19	"	A	A	none	A or hybrid	large or small
20	F_1(A ♀ ×B♂) ♀	A	A	"	"	"

ence of race A X chromosome (class 3). The rôle of the cytoplasm in the production of the differences between reciprocal crosses is evident, but it is likewise evident that the properties of the cytoplasm of an egg are determined by the chromosomes that have been present in the body of the mother in which the egg has developed. We have shown above that the sterility of interracial hybrids in *Drosophila pseudoobscura* is due to the action of complementary genes contributed by the parental races. It is possible that our conclusions might be better formulated thus: the sterility of the hybrids

in question is due to interactions between the chromosomal constitution of the hybrid itself and the properties of the cytoplasm of the
egg from which it develops, always keeping in mind that the properties of the latter are determined by the chromosomal constitution of
the mother.

Another, and very interesting, kind of maternal effect is observed
in the progenies from backcrosses of F_1 hybrid females to males of
the parental races of *D. pseudoobscura*. Deviations from the normal
$1 : 1$ sex-ratio in these progenies have been recorded by Lancefield
(1929); Dobzhansky and Sturtevant have found in addition that
the general viability of the backcross individuals is very low in
comparison both with the pure races and with the F_1 hybrids. Since
some, in fact, a majority, of the offspring of the backcrosses have
various mixtures of the chromosomes of race A and race B, it was
natural to suppose that their decreased vitality is due to the formation of unfavorable recombinations of genes of the parental races.
Although this explanation is still not excluded, further studies have
shown that it is by no means adequate to account for the whole situation. A sample of the actual data (unpublished) will illustrate the
essential features of the phenomenon.

Race A females homozygous for the sex-linked recessive genes
beaded (*bd*), yellow (*y*), short (*s*), the dominant Bare (*Ba*, second
chromosome), and the recessive purple (*pr*, third chromosome) were
crossed to race B males homozygous for the recessive orange (*or*,
third chromosome) and heterozygous for the dominant Curly (*Cy*,
fourth chromosome). In accordance with expectation, the following
results were obtained in the F_1 generation:

Ba ♀ ♀	432	845	bd y s Ba ♂ ♂	401	786
Ba Cy ♀ ♀	413		bd y s Ba Cy ♂ ♂	385	

Males are somewhat less numerous than females, which is undoubtedly the result of a slight decrease of the viability of the
former due to the sex-linked recessives *bd*, *y*, and *s*. The *Ba Cy*
hybrid females were backcrossed to race A males homozygous for
purple (*pr*) and orange (*or*). It may be noted that the females used
have every chromosome, except the small fifth, marked with at least
one mutant gene. Therefore in the backcross progeny the genetic

constitution of every male individual may be ascertained from its appearance. Disregarding the crossing over in the X and in the third chromosomes, sixteen classes of males must appear in equal numbers, carrying the combinations of the chromosomes of race A and race B represented diagramatically in Figs 20 and 21. Only eight classes of females are distinguishable (since, according to the setting

TABLE 22

(*Explanation in Text*)

CLASS NUMBER	MALES		FEMALES	
	Phenotype	Observed	Phenotype	Observed
1	*bd y s Ba pr*	2	*Ba pr*	41
2	*bd y s Ba pr Cy*	—	*Ba pr Cy*	32
3	*bd y s Ba or*	4	*Ba or*	92
4	*bd y s pr*	7	*pr*	190
5	*bd y s Ba or Cy*	1	*Ba or Cy*	89
6	*bd y s pr Cy*	7	*pr Cy*	372
7	*bd y s or*	14	*or*	140
8	*bd y s or Cy*	13	*or Cy*	336
9	*or Cy*	147		
10	*or*	143		
11	*pr Cy*	62		
12	*Ba or Cy*	17		
13	*pr*	58		
14	*Ba or*	21		
15	*Ba pr Cy*	14		
16	*Ba pr*	6		
Crossovers		121	Crossovers	311
Total		637	Total	1603

of the experiment, the sex-linked recessive genes do not manifest themselves in the female progeny). The results actually obtained are summarized in Table 22; the column marked "class number" refers to the diagrams in Fig. 21.

Table 22 shows that, contrary to expectation, males are much more scarce than females, and the representatives of the different classes are far from being equally numerous. Here mention must be made of the fact that the number of eggs deposited by an F_1 hybrid female is of the same order of magnitude as in the pure races, and yet the yield of the adult backcross individuals per mother is very small. The conclusion follows that some, perhaps a majority, of the backcross individuals die, and, as shown by the results presented in Table 22, the mortality of the representatives of the different classes is selective. An important, and at first sight paradoxical, feature is that class 1 of the males, which consists of individuals having only race A chromosomes, is almost obliterated. Now, if the decrease in viability were due to mixing the chromosomes of the two races in one individual, class 1 would be expected to be the most viable one.

A closer examination of Table 22 shows that the number of individuals of a given class recovered in this backcross is inversely proportional to the number of mutant genes this class carries. All the classes carrying *bd, y,* and *s* are much decreased in frequency. The gene *Ba* also depresses the viability greatly, *pr* follows next, while *or* and *Cy* are relatively innocuous. And yet, the same mutant genes produce no disastrous effects on viability in the pure races and in the F_1 hybrids. The results may be accounted for only if one assumes that the eggs deposited by F_1 hybrid females give rise to individuals that are afflicted with a general constitutional weakness (maternal effect); mutant genes that do not impair greatly the viability of the pure races or of F_1 hybrids act as semi-lethals in individuals developing from the eggs deposited by hybrid females.

To test the above assumption, experiments were so arranged that the class of backcross progeny which is identical in constitution with race A (corresponding to class 1 in Table 22) was free from mutant genes, and the class having hybrid autosomes (corresponding to 9 in Table 22) manifested several mutants. The result was the opposite of that observed in the first experiment: class 9 was depressed in frequency more than class 1. Experiments involving backcrosses of F_1 hybrid females to race B gave results consistent with the hypothesis; backcross individuals carrying only race B chromosomes may live or die depending upon the mutant genes they are made to carry.

It is evident that the decreased viability of the offspring of the F_1 hybrid females, entails an intensification of the isolation between the parental races. The hybrid females are in effect semi-sterile. It is important in this connection that the degree of depression of viability in the backcross progenies is a function of the strains of the parental races used in a cross. In other words, the population of either race of *Drosophila pseudoobscura* manifests hereditary variations with respect to the genetic factors producing the maternal effects. Dr. A. H. Sturtevant has observed such variations in wild strains coming from geographically different localities (unpublished).

The writer (unpublished) has found certain strains carrying mutant genes that destroy the backcross progenies entirely. The two race A strains, one homozygous for the sex-linked recessives beaded, yellow, vermilion, singed, and short (*bd y v sn s*), and the other for beaded, miniature, snapt, and sepia (*by m sp se*), are especially remarkable. The strain *bd y v sn s* has been crossed to a wild B race strain (Seattle-6), and also to a race B strain homozygous for the sex-linked gene scutellar. The reciprocal crosses A ♀ × B ♂ and B ♀ × A ♂ have been made in each case, and numerous F₁ hybrids of an apparently good viability were obtained. And yet if the hybrid females are backcrossed to *bd y v sn s* and *bd m sp se* males, respectively, no adult offspring at all are obtained. These experiments were conducted on a fairly large scale; the F₁ females deposit numerous eggs some of which give rise to young larvae, but none of the latter develop to the adult stage. The genetic constitution of the mothers is evidently such that the eggs produced are virtually inviable when fertilized by spermatozoa of certain males. The very same mothers produce, however, fairly viable offspring when crossed to males from certain other strains (the race A strain "Texas," for example).

The outcome of the cross is determined, therefore, by the genetic constitution of the maternal grandparents, as well as by that of the father. The eggs of a hybrid female give rise to a weak progeny, but the degree of weakness may vary depending upon the kind of spermatozoa that fertilize the eggs. The nature of the differences between the strains of *D. pseudoobscura* that determine the extent of the deleterious maternal effects in the backcross progenies is unknown, since the analysis has not been completed. The present hypothesis, subject to verification in further tests, is that these differences are due to the genes these strains carry. If this proves to be the case, one can visualize how an evolutionary change can occur which transforms a pair of species giving rise to hybrids that are sterile in one sex and fertile in the other to a condition when hybrids of both sexes will be sterile. Crosses between most strains of race A and race B of *D. pseudoobscura* give, as we know, sterile males and fertile females in the F_1 generation. If, however, race A would uniformly acquire the genetic constitution now present in the *bd y v sn s* strain, and race B that present in the Seattle-6 and scutellar strains, the F_1 hybrid females would be sterile when backcrossed to the race A parents. How widely maternal effects, leading to sterility of the hybrids, are distributed in nature is at present an open question.

X: SPECIES AS NATURAL UNITS

BIOLOGICAL CLASSIFICATION

THE PROCESS of evolution has two aspects, since it involves the development of diversity as well as that of discontinuity in the living world. The aspect of discontinuity should be especially emphasized, not because it is the more important of the two, but because it is the less obvious one to a superficial observer. The characteristics of discontinuity as a static phenomenon have been dealt with in Chapter I. The essential point stressed there is that the variation observed among organisms living at any given time level does not form a single probability distribution, but rather an array of discrete distributions. Moreover, the variation is hierarchical, since the small discrete arrays are grouped into larger ones, these into still larger ones, etc. The discontinuity is preserved throughout the hierarchy of arrays. The discontinuous variation of morphological and physiological characteristics of organisms is a fact given us in experience, an objective phenomenon rather than a projection on nature of some concepts created by the investigator.

The scientific classification of organisms is founded on the discontinuity and the hierarchy of variation, or, to put it more precisely, these properties of variation have been used for the purpose of making a classification. The fact that this is not the only possible kind of classification should not be lost sight of. Books in a library may be classified according to contents, name of the author, year of publication, size, color of the cover, etc.; which of these methods is selected depends on convenience. The same principle may be applied to the classification of organisms. In fact, Pliny did reject the system of animals proposed before him by Aristotle, which happened to be fairly similar to the modern one, and subdivided the animals into those living in water, on land, and in the air. One might just as well use as a basis of classification such characters as usefulness or harmfulness to man, occurrence in different climates, etc. It is amusing that at least one contemporary biologist has apparently in all serious-ness suggested that an "ecological" classification might be superior to the one now in use. The pragmatism of the existing classification is openly acknowledged by many systematists; this attitude seems not very consistent with the claims to the effect that it is also the only "natural" one, and that its naturalness must be preserved at all costs. A natural and a convenient system need not necessarily be identical.

The concept of a "natural" system is far from being always made clear. Being told that an organism belongs to the genus Drosophila we may safely predict that its body consists of segments, that it re-spires with the aid of tracheae, that it has a legless larva, is likely to have certain wing veins, bristles on the thorax, a branched arista, a short period of development, a rather low number of chromosomes, etc. The position of an organism in the system of Pliny, or one like' it, would not define so many of its characteristics. This seems to be the only reason why our system is more natural than that of Pliny. A knowledge of the position of an organism in an ideal natural system would permit the formation of a sufficient number of deductive prop-ositions for its complete description. Hence, a system based on the empirically existing discontinuities in materials to be classified, and following the hierarchical order of the discontinuous arrays, ap-proaches most closely to the ideal natural one. Every subdivision made in such a system conveys to the student the greatest possible amount of information pertaining to the objects before him. The modern classification of organisms uses the principles on which an ideal system could be built, although it would be an exaggeration to think that the two are consubstantial.

On the other hand, since the time of Darwin and his immediate followers the term "natural classification" has meant in biology one based on the hypothetical common descent of organisms. The forms united together in a species, genus, class, or phylum were supposed to have descended from a single common ancestor, or from a group of very similar ancestors. The lines of separation between the system-atic categories were, hence, adjusted, at least in theory, not so much to the discontinuities in the observed variations as to the branching of real or assumed phylogenetic trees. And yet the classification has continued to be based chiefly on morphological studies of the existing organisms rather than of the phylogenetic series of fossils. The logical difficulty thus incurred is circumvented with the aid of a hypothesis according to which the similarity between the organisms is a function of their descent. In other words, it is believed that one may safely base the classification on studies on the structures and functions of the organisms existing at our time level, in the assurance that if such studies are made complete enough, a picture of the phylogeny will emerge automatically. This comfortably complacent theory has received some rude shocks from certain palaeontological data that cast a grave doubt on the proposition that similarity is always a function of descent. Now, if similar organisms may, however rarely, develop from dissimilar ancestors, a phylogenetic classification must sometimes unite dissimilar, and separate similar, forms. The resulting system will be, at least in some of its parts, neither natural in the sense defined above nor convenient for practical purposes.

Fortunately, the difficulty just stated is more abstract than real. The fact is that the classification of organisms that existed before the advent of evolutionary theories has undergone surprisingly little change in the times following it, and whatever changes have been made depended only to a trifling extent on the elucidation of the actual phylogenetic relationships through palaeontological evidence. The phylogenetic interpretation has been simply superimposed on the existing classification; a rejection of the former fails to do any violence to the latter. The subdivisions of the animal and plant kingdoms established by Linnaeus are, with few exceptions, retained in the modern classification, and this despite the enormous number of new forms discovered since then. These new forms were either included in the Linnaean groups, or else new groups were created to accommodate them. There has been no necessity for a basic change in the classification. This fact is taken for granted by most systematists, and all too frequently overlooked by the representatives of other biological disciplines. Its connotations are worth considering. For the only inference that can be drawn from it is that the classification now adopted is not an arbitrary but a natural one, reflecting the objective state of things.

To avoid misunderstanding, it is necessary to define in what sense the classification may be said to have remained constant. The system of Linnaeus recognized only four hierarchical ranks: beginning with the lowest, they were species, genus, order, and class. Two new categories were added very early: the family and the phylum. The number of categories now used is large indeed: species are subdivided into subspecies and races of various rank, genera split into subgenera and sections, and grouped into tribes, subfamilies, families, superfamilies, suborders, etc. Among insects, for example, most of the Linnaean genera are now treated as families and some as even higher categories. What has remained virtually unaltered through all these metamorphoses is the recognition that a given complex of forms represent a natural group, different from other similar complexes or groups. The rank ascribed to a group has been changed repeatedly, and individual authorities are quite likely to be at odds in their opinions on such matters, but the delimitation of the groups is very much less frequently a subject of contention. The evaluation of a group as a genus, tribe, subfamily, or a family is determined purely by convenience; an investigator is, within limits, free to exercise his choice. The number of discontinuities of different orders in the organic world is so large that more and more categories can be created to describe them, just as branches of a tree can be classified only into major and minor ones, or else into primary, secondary, tertiary, etc.

There is, however, a single systematic category which, in contrast to others, has withstood all the changes in the nomenclature with an amazing tenacity. This is the category of species. To be sure, some of the species described by Linnaeus have been split into two or more new ones, and such splitting of species is in general not infrequent. And yet, a majority of the Linnaean species still are treated as species, not as subgenera, genera, or anything else. In most animal and plant groups, except in the so-called difficult ones, the delimitation of species is subject to no dispute at all. Despite all refinements in the techniques of investigation, and notwithstanding the differences in the individual tastes and preferences, the agreement on which forms are to be included in a species is as a rule universal. On the other hand, some groups are "difficult," and here an appalling chaos of individual judgments on the limits of species reigns supreme. The situation encountered in these "difficult" groups is apt to convey the false impression that species in general are arbitrary units like the rest of systematic categories. Some biologists, lacking familiarity with the subject, have, in fact, fallen into this error. In reality, no category is arbitrary so long as its limits are made to coincide with those of the discontinuously varying arrays of living forms. Furthermore, the category of species has certain attributes peculiar to itself that restrict the freedom of its usage, and consequently make it methodologically more valuable than the rest.

GENETIC BASIS OF CLASSIFICATION

Before proceeding further with our discussion of species as natural units, it may be useful to make a short digression to consider certain prolegomena on which this discussion must be based. Let us examine first an imaginary situation, a living world in which all possible gene combinations are represented by equal numbers of individuals. Under such conditions no discrete groups of forms and no hierarchy of groups could occur, since the single gene differences producing striking phenotypical effects, like some of the mutations in Drosophila, would be the sole remaining source of discontinuity. Disregarding these, the variability would become a perfect continuum. The most "natural," although not the only possible, classification would be a sort of a multi-dimensional periodic system, with a number of dimensions equal to that of the variable genes.

Clearly, the existing organic world is unlike the above imaginary one. As pointed out in Chapters VI and VIII, only an infinitesimal fraction of the possible gene combinations is realized among the living individuals, or has ever been realized. According to a conservative estimate given by Wright (1932), the number of possible combinations of genes is of the order of 10^{1000}, while the estimate of the number of electrons in the visible universe is of the order of 10^{100}. Furthermore, the existing gene combinations are by no means scattered at random through the entire field of the possible ones. On the contrary, the gene combinations are grouped together into more or less compact arrays, each array being attached, to use the symbolic picture of Wright, to one or to several related "adaptive peaks" in the field. The arrays are therefore complexes of fairly similar gene combinations that make their carriers fit to survive in the environments that are encountered in nature. The "adaptive valleys" intervening between the peaks correspond to discordant gene combinations, most of which would be nearly or completely inviable. A promiscuous formation of gene combination would give mainly a mass of freaks, something like the primeval monsters in the poetic myths of Empedocles and Lucretius.

The discontinuous variation in the organic world is therefore not merely a superficial appearance, but the consequence of a fundamental discontinuity in the gametic make-up of organisms. The discontinuity and the hierarchical character of the empirically observed variation may be viewed as a corollary of the particulate structure of the hereditary materials, and as a response of living matter to the pressure of the secular environment. Each race, species, genus, or any other group embraces a certain array of gene combinations attached to an "adaptive peak," or to several neighboring peaks. The fact that one group may be distinguished from the related ones necessarily implies that the gene combinations lying in the field between the peaks are formed only rarely or not at all. Now, if the representatives of the different groups interbred at random, all the gene combinations that are now rare or absent would be produced, given a sufficient number of individuals, within a few generations from the start of random breeding. That would mean a breakdown of the separation of the groups, and an emergence of a continuous variability over a part of the field. If all the organisms were to interbreed freely, a perfect continuum postuated above would result.

The conclusion that is inexorably forced on us is that the discontinuous variation encountered in nature, except that based on single gene differences, is maintained by means of preventing the random interbreeding of the representatives of the now discrete groups. This conclusion is evidently applicable to discrete groups of any rank whatever, beginning with minor races of a species and up to and including classes and phyla. The development of isolating mechanisms is therefore a *conditio sine qua non* for emergence of discrete groups of forms in evolution (cf. Chapter VIII).

The above conclusion is certainly not vitiated by the well-known fact that the isolation between groups may be complete or only partial. An occasional exchange of genes, not attaining to the frequency of a random interbreeding, results in the production of some intergrades, without, however, swamping the differences between the groups entirely. On the whole, the degree of isolation is proportional to the remoteness of the groups concerned, although no strict rules can be formulated governing this phenomenon. The fact that representatives of different phyla, classes, and orders do not interbreed is generally taken for granted. Interbreeding of families probably also never takes place, and that of genera is rare enough. Interbreeding of species and races has been recorded in many instances.

SPECIES IN SEXUALLY REPRODUCING ORGANISMS

In the pre-Darwinian days the relative ease of separating species, and the greater difficulty of combining them into higher categories or subdividing them into lower ones, seemed to offer no great difficulties for explanation. Indeed, if each species has arisen owing to a separate act of creation it must be regarded a fundamental unit of classification, and the task of the investigator is merely to learn to discriminate between these primordial entities and their secondary groupings and subdivisions. The situation has changed completely in the light of the evolution theories, for now such concepts as race, species, genus, family, etc., have come to be understood as connoting nothing more than degrees of separation in the process of a gradual phylogenetic divergence. Yet, despite all the difficulties encountered in classifying species in certain exceptional groups of organisms, biologists have continued to feel that there is something about species that makes them more definite entities than all other categories. W. Bateson has expressed this vague feeling quite concisely: "Though we cannot strictly define species, they yet have properties which varieties have not, and . . . the distinction is not merely a matter of degree."

There was no shortage of attempts to invent methods whereby separate species could be distinguished from groups that have attained to a racial rank only. The lack of intergrades between species and their presence between races has been frequently depended upon, but there are some very distinct species the intergrades between which occur in exceptional individuals, and there are obviously closely related races and variations produced by a single Mendelian gene that are discrete. Among insects, the differences in the genitalia were assumed to mark species, but such differences may be present in races and absent in species (cf. Chapter VIII). The geographical distributions of species frequently overlap, while those of races do not—but again there are numerous exceptions. Experimental biologists have naturally looked in a somewhat different direction for criteria of species distinction. The sterility of hybrids has been frequently supposed to be confined to species crosses. Thus Standfuss (1896) gives the following definition: "Arten sind Gruppen von Individuen, die sich in ihren geschlechtlich entwickelten Formen nicht mehr dergestalt kreuzen können, dass sich die aus dieser Kreuzung hervorgehenden vollkommen ausgebildeten Tiere unbeschränkt miteinander fortzupflanzen vermögen." Yet hybrids between some apparently "good" species seem to be fully fertile. Writers inclined toward eclecticism prefer to believe that none of the above criteria are sufficient when taken singly, but that a satisfactory result may be obtained by combining them. Of late, the futility of attempts to find a universally valid criterion for distinguishing species has come to be fairly generally, if reluctantly, recognized. This diffidence has prompted an affable systematist to propose something like the following definition of species: "a species is what a competent systematist considers to be a species."

The cause of this truly amazing situation—a failure to define species which is supposedly one of the basic biological units—is not too difficult to fathom. All of the attempts mentioned above have strived to accomplish a patently impossible task, namely to produce a definition that would make it possible to decide in any given case whether two given complexes of forms are already separate species or are still only races of a single species. Such a task might be practicable either if species were separate acts of creation, or else if species would arise from one another by a sudden, catastrophic, change, like a single mutational step. The first of the above alternatives finds however no sympathy in modern science; as to the second, it is apparently realized only in some groups of organisms, notably among certain higher plants (origin of species through allopolyploidy, cf. Chapter VII). The far more general method of species formation, believed to be encountered in all groups of organisms and in some groups being apparently the only one, is through a slow process of accumulation of genetic changes of the type of gene mutations and chromosomal reconstructions. This premise being granted, it follows that instances must be found in nature when two or more races have become so distinct as to approach, but not to attain completely, the species rank. The decision of a systematist in such instances can not but be an arbitrary one.

These difficulties need not however deter biologists from attempting to elucidate the nature of speces, provided it is clearly realized at the start that the outcome of these attempts will not be the emergence of a rigid standard of species distinction. An interesting trial of this kind has been made by Lotsy (1931). Starting from premises somewhat like those formulated by us above, namely that the existence of discrete groups of organisms presupposes a prevention of the random interbreeding between their representatives, Lotsy concludes that the fundamental unit among systematic categories is a "syngameon," which he seems to hold to be equivalent to a species. A syngameon is "an habitually interbreeding community of individuals."

Lotsy's attempt to clarify the species concept seems sound in principle. Nevertheless, it can hardly be accepted as adequate in its original form, since it fails to take into account that almost any widely distributed species is in all probability broken up into numerous more or less isolated colonies (Chapter V). The colonies, and not species as wholes, are the elementary "habitually interbreeding communities." The stress should be placed on the nature of the causes preventing the interbreeding of populations rather than on the fact itself. We have seen (Chapter VIII) that the isolating mechanisms may be divided into two fairly sharply distinct classes, namely, geographical isolation and physiological isolating mechanisms. As far as isolation (that is, the prevention of the exchange of genes between populations) is concerned, geographical separation may be just as effective as a physiological one. The rôle of geographical isolation in the molding of the genetic variability in the racial and specific complexes has been duly emphasized above. It is, however, self-evident that geographical isolation can be effective only so long as it lasts and no longer. In other words, if the discreteness of groups of organisms were guarded by geographical isolation alone, every locality could be inhabited by one and only one kind of living beings. For, as soon as two once discrete groups would, in the process of expanding their distribution regions, meet in a section of territory, the interchange of genes would begin. A progressive infiltration of the territory of one group by the other can have no other result but their fusion. How greatly the possibilities of the evolutionary process would be curtailed under such conditions is clear enough.

In reality, discrete groups of organisms frequently coexist in the same territory without losing their discreteness, because their interbreeding is prevented through one, or a combination of several, physiological isolating mechanisms (Chapters VIII and IX). The development of the latter causes a more or less permanent fixation of the organic discontinuity (the fixation need not be absolutely permanent in every case since the physiological isolation may be incomplete). The stage of the evolutionary process at which this fixation takes place is fundamentally important, and the attainment of this stage by a group of organisms signifies the advent of species distinction. The present writer has therefore proposed (Dobzhansky 1935e) to define species as that stage of evolutionary process, "at which the once actually or potentially interbreeding array of forms becomes segregated in two or more separate arrays which are physiologically incapable of interbreeding."

The definition of species just quoted differs from those hitherto proposed in that it lays emphasis on the dynamic nature of the species concept. Species is a stage in a process, not a static unit. This difference is important, for it frees the definition of the logical difficulties inherent in any static one. At the same time, our definition can not pretend to offer to a systematist a fixed yardstick with the aid of which he could decide in any given case whether two or more groups of forms have or have not reached the species rank. This drawback is unavoidable. A systematist is forced to describe the changing patterns of life in terms of abstract static conceptions. His task is first of all a practical one, viz., to classify and to systematize. To put it crudely, he wishes to be able to write a determination label under every specimen. Instances where groups of forms are caught at our time level in the transition stage between races and species are interesting and at the same time are a hindrance in practice. And yet systematists have intuitively grasped the existence of species as natural units. Hence, a dynamic definition may be of value also for systematists, if it substitutes an analytical judgment for the less communicable judgment of intuition.

The stage when physiological isolating mechanisms develop, and at which the genetic discontinuity reaches a state of fixation, undoubtedly occurs in evolution. Therefore, there is no doubt that our definition of species refers to a real and important phenomenon in nature. If this phenomenon can not be called "species" it must be called something else and a new term should be invented for this purpose. It is pertinent to inquire to what extent "species," so defined, correspond to the species established by nearly two centuries of usage of this term in descriptive biology. Fortunately, there is enough evidence to show that the correspondence is rather far reaching. Although in separating species the systematists, with rare exceptions, have no direct information on the ability of the forms concerned to interbreed, the criteria used by them are capable of producing indirect evidence bearing on this point. Some of these criteria are reviewed below.

Separate species possess different cycles of variability in morphological and physiological characters. The variations in at least some characters usually do not overlap, so that a hiatus is formed. Varieties, especially those that are encountered in the same geographical region with the "type" form, differ from each other mostly in single, although sometimes striking, characters. Geographical races, and especially species, differ in complexes of traits, some of which may be less striking to the eye than varietal differences. As might be expected theoretically, and is actually found to be the case in the few instances studied, the differences between geographical races and species are due to the cooperation of numerous genes. The differences between non-geographical variations are more frequently monogenic. Now, the existence of discrete forms differing in complexes of genes is possible only provided they are debarred from interbreeding. The reverse is true for the non-geographical varieties, since there the interbreeding does not result in any decrease in the extent of the difference.

The geographical criterion of species distinction is regarded by some systematists as no less decisive than the presence of a hiatus in morphological and physiological characteristics (Semenov-Tian-Shansky 1910). The geographical distributions of races (subspecies) as a rule do not overlap. In species with a continuous distribution the races may merge into each other by a series of almost imperceptible gradations (Rensch 1929); or else races may be fairly definite entities separated from each other by more or less pronounced physical barriers; intermediate situations are likewise encountered, where the intergradations are localized in a narrow stretch of territory in which races come in contact. On the other hand, the geographical distributions of species overlap very frequently, without however hybrids or intermediates being formed in the territory common to two or more species. In fact, the area of one species may be included in that of another without in the least decreasing the sharpness of the separation. The presence of physiological isolating mechanisms is the only possible explanation of the preservation of differences between species with overlapping distributions.

The formation of interspecific hybrids in the territories coinhabited by two or several closely related species presents a number of most interesting, though exceptional, situations. These exceptions may be truly said to prove the rule. For example, the distributions of two snail species, *Cepaea nemoralis* and *C. hortensis*, overlap in France. Hybrids between them have been observed in small numbers in several localities, but they are semi-sterile and have nowhere become established as a separate type competing with the parental species (Boettger 1922). Two water snails, *Viviparus ater* and *V. pyramidalis*, form a hybrid population in Lake Garda; in this lake, or at least in a part of it, the two species may be said to have fused into one (Franz 1928).

Instances of this kind seem to be more frequent in plants than among animals, though perhaps their apparent greater frequency in the former is simply due to the fact that botanists have paid more attention to this problem than zoologists. Du Rietz (1930) quotes a large number of examples from different genera and families. Hybrids may appear as exceptional individuals, so that the limits of species are only slightly blurred. Or else the territory where two species meet is populated with "hybrid swarms"; the parental species are almost completely lost in a mass of intermediates and of "new" types that presumably arise through recombination of the genes of two species. The "hybrid swarms" remind one of the complex segregations obtained in the F_2 and further generations of experimental interspecific hybrids (cf. Chapter III). The population of Salix inhabiting Greenland is described as consisting almost exclusively of hybrids between several species, while the putative parents are encountered rarely or not at all. This situation is exceptional even among plants, and the suspicion may arise that it is in general not justifiable to speak here about hybrids; we may be dealing with a primitive highly variable population from which separate species may differentiate through elimination of certain gene combinations. In some instances hybrids between species are observed mainly in localities modified by man's activity (Wiegand 1935). The physiological isolating mechanisms that have kept these species separate (presumably ecological isolation) prove inadequate when the conditions have changed due to the introduction of new factors. So long as a complete physiological isolation has not developed, the process of evolution is partly reversible: once discrete groups may become fused again into a single one.

Notice should also be taken of the most interesting rings of races that have been described in a number of animals (see Meise 1936). Two apparently unquestionably different species may coexist in the same territory without giving rise to intergrades, and yet they may be united by a series of races inhabiting other localities and imperceptibly grading into each other and into the extreme members of the series. A situation of this kind appears paradoxical unless one adopts a dynamic species concept: the extreme forms have already developed physiological isolating mechanisms that keep them separate from each other, and yet each of them has not become isolated from certain other forms that unite them together. Experimental work on such rings of races might produce much valuable information for elucidation of the mechanisms of species formation.

We may conclude that if the species separation is defined as a stage of the evolutionary process at which physiological isolating mechanisms become developed, the species so defined and the species of systematists will largely coincide. Discrepancies will occur only in those, relatively rare, groups where the appearance of physiological isolating mechanisms is not accompanied by a divergence in visible morphological traits. "Biological species" or "biological races" of this kind are fairly well known, especially among insects (cf. Thorpe 1930), where their separation has sometimes a great practical value (insect pests). The example of "race A" and "race B" of *Drosophila pseudoobscura* has been referred to repeatedly in this book. These "races" have distinctive geographical distributions, that however broadly overlap without the formation of hybrids in nature; the ecological preferences are somewhat different; a pronounced, though incomplete, sexual isolation is observed; the F_1 hybrid males are absolutely sterile; the viability of the backcross products is decreased. And yet the "races" are not distinguishable morphologically, this being the only reason why they are not classified as distinct species. The terms "biological species" or "biological races" are not particularly happy ones, since they convey a wrong impression, as though there may exist a fundamental difference between morphological and biological criteria of species separation. Some systematists are loath to accept the possibility that species may not be determinable in dead and preserved specimens, contending that a classification based on such an assumption will cease to be convenient. A discussion of this topic is hardly necessary here; we may only remark that the decision is here contingent on whether one is or is not satisfied with a classification that serves no other purpose than that of a well-ordered catalogue.

"SPECIES" IN ASEXUAL ORGANISMS

In the above discussion it has been assumed that the organisms concerned reproduce by cross-fertilization, through a union of gametes contributed by two individuals. In such organisms an interbreeding community (colony, race, species) may be said to possess a collective genotype. The genes present in the individuals composing the community are recombined in every generation; any combination of genes may occur from time to time, although a constellation of genes present in a given individual is very likely not to reappear in its immediate, and sometimes not even in its remote offspring. Fixed gene constellations differing from each other in two or more genes may become established only with the aid of isolation, that is through a splitting of the once integrated breeding community into two or more independent ones. It must be noted however, that some plants, among which Oenothera is the best known one, have developed an extraordinary genetic mechanism that permits coexistence of discrete gene combinations not only in the same population but in fact in the same individual. Here a gene combination is inherited as though it were a single gene. This situation, however, must be regarded as exceptional in the living world at large.

Cross-fertilization though by far the commonest, is not, however, the only known method of reproduction. Some organisms reproduce asexually without formation of specialized germ cells: simple fission, budding, adventitious sprouts, etc. Others preserve the specialized germ cells, but the eggs or ovules develop without fertilization (parthenogenesis, apomixis). Still others retain the sexual process, the union of two gametes takes place in every generation, but both gametes come from the same individual (self-fertilization). The obvious genetic consequence of asexual reproduction is that, barring the occurrence of mutation, the entire progeny of an individual possesses exactly the same genotype as its ancestor. Clones of genetically similar individuals are thus formed, every clone being isolated from every other simply by its mode of reproduction. Self-fertilization leads eventually to the same situation as asexual reproduction. Starting with heterozygous individuals, the proportion of homozygotes in a population increases with every generation, until the population becomes segregated into separate biotypes (pure lines). The offspring of a homozygote is, again barring mutation, genetically homogeneous and similar to its ancestor.

In some organisms, asexual reproduction alternates at more or less regular intervals with the sexual one. This is the case of most plants, in coelenterates, rotifers, cladocerans, and in some insects. Likewise, some organisms may produce offspring either through self- or through cross-fertilization (many plants, some hermaphroditic molluscs). Unless the cross-fertilization takes place very rarely, such organisms are genetically similar to those in which the cross-breeding is an invariable rule. Many individuals carrying the same genotype may recur in several successive generations, but sooner or later cross-fertilization, and the consequent recombination of genes and a mass production of new genotypes are bound to occur. Selection or elimination of individual carriers of certain genotypes is replaced in the asexual generations by a selection of clones, but the advent of the sexual generation again merges the genotypes of the clones into a single population genotype.

Groups of organisms are known, however, in which asexual reproduction or self-fertilization are not facultative but obligatory. In very primitive forms, such as bacteria and some lower fungi, the sexual process has perhaps never taken place in the phylogenetic history. Others apparently have lost the sexual reproduction secondarily, since their close relatives reproduce sexually. The genetic mechanisms of the evolutionary process in such organisms are bound to differ fairly widely from those with which we are familiar in cross-fertilizing forms. As pointed out by many writers since Weismann, the most important genetic consequence of amphimixis is the production of a variety of gene combinations, and hence an increase of the field of the hereditary variability. Now, the process of evolution entails changes not only in individual genes but also in gene systems; species and even races are known to differ in complexes of many genes, hence any satisfactory account of race and species formation must include not only a description of how the gene changes originate, but also how they are combined to give rise to the specific "radicals." A gene which in a given environment produces an unfavorable effect on the viability of its carrier may become favorable in combinations with other genes, or vice versa. Since every clone or pure line is isolated permanently from others, the possibility of exchanging genes, to discover by trial and error the most favorable gene combinations, is done away with in obligatorily asexual and self-fertilizing organisms.

The only way in which a genetic change may occur in a clone or a pure line is through a gene mutation or a chromosomal alteration. A new, altered clone or pure line arises in this manner. If the effects of the mutation are favorable, the new clone will successfully compete, and may eventually supersede the old one; otherwise it will be eliminated by natural selection. The formation of favorable gene combinations is made extremely difficult, however, for this requires the occurrence of a series of mutations in the same line of descent.

If each mutation is favorable per se, the task may be finally accomplished, although at the expense of a great loss of time; otherwise the potentially possible evolutionary change may never become realized.

These arguments seem to prove too much, since obligatorily asexual and self-fertilizing organisms exist, and are sometimes quite successful in the struggle for existence. The solution of the paradox seems to lie in the fact that having dispensed with the advantages of the process of apomixis these forms have acquired certain, at least partly compensating, advantages. Under cross-fertilization a favorable gene combination once formed is compelled to run the gauntlet of unlimited crossing with others. Without isolation, the mechanism of Mendelian recombination is just as efficient in dismantling gene combinations as it is in forming them. A favorable gene combination is probably formed in nature many times before it becomes established. In asexual or self-fertilizing forms any gene combination is at once fixed and isolated from others, and is ready to undergo the process of testing by natural selection. Moreover, certain genetic systems that are essentially unstable with cross-fertilization may be retained intact indefinitely in its absence. Here belong the autopolyploidy and various unstable chromosomal conditions, such as possession of odd numbers of sets of chromosomes (triploidy, pentaploidy, etc.), presence of extra chromosomes and absence of single members of a chromosome pair (trisomics, monosomics, aneuploidy). All these conditions are encountered among the asexual and apogamic plants much more frequently than among the cross-fertilizing ones (Rosenberg 1930, Darlington 1932a). The corresponding groups of animals have not been studied sufficiently to permit any generalization regarding them.

The modification of the evolutionary patterns wrought by the obligatory asexual reproduction and self-fertilization manifests itself in the absence of a definite species category in such organisms. It is not surprising that the groups of forms which are recognized as being uncommonly "difficult" from the standpoint of delimiting species have proved to be mainly those in which asexual reproduction or self-fertilization are the only, or the predominant, modes of propagation. The standard examples of such "difficult" groups are the plant genera

Hieracium and Rubus. The opinions of different authorities on what constitutes a species in these genera vary so widely that it is not uncommon to find that one investigator unites under a single specific name a complex of forms that is divided by others into numerous "species." Crow (1924) in his work on Cyanophyceae (algae) emphasizes that in this asexually reproducing group "the differences between closely related species rarely show that discontinuity which is apparent in many other groups of organisms," and that "individuals transitional between species are exceedingly common among the Cyanophyceae." The subdivision of the mass of clones into the species *Bacterium coli, B. typhi,* and *B. enteridis* is purely a matter of taste; one might just as well regard all of them as a single species (Baur 1930). The same is true for the lichen genus Cladonia and the related ones, in which a clear separation of species is impossible (but in which, nevertheless, a tremendous number of "species" have been described). Baur rightly points out, however, that such genera, constituting a "crux et scandalum botanicorum," are found mainly among the asexually reproducing forms.

The above statements should not be misunderstood as implying that the variation in asexually reproducing groups is absolutely continuous. On the contrary, we find there aggregations of numerous more or less clearly distinct biotypes, each of which is constant and reproduces its like if allowed to breed. These constant biotypes are sometimes called elementary species, but they are not united into integrated groups that are known as species in the cross-fertilizing forms. The term "elementary species" is therefore misleading and should be discarded. The existing biotypes obviously do not embody all the potentially possible combinations of genes. As in cross-fertilizing organisms, the biotypes in the asexual ones are clustered around some of the "adaptive peaks" in the field of gene combinations, while the "adaptive valleys" remain more or less uninhabited. Furthermore, the clusters are arranged in a hierarchical order, in a way which is again analogous to that encountered in sexual forms. The different clusters may, then, be designated some as species, others as subgenera, still others as genera, etc. Which one of these ranks is ascribed to a given cluster is, however, decided by considerations of convenience, and the decision is in this sense purely arbitrary. In other words, the species as a category which is more fixed, and therefore less arbitrary than the rest, is lacking in asexual and obligatorily self-fertilizing organisms. All the criteria of species distinction (see above) utterly break down in such forms.

The binominal system of nomenclature, which is applied universally to all living beings, has forced systematists to describe "species" in the sexual as well as in the asexual organisms. Two centuries have rooted this habit so firmly that any thorough reform will meet with a determined opposition. Nevertheless, systematists themselves have come to the conclusion that sexual species and "asexual species" must be distinguished (Du Rietz 1930). In the opinion of the writer, all that is saved by this method is the word "species." A realization of the fundamental difference between the two kinds of "species" can make the species concept methodologically more valuable than it has been.

LITERATURE

LITERATURE

The following abbreviations of the names of periodicals are used below.

A.E. — Roux' Archiv für Entwicklungs-mechanik der Organismen
A.N. — American Naturalist
B.B. — Biological Bulletin (Woods Hole)
B.Z. — Biologisches Zentralblatt
B.Zh. — Biologichesky Zhurnal (Moscow)
C. — Cytologia
G. — Genetics
H. — Hereditas
J.E.Z. — Journal of Experimental Zoology
J.G. — Journal of Genetics
J.H. — Journal of Heredity
P.N.A.S. — Proceedings of the National Academy of Sciences (U.S.A.)
P.VI.C.G. — Proceedings of the VI International Congress of Genetics (Ithaca)
S. — Science
U.C.P.B. — University of California Publications, Botany
V.K.V. — Verhandlungen des V. Internationalen Kongresses für Vererbungswissenschaft (Berlin)
Z.i.A.V. — Zeitschrift für induktive Abstammungs- und Vererbungslehre
Z.Z.m.A. — Zeitschrift für Zellforschung und mikroskopische Anatomie

Papers that have not been seen by the writer in the original
are marked by an asterisk.

Alpatov, W. W. 1932. Egg production in *Drosophila melanogaster* and some factors which influence it. J.E.Z., 63:85-111.

Anderson, Edgar. 1924. Studies on self-sterility. VI. The genetic basis of cross-sterility in Nicotiana. G., 9:13-40.

———— 1936. The species problem in Iris. Ann. Missouri Bot. Garden, 23:457-509.

———— and K. Sax. 1936. A cytological monograph of the American species of Tradescantia. Bot. Gazette, 97:433-476.

———— and R. E. Woodson. 1935. The species of Tradescantia indigenous to the United States. Contr. Arnold Arboretum, 9:1-132.

Arnason, T. J. 1936. Cytogenetics of hybrids between *Zea mays* and *Euchlaena mexicana*. G., 21:40-60.

Artom, Ch. 1931. L'origine e l'evoluzione della partenogenesi attraverso i differenti biotipi di una specie collettiva (*Artemia salina L.*) con speciale riferimento al biotipo diploide partenogenetico di Sète. Mem. Reale Accad. Italia, Fis. Mat. Nat., 2:1-57.

Averinzev, S. 1923-30. Herring of the White Sea. Wiss. Meeresunters., N.F., Abt. Helgoland, 15, No. 18:1-24.

Babcock, E. B., and M. Navashin. 1930. The genus Crepis. Bibliogr. Genetica, 6:1-90.

Backhouse, W. O. 1916. Note on the inheritance of "crossability." J.G., 6:91-94.

Balkaschina, E. I., and D. D. Romaschoff. 1935. Genetische Struktur der Drosophila Populationen. I. Swenigoroder Populationen von *D. phalerata Meig., transversa Fall.* und *vibrissina Duda.* B.Zh., 4:81-106.

Banta, A. M., and T. R. Wood. 1927. A thermal race of Cladocera originating by mutation. V.K.V., 1:397-398.

Bateson, W. 1922. Evolutionary faith and modern doubts. S., 55:55-61.

Bauer, H. 1936. Beiträge zur vergleichenden Morphologie der Speichel-drüsen-chromosomen. Zool. Jahrb. allg. Zool. Physiol., 56:239-276.

Baur, E. 1924. Untersuchungen über das Wesen, die Entstehung und die Vererbung von Rassenunterschieden bei Antirrhinum majus. Bibliotheca Genetica, 4:1-170.

———— 1925. Die Bedeutung der Mutationen für das Evolutionsproblem. Z.i.A.V., 37:107-115.

———— 1930. Einführung in die Vererbungslehre. 7-11 Aufl., Borntraeger, Berlin.

———— 1932. Artumgrenzung und Artbildung in der Gattung Antirrhinum, Sektion Antirrhinastrum. Z.i.A.V., 63:256-302.

Beadle, G. W. 1930. Genetical and cytological studies of Mendelian asynapsis in *Zea mays.* Cornell Univ. Agr. Exp. Sta., 129:1-23.

———— 1931. A gene in maize for supernumerary cell divisions following meiosis. Cornell Univ. Agr. Exp. Sta., 135:1-12.

———— 1932a. A gene for sticky chromosomes in *Zea mays.* Z.i.A.V., 63: 195-217.

———— 1932b. Genes in maize for pollen sterility. G., 17:413-431.

———— 1932c. A gene in *Zea mays* for failure of cytokinesis during meiosis. C., 3:142-155.

———— 1932d. A possible influence of the spindle fibre on crossing-over in Drosophila. P.N.A.S., 18:160-165.

———— 1933. Further studies of asynaptic maize. C., 3:269-287.

Beljajeff, M. M. 1927. Ein Experiment über die Bedeutung der Schutzfärbung. B.Z., 47:107-113.

———— 1930. Die Chromosomenkomplexe und ihre Beziehung zur Phylogenie bei den Lepidopteren. Z.i.A.V., 54:369-399.

Bellamy, A. W. 1922. Breeding experiments with the viviparous teleosts *Xiphophorus helleri* and *Platypoecilus maculatus.* Anat. Rec., 23:98-99.

Belling, J. 1927a. Configurations of bivalents of Hyacinthus with regard to segmental interchange. B.B., 52:480-487.

———— 1927b. The attachment of chromosomes at the reduction division in flowering plants. J.G., 18:177-205.

———— 1928. The ultimate chromomeres of Lilium and Aloe with regard to the number of genes. U.C.P.B., 14:307-318.

———— 1931. Chromomeres of Liliaceous plants. U.C.P.B., 16:153-170.

———— and A. F. Blakeslee. 1926. On the attachment of non-homologous chromosomes at the reduction division in certain 25-chromosome Daturas. P.N.A.S., 12:7-11.

Bergner, A. D., and A. F. Blakeslee. 1932. Cytology of the ferox-quercifolia-stramonium triangle in Datura. P.N.A.S., 18:151-159.

———— 1935. Chromosome ends in Datura discolor. P.N.A.S., 21:369-374.

———— J. L. Cartledge, and A. F. Blakeslee. 1934. Chromosome behavior due to a gene which prevents metaphase pairing in Datura. C., 6:19-37.

———— S. Satina, and A. F. Blakeslee, 1933. Prime types in Datura. P.N.A.S., 19:103-115.

Bernstein, F. 1925a. Beiträge zur mendelistischen Anthropologie I. Quantitative Rassenanalyse auf Grund von statistischen Beobachtungen über den Klangcharakter der Singstimme. Sitzungsb. Preussischen Akad. Wiss., 5:61-70.

———— 1925b. Beiträge zur mendelistischen Anthropologie II. Quantitative Rassenanalyse auf Grund von statistischen Beobachtungen über den Drehsinn des Kopfhaarwibels. Sitzungsb. Preussischen Acad. Wiss., 5: 71-82.

Blackburn, K. B., and J. W. H. Harrison. 1924. Genetical and cytological studies in hybrid roses I. J. Exp. Biol., 1:557-570.

Blakeslee, A. F. 1922. Variations in Datura due to changes in chromosome number. A.N., 56:16-31.

———— 1929. Cryptic types in Datura due to chromosomal interchange and their geographical distribution. J.H., 20:177-190.

———— 1932. The species problem in Datura. P.VI.C.G., 1:104-120.

———— and J. Belling. 1924. Chromosomal mutations in the Jimson weed, *Datura stramonium*. J.H., 15:194-206.

———— J. Belling, and M. E. Farnham. 1923. Inheritance of tetraploid Daturas. Bot. Gazette, 76:329-373.

———— and R. E. Cleland, 1930. Circle formation in Datura and Oenothera. P.N.A.S., 16:177-189.

———— G. Morrison, and A. G. Avery. 1927. Mutations in a haploid Datura. J.H., 18:193-199.

Bleier, H. 1928. Genetik und Cytologie teilweise und ganz steriler Getreidebastarde. Bibliogr. Genetica, 4:321-400.

———— 1933. Die meiosis von Haplodiplonten. Genetica, 15:129-176.

———— 1934. Bastardkaryologie. Bibliogr. Genetica, 11:393-485.

Boettger, C. 1922. Über freilebende Hybriden der Landschnecken *Cepaea nemoralis L.* und *Cepaea hortensis Müll.* Zool. Jahrb. Systematik, 44: 297-336.

Bonnier, G. 1924. Contributions to the knowledge of intra- and interspecific relationship in Drosophila. Acta Zool., 5:1-122.

———— 1927. Species-differences and gene-differences. H., 9:137-144.

328 LITERATURE

Bridges, C. B. 1917. Deficiency. G., 2:445-465.
—— 1919. Duplications. Anat. Record, 15:357.
—— 1923. The translocation of a section of chromosome II upon chromosome III in Drosophila. Anat. Record, 24:426-427.
—— 1935. Salivary chromosome maps. J.H., 26:60-64.
—— 1936. The bar "gene" a duplication. S., 83:210-211.
Brieger, F. 1928. Über die Verdoppelung der Chromosomenzahl bei Nicotiana Artbastarden. Z.i.A.V., 47:1-53.
—— 1930. Selbststerilität und Kreuzungssterilität im Pflanzenreich und Tierreich. J. Springer, Berlin.
—— 1935. Genetic analysis of the cross between the self-fertile *Nicotiana langsdorfii* and the self-sterile *N. sanderae*. J.G., 30:79-100.
Brink, R. A. 1929. The occurrence of semi-sterility in maize. J.H., 20: 266-269.
—— and C. R. Burnham. 1929. Inheritance of semi-sterility in maize. A.N., 43:301-316.
*Bristowe, W. S., and G. H. Locket. 1929. The courtship of British Lycosid spiders, and its probable significance. Proc. Zool. Soc. London: 317-347.
Buchholz, J. T., L. F. Williams, and A. F. Blakeslee. 1935. Pollen-tube growth of ten species of Datura in interspecific pollinations. P.N.A.S., 21:651-656.
Burgeff, H. 1928. Variabilität, Vererbung und Mutation bei Phycomyces blakesleeanus. Z.i.A.V., 49:228-242.
Burnham, C. R. 1930. Genetical and cytological studies of semisterility and related phenomena in maize. P.N.A.S., 16:269-277.
—— 1932. An interchange in maize giving low sterility and chain configurations. P.N.A.S., 18:434-440.
Buxton, B. H., and W. C. F. Newton. 1928. Hybrids of *Digitalis ambigua* and *Digitalis purpurea*, their fertility and cytology. J.G., 19:269-279.
—— and C. D. Darlington. 1932. Behaviour of a new species, *Digitalis mertonensis*. New Phytol., 31:225-240.
Bytinski-Saltz, H. 1933. Untersuchungen an Lepidopterenhybriden. II. Entwicklungsphysiologische Experimente über die Wirkung der disharmonischen Chromosomenkombinationen. A.E., 129:356-378.
Carothers, E. 1917. The segregation and recombination of homologous chromosomes as found in two genera of Acrididae (Orthoptera). J. Morphology, 28:445-521.
—— 1931. The maturation divisions and segregation of heteromorphic homologous chromosomes in Acrididae (Orthoptera). B.B., 61:324-349.
Catcheside, D. G. 1932. The chromosomes of a new haploid Oenothera. C., 4:68-113.
*di Cesnola, A. P. 1904. Preliminary note on the protective value of color in *Mantis religiosa*. Biometrica, 3:58-59.

Chipman, R. H., and T. H. Goodspeed. 1927. Inheritance in *Nicotiana tabacum* VIII. Cytological features of purpurea haploid. U.C.P.B., 11: 141-158.

Chopard, L. and R. Bellecroix. 1928. Dimorphism alaire chez les Gryllides; répartition géographique des forms macroptères et brachyptères. Bull. Biol. France Belgique, 62:157-163.

Clausen, J. 1927. Chromosome number and the relationship of species in the genus Viola. Ann. Bot., 41:677-714.

———— 1931a. Cyto-genetic and taxonomic investigations on Melanium violets. H., 15:219-308.

———— 1931b. Genetic studies on Polemonium. III. Preliminary account on the cytology of species and species hybrids. H., 15:62-66.

———— 1933. Cytological evidence for the hybrid origin of *Penstemon neotericus Keck*. H., 18:65-76.

Clausen, R. E. 1928a. Interspecific hybridization in Nicotiana. VII. The cytology of hybrids of the synthetic species, *digluta*, with its parents, *glutinosa* and *tabacum*. U.C.P.B., 11:177-211.

———— 1928b. Interspecific hybridization and the origin of species in Nicotiana. V.K.V., 1:547-553.

———— and T. H. Goodspeed. 1925. Interspecific hybridization in Nicotiana. II. A tetraploid *glutinosa-tabacum* hybrid, an experimental verification of Winge's hypothesis. G., 10:278-284.

Cleland, R. E., and A. F. Blakeslee. 1931. Segmental interchange, the basis of chromosomal attachments in Oenothera. C., 2:175-233.

Collins, J. L., L. Hollingshead and P. Avery. 1929. Interspecific hybrids in Crepis. III. Constant fertile forms containing chromosomes derived from two species. G., 14:305-320.

*Correns, C. 1902. Über Bastardirungsversuche mit Mirabilis-Sippen. Ber. Deutsch. Bot. Ges., 20:549-608.

———— 1928. Über nichtmendelnde Vererbung. V.K.V., 1:131-168.

Cousin, G. 1934. Sur la fécondité normale et les caractères des hybrides issues du croisement de deux éspèces de Gryllides. C.R. Acad. Sci. Paris, 198:853-855.

Crampton, H. E. 1916. Studies on the variation, distribution, and evolution of the genus Partula. The species inhabiting Tahiti. Carnegie Inst. Washington, Publ., 228:1-311.

———— 1932. Studies on the variation, distribution, and evolution of the genus Partula. The species inhabiting Moorea. Carnegie Inst. Washington, Publ. 410:1-335.

Cretschmar, M. 1928. Das Verhalten der Chromosomen bei der Spermatogenese von *Orgyia thyellina* Btl. und *antiqua* L., sowie eines ihrer Bastarde. Z.Z.m.A., 7:290-399.

Crew, F. A. E., and P. C. Koller. 1936. Genetical and cytological studies of the intergeneric hybrid of *Cairina moschata* and *Anas platyrhyncha platyrhyncha*. Proc. R. Soc. Edinburgh, 56, III:210-241.

Crew, F. A. E., and R. Lamy. 1935. Linkage groups in *Drosophila pseudoobscura*. J.G., 30:15-29.

Crow, W. B. 1924. Variation and species in Cyanophyceae. J.G., 14:397-424.

Cuénot, L. 1933. La seiche commune de la Méditerranée; étude sur la naissance d'une éspèce. Arch. Zool. Exp. Gén. 75:319-330.

Darlington, C. D. 1928. Studies in Prunus, I and II. J.G., 19:213-256.

——— 1929a. Chromosome behaviour and structural hybridity in the Tradescantiae. J.G., 21:207-286.

———1929b. Ring-formation in Oenothera and other genera. J.G., 20:345-363.

———1931. Meiosis in diploid and tetraploid *Primula sinensis*. J.G., 24:65-96.

——— 1932a. Recent advances in cytology. Blakiston's, Philadelphia.

——— 1932b. The control of the chromosomes by the genotype and its bearing on some evolutionary problems. A.N., 66:25-51.

——— 1936. Crossing over and its mechanical relationships in Chorthippus and Stauroderus. J.G., 33:465-500.

De Buck, A. 1935. Beitrag zur Rassenfrage bei *Culex pipiens*. Zeits. angew. Entomol., 22:242-252.

——— E. Schoute, and N. H. Swellengrebel. 1934. Cross-breeding experiments with Dutch and foreign races of *Anopheles maculipennis*. Riv. Malariologia, 13:237-263.

——— and N. H. Swellengrebel. 1931. Das Vorkommen von zwei verschiedenen Rassen des *Anopheles maculipennis*, als Erklärung des Anophelismus sine Malaria in Niederland. Zool. Anz., Suppl., 5:225-230.

——— G. v. d. Torren, and N. H. Swellengrebel. 1933. Report for the year 1932 on investigations into the racial composition of *Anopheles maculipennis* in Holland. Riv. Malariol., 12:265-280.

Demerec, M. 1929a. Genetic factors stimulating mutability of the miniature-gamma wing character of *Drosophila virilis*. P.N.A.S., 15:834-838.

——— 1929b. Cross sterility in maize. Z.i.A.V., 50:281-291.

——— 1933. What is a gene? J.H., 24:369-378.

——— 1934. Biological action of small deficiencies of X-chromosome of *Drosophila melanogaster*. P.N.A.S., 20:354-359.

——— 1935. Unstable genes. Bot. Reviews, 1:233-248.

——— 1937. A mutability stimulating factor in the Florida stock of *Drosophila melanogaster*. G., 22:190.

——— and M. E. Hoover. 1936. Three related X-chromosome deficiencies in Drosophila. J.H., 27:207-212.

Detlefsen, J. A. 1914. Genetic studies on a cavy species cross. Carnegie Inst. Washington, Publ. 205:5-134.

Dice, L. C. 1931. The occurrence of two subspecies of the same species in the same area. J. Mammalogy, 12:210-213.

———— 1933. Fertility relationships between some of the species and subspecies of mice in the genus Peromyscus. J. Mammalogy, 14:298-305.

Digby, L. 1912. The cytology of Primula kewensis and of other related Primula hybrids. Ann. Bot., 26:357-388.

Diver, C., A. E. Boycott, and S. Garstang. 1925. The inheritance of inverse symmetry in *Limnaea peregra*. J.G., 15:113-200.

Dobzhansky, Th. 1927. Studies on manifold effect of certain genes in *Drosophila melanogaster*. Z.i.A.V., 43:330-388.

———— 1930a. The manifold effects of the genes Stubble and stubbloid in *Drosophila melanogaster*. Z.i.A.V., 54:427-457.

———— 1930b. Translocations involving the third and the fourth chromosomes of *Drosophila melanogaster*. G., 15:347-399.

———— 1931. The decrease of crossing over observed in translocations, and its probable explanation. A.N., 65:214-232.

———— 1932. Studies on chromosome conjugation. I. Translocations involving the second and the Y-chromosome of *Drosophila melanogaster*. Z.i.A.V., 60:235-286.

———— 1933a. Studies on chromosome conjugation. II. The relation between crossing over and disjunction of chromosomes. Z.i.A.V., 64:269-309.

———— 1933b. On the sterility of the interracial hybrids in *Drosophila pseudoobscura*. P.N.A.S., 19:397-403.

———— 1933c. Geographical variation in lady-beetles. A.N., 67:97-126.

———— 1933d. Role of the autosomes in the *Drosophila pseudoobscura* hybrids. P.N.A.S., 11:950-953.

———— 1934a. Studies on chromosome conjugation. III. Behavior of duplicating fragments. Z.i.A.V., 68:134-162.

———— 1934b. Studies on hybrid sterility I. Spermatogenesis in pure and hybrid *Drosophila pseudoobscura*. Z.Z.m.A., 21:169-223.

———— 1935a. *Drosophila miranda*, a new species. G., 21:377-391.

———— 1935b. The Y-chromosome of *Drosophila pseudoobscura*. G., 20:366-376.

———— 1935c. Maternal effects as a cause of the difference between the reciprocal crosses in *Drosophila pseudoobscura*. P.N.A.S., 21:443-446.

———— 1935d. Fecundity in *Drosophila pseudoobscura* at different temperatures. J.E.Z., 71:449-464.

———— 1935e. A critique of the species concept in biology. Philosophy of Science, 2:344-355.

————1936a. Induced chromosomal aberrations in animals. Biol. Effects of Radiation, 2:1167-1208.

———— 1936b. Studies on hybrid sterility II. Localization of sterility factors in *Drosophila pseudoobscura* hybrids. G., 21:113-135.

———— 1936c. Position effects of genes. Biol. Reviews, 11:364-384.

Dobzhansky, Th. 1936d. L'effet de position et la théorie de l'hérédité. Hermann, Paris.

———— 1936e. The persistence of the chromosome pattern in successive cell divisions in *Drosophila pseudoobscura*. J.E.Z., 74:119-135.

———— 1937a. Genetic nature of species differences. A.N., 71:404-420.

———— 1937b. Further data on *Drosophila miranda* and its hybrids with *Drosophila pseudoobscura*. J.G., 34:135-151.

———— 1937c. Further data on the variation of the Y-chromosome in *Drosophila pseudoobscura*. G., 22:340-346.

———— and G. W. Beadle. 1936. Studies on hybrid sterility IV. Transplanted testes in *Drosophila pseudoobscura*. G., 21:832-840.

———— and R. D. Boche. 1933. Intersterile races of *Drosophila pseudoobscura* Frol. B.Z., 54:314-330.

———— and A. H. Sturtevant. 1931. Translocations between the second and third chromosomes of Drosophila and their bearing on Oenothera problems. Carnegie Inst. Washington, Publ., 421:29-59.

———— 1935. Further data on maternal effects in *Drosophila pseudoobscura* hybrids. P.N.A.S., 21:566-570.

———— and C. C. Tan. 1936. Studies on hybrid sterility III. A comparison of the gene arrangement in two species, *Drosophila pseudoobscura* and *Drosophila miranda*. Z.i.A.V., 72:88-114.

Dodge, B. O. 1936. Reproduction and inheritance in Ascomycetes. S., 83: 169-175.

Donald, H. P. 1936. On the genetical constitution of *Drosophila pseudoobscura*, race A. J.G., 33:103-122.

Dubinin, N. P. 1930. On the origin of deleted X-chromosomes. J. Exp. Biol. (Russian), 6:365-368.

———— 1931. Genetico-automatical processes and their bearing on the mechanism of organic evolution. J. Exp. Biol. (Russian), 7:463-479.

———— 1934. Experimental reduction of the number of chromosome pairs in *Drosophila melanogaster*. B.Zh., 3:719-736.

———— 1936. Experimental alteration of the number of chromosome pairs in *Drosophila melanogaster*. B.Zh., 5:833-850.

———— and D. D. Romaschoff. 1932. Die genetische Struktur der Art und ihre Evolution. B.Zh., 1:52-95.

———— and B. N. Sidorov. 1935. The position effect of the hairy gene. B.Zh., 4:555-568.

———— N. N. Sokolov, and G. G. Tiniakov. 1936. Occurrence and distribution of chromosome aberrations in nature. Nature, 138:1035-1036.

———— N. N. Sokolov, G. G. Tiniakov, and W. W. Sacharov. 1935. On the problem of chromosome conjugation. B.Zh., 4:175-204.

———— and fourteen collaborators. 1934. Experimental study of the ecogenotypes of *Drosophila melanogaster*. B.Zh., 3:166-216.

Dubovskij, N. V. 1935. On the question of the comparative mutability of stocks of *Drosophila melanogaster* of different origin. C.R. Acad. Sci. U.R.S.S., 4:95-97.

Dunn, L. C. 1921. Unit character variation in rodents. J. Mammalogy, 2: 125-140.

Du Rietz, G. E. 1930. The fundamental units of botanical taxonomy. Svensk. Bot. Tidskrift, 24:333-428.

East, E. M. 1916. Inheritance in crosses between *Nicotiana langsdorfii* and *Nicotiana alata*. G., 1:311-333.

–––––– 1921. A study of partial sterility in certain hybrids. G., 6:311-365.

Eisentraut, M. 1934. Markierungsversuche bei Fledermäusen. Zeits. Morph. Ökol. Tiere, 28:553-560.

Elton, C. S. 1924. Periodic fluctuations in the numbers of animals: their causes and effects. Brit. J. Exp. Biol., 2:119-163.

––––––1927. Animal ecology. Macmillan, New York.

Emerson, S. 1929. The reduction division in a haploid Oenothera. La Cellule, 39:159-165.

–––––– and A. H. Sturtevant. 1931. Genetic and cytological studies on Oenothera. III. The translocation interpretation. Z.i.A.V., 59:395-419.

Ephrussi, B., et G. W. Beadle. 1935. La transplantation des ovaires chez la Drosophile. Bull. Biol. France, 69:492-502.

Federley, H. 1913. Das Verhalten der Chromosomen bei der Spermatogenese der Schmetterlinge *Pygaera anachoreta*, *curtula* und *pigra* sowie einiger ihrer Bastarde. Z.i.A.V., 9:1-110.

–––––– 1914. Ein Beitrag zur Kenntnis der Spermatogenese bei Mischlingen zwischen Eltern verschiedener systematischer Verwandtschaft. Öfv. Finska Veten. Soc. Förhandl., 56:1-28.

–––––– 1915a. Chromosomenstudien an Mischlingen. I. Die Chromosomenkonjugation bei der Gametogenese von *Smerinthus populi* var. *austauti* × *populi*. Öfv. Finska Veten. Soc. Förhandl., 57, No. 26:1-36.

–––––– 1915b. Chromosomenstudien an Mischlingen. II. Die Spermatogenese des Bastards *Dicranura erminea* ♀ × *D. vinula* ♂. Öfv. Finska Veten. Soc. Förhandl., 57, No. 30:1-26.

–––––– 1916. Chromosomenstudien an Mischlingen. III. Die Spermatogenese des Bastards *Chaerocampa porcellus* ♀ × *elpenor* ♂. Öfv. Finska Veten. Soc. Förhandl., 58, No. 12:1-17.

–––––– 1928. Das Inzuchtsproblem. Handbuch der Vererbungswissenschaft, Borntraeger, Berlin.

–––––– 1929a. Über subletale und disharmonische Chromosomenkombinationen. H., 12:271-293.

–––––– 1929b. Metoden zur Erforschung der Vererbung bei den Lepidopteren. Abderhalden Handb. biol. Arbeitsmethoden, Abt. 9, 3:637-390.

–––––– 1931. Chromosomenanalyse der reziproken Bastarde zwischen *Pygaera pigra* und *P. curtula* sowie ihrer Rückkreuzungsbastarde. Z.Z.m.A., 12:772-816.

Federley, H. 1932. Die Bedeutung der Kreuzung für die Evolution. Jena-ische Zeits. Naturwiss., 67:364-386.

Fisher, R. A. 1922. On the dominance ratio. Proc. R. Soc. Edinburgh, 42: 321-341.

———— 1928. The possible modification of the response of the wild type to recurrent mutations. A.N., 62:115-126.

———— 1930. The genetical theory of natural selection. Clarendon Press, Oxford.

———— 1931. The evolution of dominance. Biol. Reviews, 6:345-368.

———— 1932. The evolutionary modification of genetic phenomena. P.VI.C.G., 1:165-172.

———— 1936. The measurement of selective intensity. Proc. Royal Soc. London, Serie B, 121:58-62.

Ford, E. B. 1930. The theory of dominance. A.N., 64:560-566.

Franz, V. 1928. Über Bastardpopulation in der Gattung Paludina (recte: Viviparus). B.Z., 48:79-93.

Friesen, H. 1936. Röntgenomorphosen bei Drosophila. A.E., 134:147-165.

Frolova, S. L. 1936. Several spontaneous chromosome aberrations in Dro-sophila. Nature, 138:204-205.

Gaines, E. F., and H. C. Aase. 1926. A haploid wheat plant. Amer. J. Bot., 13:373-385.

Gairdner, A. E. and C. D. Darlington. 1931. Ring formation in diploid and polyploid Campanula persicifolia. Genetica, 13:113-150.

Gaisinovich. 1928. A study on the phenomenon of malelessness in Drosophila falerata Meig. J. Exp. Biol. (Russian), 4:233-250.

Galtsoff, P. S. 1930. The role of chemical stimulation in the spawning reac-tions of Ostrea virginica and Ostrea gigas. P.N.A.S., 16:555-559.

Gershenson, S. 1934. Mutant genes in a wild population of Drosophila ob-scura Fall. A.N., 68:569-571.

Glass, H. B. 1935. A study of factors influencing chromosomal segregation in translocations of Drosophila melanogaster. Univ. Missouri Agric. Exp. Sta., Res. Bull., 231:1-28.

Godlewski, E. 1926. L'inhibition réciproque de l'aptitude à féconder de spermes d'espèces éloignées comme conséquence de l'agglutination des spermatozoides. Arch. Biologie, 36:311-350.

Goldschmidt, R. 1921. Erblichkeitsstudien an Schmetterlingen III. Der Melanismus der Nonne, Lymantria monacha L. Z.i.A.V., 25:89-163.

———— 1924. Erblichkeitsstudien an Schmetterlingen IV. Weitere Unter-suchungen über die Vererbung des Melanismus. Z.i.A.V., 34:229-244.

———— 1929a. Untersuchungen zur Genetik der geographischen Variation II. A.E., 116:136-201.

———— 1929b. Experimentelle Mutationen und das Problem der sogenann-ten Parallel-Induktion. B.Z., 49:437-448.

———— 1931. Die sexuellen Zwischenstufen. J. Springer, Berlin.

Goldschmidt, R. 1932a. Untersuchungen zur Genetik der geographischen Variation III. Abschliessendes über die Geschlechtsrassen von *Lymantria dispar.* A.E., 126:277-324.

——— 1932b. Untersuchungen zur Genetik der geographischen Variation. IV. Cytologisches. A.E., 126:591-612.

———1932c. Untersuchungen zur Genetik der geographischen Variation. V. Analyse der Überwinterungszeit als Anpassungscharakter. A.E., 126: 674-768.

——— 1933a. Untersuchungen zur Genetik der geographischen Variation. VI. Die geographische Variation der Entwicklungsgeschwindigkeit und des Grössenwachstums. A.E., 130:266-339.

——— 1933b. Untersuchungen zur Genetik der geographischen Variation. VII. A.E., 130:562-615.

——— 1933c. Some aspects of evolution. S., 78:539-547.

——— 1934a. Die Genetik der geographischen Variation. P.VI.C.G., 1: 173-184.

——— 1934b. Lymantria. Bibliogr. Genetica, 11:1-186.

——— 1935. Gen und Ausseneigenschaft. Z.i.A.V., 69:38-131.

——— J. Seiler, and H. Poppelbaum. 1924. Untersuchungen sur Genetik der geographischen Variation I. A.E., 101:92-337.

Goodspeed, T. H., and R. E. Clausen. 1928. Interspecific hybridization in Nicotiana. VIII. The *sylvestris-tomentosa-tabacum* hybrid triangle and its bearing on the origin of tabacum. U.C.P.B., 11:245-256.

Gordon, C. 1936. The frequency of heterozygosis in free-living populations of *Drosophila subobscura.* J.G., 33:25-60.

Gowen, J. W. 1931. Genetic non-disjunctional forms in Drosophila. A.N., 65:193-213.

Gregor, J. W., and F. W. Sansome. 1930. Genetics of wild populations. II. *Phleum pratense L.* × *P. Alpinum L.* J.G., 22:373-387.

Gregory, R. P. 1914. On the genetics of tetraploid plants in *Primula sinensis.* Proc. Roy. Soc., B, 87:484-492.

Gross, F. 1932. Untersuchungen über die Polyploidy und die Variabilität bei Artemia salina. Naturwissenschaften, 20:962-967.

Gulick, J. T., 1905. Evolution, racial and habitudinal. Carnegie Inst. Washington, Publ., 25:1-269.

Guyénot, E., and Duszynska-Wietrzykowska. 1935. Stérilité et virilisme chez des femelles de cobayes issues d'un croisment interspécifique. Revue Suisse Zool., 42:341-388.

Haase-Bessel, G. 1921. Digitalisstudien II. Z.i.A.V., 27:1-26.

Hackett, L. W. 1934. The present status of our knowledge of the sub-species of *Anopheles maculipennis.* Trans. Roy. Soc. Tropical Med. Hyg., 28: 109-128.

——— E. Martini, and A. Missiroli. The races of A. maculipennis. Amer. Journ. Hygiene, 16:137-162.

Hagedoorn, A. L., and A. C. Hagedoorn. 1921. The relative value of the processes causing evolution. Martius Nijhoff, The Hague.

Hagerup, O. 1931. Über Polyploidy in Beziehung zu Klima, Ökologie und Phylogenie. H., 16:19-40.

Håkansson, A. 1929a. Chromosomenringe in Pisum und ihre mutmässliche genetische Bedeutung. H., 12:1-10.

—————— 1929b. Die Chromosomen in der Kreuzung *Salix viminalis* × *caprea* von Heribert Nilsson. H., 12:1-52.

—————— 1931a. Über Chromosmenverkettung in Pisum. H., 15:17-61.

—————— 1931b. Chromosomenverkettung bei Godetia und Clarkia. Ber. Deut. Bot. Ges., 49:228-234.

—————— 1933. Die Konjugation der Chromosomen bei einigen Salix-Bastarden. H., 18:199-214.

—————— 1934. Chromosomenbindungen in einigen Kreuzungen zwischen halbsterilen Erbsen. H., 19:341-358.

—————— 1935. Die Reduktionsteilung in einigen Artbastarden von Pisum. H., 21:215-222.

Haldane, J. B. S. 1922. Sex-ratio and unisexual sterility in hybrid animals. J.G., 12:101-109.

—————— *1924-1932. A mathematical theory of natural and artificial selection. Proc. Cambridge Phil. Soc., 23:19-41, 158-163, 363-372, 607-615, 838-844; 26:220-230; 27:131-142; 28:244-248.

—————— 1930. A note on Fisher's theory of the origin of dominance, and on a correlation between dominance and linkage. A.N., 64:87-90.

—————— 1932. The causes of evolution. Harper & Bros., New York and London.

Hardy, G. H. 1908. Mendelian proportions in a mixed population. S., 28:49-50.

Harland, S. C. 1932a. The genetics of Gossypium. Bibliogr. Genetica, 9:107-182.

—————— 1932b. The genetics of cotton. V. Reversal of dominance in the interspecific cross *G. Barbadense* Linn. *G. hirsutum* Linn. and its bearing on Fisher's theory of dominance. J.G., 25:261-270.

—————— 1933. The genetics of cotton. IX. Further experiments on the inheritance of the crinkled dwarf mutant of *G. Barbadense* L. in interspecific crosses and their bearing on the Fisher's theory of dominance. J.G., 28:315-325.

—————— 1935. The genetics of cotton. XII. Homologous genes for anthocyanin pigmentation in new and old world cotton. J.G., 30:465-476.

—————— 1936. The genetical conception of the species. Biol. Reviews, 11:83-112.

Harrison, J. W. H. 1920. Genetical studies in the moths of the geometrid genus Oporabia (Oporinia) with a special consideration of melanism in the Lepidoptera. J.G., 9:195-280.

Harrison, J. W. H., and F. Garrett. 1926. The induction of melanism in the Lepidoptera. Proc. Royal Soc. London, B, 99:241-263.

Hartmann, M. 1933. Die methodologischen Grundlagen der Biologie. Felix Meiner, Leipzig.

Hasebroek, K. 1934. Industrie und Grosstadt als Ursache des neuzeitlichen vererblichen Melanismus der Schmetterlinge in England und Deutschland. Zool. Jahrb., alg. Zool. Phys. 53:411-460.

Heberer, G. 1924. Die Spermatogenese der Copepoden. I. Die Spermatogenese der Centropagiden nebst Anhang über die Oogenes von *Diaptomus castor*. Z. Wiss. Zool., 123:555-646.

Heikertinger, F. 1933-36. Kritik der Schmetterlingsmimikry I-V. B.Z., 53:561-590, 54:365-389, 55:461-483, 56:151-166, 463-494.

Heincke, F. 1898. Die Naturgeschichte des Herings. I. Abh. Deut. Seefischerei Vereins, 2:1-178.

Heitz, E. 1933. Die somatische Heteropyknose bei *Drosophila melanogaster* und ihre genetische Bedeutung. Z.Z.m.A., 20:237-287.

Heitz, E., und H. Bauer. 1933. Beweise für die Chromosomennatur der Kernschleifen in den Knäuelkernen von *Bibio hortulanus*. Z.Z.m.A., 17:67-83.

Helwig, E. R. 1929. Chromosomal variations correlated with geographical distribution in *Circotettix verruculatus* (Orthoptera). J. Morphology, 47:1-36.

*Heribert-Nilsson, N. 1918. Experimentelle Studien über Variabilitat, Spaltung, Artbildung und Evolution in der Gattung Salix. Lund Univr. Aarskr. 14:1-145.

Hertwig, P. 1936. Artbastarde bei Tieren. Handbuch Vererbungswiss., 21:1-140.

Hollingshead, L. 1930a. A lethal factor in Crepis effective only in interspecific hybrid. G., 15:114-140.

—— 1930b. Cytological investigations of hybrids and hybrid derivatives of *Crepis capillaris* and *Crepis tectorum*. Univ. California Publ. Agr. Sci., 6:55-94.

Honing, J. A. 1923. Canna crosses I. Mededeelingen Landbouwhoofeschool Wageningen, 26:1-56.

—— 1928. Canna crosses II. Mededeelingen Landbouwhoofeschool Wageningen, 32:1-14.

Huskins, C. L. 1931. The origin of Spartina townsendii. Genetica 12:531-538.

—— and E. M. Hearne. 1933. Meiosis in asynaptic dwarf oats and wheat. J. Royal Micr. Soc., 53:109-117.

—— and S. G. Smith. 1934. A cytological study of the genus *Sorghum Ters*. II. The meiotic chromosomes. J.G., 28:387-395.

Hutchinson, J. B. 1934. The genetics of cotton. X. The inheritance of leaf shape in Asiatic Gossypiums. J.G., 28:437-513.

Irwin, M. R., and L. J. Cole. 1936a. Immunogenetic studies of species and species hybrids in doves and the separation of species-specific substances in the back-cross. J.E.Z., 73:85-108.

———— 1936b. Immunogenetic studies of species and species hybrids from the cross *Columba livia* and *Streptopelia risoria*. J.E.Z., 73:309-318.

———— L. J. Cole, and C. D. Gordon. 1936. Immunogenetic studies of species and species hybrids in pigeons, and the separation of species-specific characters in back-cross generations. J.E.Z., 73:285-308.

Jenkin, T. J. 1933. Interspecific and intergeneric hybrids in herbage grasses. J.G., 28:205-264.

Jollos, V. 1931. Genetik und Evolutionsproblem. Verh. Deuts. Zool. Ges., 252-295.

———— 1934. Inherited changes produced by heat-treatment in *Drosophila melanogaster*. Genetica, 16:476-494.

———— 1935. Studien zum Evolutionsproblem. B.Z., 55:390-436.

Jones, F. M., 1932. Insect coloration and the relative acceptability of insects to birds. Trans. Entom. Soc. London, 80:345-385.

Jordan, D. S. 1905. The origin of species through isolation. S., 22:545-562.

*Jordan, K. 1905. Der Gegensatz zwischen geographischer und nichtgeographischer Variation. Zeits. wiss. Zool., 83.

Jorgensen, C. A. 1928. The experimental formation of heteroploid plants in the genus Solanum. J.G., 19:133-211.

———— and M. B. Crane. 1927. Formation and morphology of *Solanum chimaeras*. J.G., 18:247-273.

Karpechenko, G. D. 1927a. The production of polyploid gametes in hybrids. H., 9:349-368.

———— 1927b. Polyploid hybrids of *Raphanus sativus* L. × *Brassica oleracea* L. Bull. Appl. Botany, 17:305-408.

———— 1928. Polyploid hybrids of *Raphanus sativus* L. × *Brassica oleracea* L. Z.i.A.V., 48:1-85.

———— 1935. Theory of remote hybridization. Moscow-Leningrad.

———— and Shchavinskaia. 1929. On sexual incompatibility of tetraploid hybrids *Raphanus Brassica*. Proc. U.S.S.R. Congr. Genetics, 2:267-276.

Katterman, G. 1931. Über die Bildung palyvalenter Chromosomenverbände bei einigen Gramineen. Planta, 12:732-744.

Kaufmann, B. P. 1936. A terminal inversion in *Drosophila ananassae*. P.N.A.S., 22:591-594.

———— 1936b. The chromosomes of *Drosophila ananassae*. S., 83:39.

Kawaguchi, E. 1928. Zytologische Untersuchungen am Seidenspinner und seine Verwandten. Z.Z.m.A., 7:519-552.

Kerkis, J. 1931. Vergleichende Studien über die Variabilität der Merkmale des Geschlechtsapparats und der äusseren Merkmale bei Eurygaster integriceps Put. Zool. Anz., 93:129-143.

Kerkis, J. 1933. Development of gonads in hybrids between *Drosophila melanogaster* and *Drosophila simulans*. J.E.Z., 66:477-509.

―――― 1936. Chromosome configuration in hybrids between *Drosophila melanogaster* and *Drosophila simulans*. A.N., 70:81-86.

Kihara, H. 1919. Über cytologische Studien bei einigen Getreidearten I. Bot. Mag. Tokyo., 32:17-38.

―――― 1924. Cytologische und genetische Studien bei wichtigen Getreidearten mit besonderer Rücksicht auf das Verhalten der Chromosomen und die Sterilität in den Bastarden. Mem. Coll. Sci. Kyoto Imp. Univ. 1:1-200.

―――― and F. Lilienfeld. 1932. Genomanalyse bei Triticum and Aegilops IV. C., 3:384-456.

―――― 1935. Genomanalyse bei Triticum und Aegilops VI. C., 6:195-216.

―――― and J. Nishiyama. 1930. Genomanalyse bei Triticum und Aegilops I. C., 1:263-284.

Kinsey, A. C. 1936. The origin of higher categories in Cynips. Indiana Univ. Publ., Science Series, 4:1-334.

―――― 1937. An evolutionary analysis of insular and continental species. P.N.A.S., 23:5-11.

Knuth, P., and J. R. Ainsworth Davis. 1906-9. Handbook of flower pollination. Clarendon, Oxford, 3 volumes.

Koller, P. Ch. 1932. The relation of fertility factors to crossing over in the *Drosophila obscura* hybrids. Z.i.A.V., 60:137-151.

―――― 1936. Structural hybridity in *Drosophila pseudoobscura*. J.G., 32:79-102.

Kosharin, Th. S. and G. W. Samochwalowa. 1933. Einige Blutelemente bei der Hybridierung des Jaks mit dem örtlichen Hornvieh. B.Zh., 3:513-532.

Kosswig, G. 1929a. Über die veränderte Wirkung von Farbgenen des Platypoecilus in der Gattungskreuzung mit Xiphophorus. Z.i.A.V., 50:63-73.

―――― 1929b. Zur Frage der Geschwulstbildung bei Gattungsbastarden der Zahnkarpfen Xiphophorus und Platypoecilus. Z.i.A.V., 52:114-120.

Kostoff, D. 1936. Polyploid hybrids *Nicotiana rustica var. texana* L. × *Nicotiana glauca Grah.* Bull. Appl. Bot., Ser. 2, 9:153-162.

―――― and I. A. Axamitnaja. 1935. Studies on polyploid plants. C. R. Acad. Aci. URSS, 1:325-329; 2:293-297.

Kozhevnikov, B. Th. 1933. Partial non-homology of the sex chromosomes in *Drosophila melanogaster* and *Drosophila simulans*. B.Zh., 3:585-601.

―――― 1936. Experimentally produced karyotypical isolation. B.Zh., 5:727-752.

Krumbiegel, I. 1932. Untersuchungen über physiologische Rassenbildung. Zool. Jahrbücher, Syst., 63:183-280.

Kühn., 1932. Entwicklungsphysiologische Wirkung einiger Gene von Ephestia kuhniella. Naturwissenschaften, 20:947-977.

Laibach, F. 1925. Das Taubwerden von Bastardsamen und die Künstliche

Aufzucht früh absterbender Bastardembryonen. Zeits. Botanik, 17:417-459.

Lamm, R. 1936. Cytological studies on inbred rye. H., 22:217-240.

Lammerts, W. E. 1931. Interspecific hybridization in Nicotiana XII. The amphidiploid *rustica-paniculata* hybrid; the origin and cytogenetic behavior. G., 16:191-211.

―――― 1934a. Derivative types obtained by back-crossing *Nicotiana rustica-paniculata* to *N. paniculata*. J.G., 29:355-366.

―――― 1934b. On the nature of chromosome association in *Nicotiana tabacum* haploid. C., 6:38-50.

Lancefield, D. E. 1929. A genetic study of two races or physiological species in *Drosophila obscura*. Z.i.A.V., 52:287-317.

*Lehmann, E., and J. Schwemmle. 1927. Genetische Untersuchungen in der Gattung Epilobium. Bibl. Bot., 95.

Leiner, M. 1934. Die drei europäische Stichlinge (*Gasterosteus aculeatus L., Gasterosteus pungitius L.* and *Gasterosteus spinachia L.*) und ihre Kreuzungsprodukte. Z. Morph. Ökol. Tiere, 28:107-154.

Lenz, F. 1928. Ein weiterer mendelnder Artbastard *Epicnaptera tremulifolia* × *ilicifolia*. V.K.V., 2:984-986.

Lesley, M. M., and J. W. Lesley. 1930. The mode of origin and chromosome behavior in pollen mother cells of a tetraploid seedling tomato. J.G., 22:419-425.

Levan, A. 1935a. Cytological studies in Allium VI. The chromosome morphology of some diploid species of Allium. H. 20:289-330.

―――― 1935b. Die Zytologie von *Allium cepa* × *fistulosum*. H., 21:195-214.

Lilienfeld, F., and H. Kihara. 1934. Genomanalyse bei Triticum and Aegilops. C., 6:87-122.

Liljefors, A. 1936. Zytologische Studien über den F₁ Bastard *Triticum turgidum* × *Secale cereale*. H., 21:240-262.

Lillie, F. R. 1921. Studies of fertilization. VIII. On the measure of specificity in fertilization between two associated species of the sea-urchin genus Strongilocentrotus. B.B., 40:1-22.

Ljungdahl, H. 1924. Über die Herkunft der in der Meiosis konjugierenden Chromosomen bei Papaver-Hybriden. Svensk Bot. Tidsk., 18:279-291.

Lobashov, M. 1935 Über die Wirkung der Asfiktion auf den Mutationsprozess bei *Drosophila melanogaster*. Bull. Soc. Natur. Leningrad, 63:371-378.

―――― and F. Smirnov. 1934. On the nature of the action of chemical agents on the mutational process in *Drosophila melanogaster*. II. The effect of ammonia on the occurrence of lethal transgenations. C.R. Acad. Sci. U.S.S.R., 3:174-176.

Lotsy, J. P. 1911. Hybrides entre espèces d'Antirrhinum. C.R. IV Confér. Internat. Génétique: 416-428.

Lotsy, J. P. 1916. Evolution by means of hybridization. M. Nijhoff, Hague.
———— 1931. On the species of the taxonomist in its relation to evolution. Genetica, 13:1-16.

Lutkow, A. N. 1930. Interspecific hybrids of *Pisum humile Boiss.* × *Pisum sativum L.* Proc. U.S.S.R. Congr. Genetics. 2:353-367.

Lutz, F. E. 1911. Experiments with *Drosophila ampelophila* concerning evolution. Carnegie Inst. Washington, Publ. 143:1-35.

Lynch, C. J. 1919. An analysis of certain cases of intraspecific sterility. G., 4:501-533.

McAtee, W. L. 1932. Effectiveness in nature of the so-called protective adaptations in the animal kingdom, chiefly as illustrated by the food habits of Nearctic birds. Smithsonian Misc. Coll., 85:1-201.

McClintock, B. 1931. A cytological demonstration of the location of an interchange between the non-homologous chromosomes of *Zea mays.* P.N.A.S., 16:791-796.

———— 1932. Cytological observations in Zea on the intimate association of non-homologous parts of chromosomes in the mid-prophase of meiosis and its relation to diakinesis configurations. P. VI. C.G., 2:126-128.

———— 1933. The association of non-homologous parts of chromosomes in the mid-prophase of meiosis in *Zea mays.* Z.Z.m.A., 19:191-237.

———— 1934. The relation of a particular chromosomal element to the development of the nucleoli in *Zea mays.* Z.Z.m.A., 21:294-328.

McClung, C. E. 1917. The multiple chromosomes of Hesperotettix and Mermiria. J. Morphology, 29:519-605.

McCray, F. A. 1933. Embryo development in Nicotiana species hybrids. G., 18:95-110.

Mangelsdorf, P. C., and D. F. Jones. 1926. The expression of Mendelian factors in the gametophyte of maize. G., 11:423-455.

Mann-Lesley, M., and H. B. Frost. 1927. Mendelian inheritance of chromosome shape in Mathiola. G., 12:449-460.

Manton, I. 1934. The problem of *Biscutella laevigata L.* Z.i.A.V., 67:41-57.

Marlatt, C. L. 1907. The periodical Cicada. U. S. Dept. Agric. Entom. Bull., 71:1-181.

Marshall, W. W., and H. J. Muller. 1917. The effect of long-continued heterozygosis on a variable character in Drosophila. J.E.Z., 22:457-470.

Mather, R. 1935. The behavior of meiotic chromosomes after X-radiation. H., 19:302-322.

———— 1935. Chromosome behavior in a triploid wheat hybrid. Z.Z.m.A., 23:117-138.

———— and L. H. A. Stone. 1933. The effect of X-radiation upon somatic chromosomes. J.G., 28:1-24.

Matsuura, H. 1933. A bibliographical monograph of plant genetics (genic analysis). Hokkaido university, Sapporo.

Mayr, E. 1932. Birds collected during the Whitney South Sea expedition. Amer. Museum. Novitates, 20:1-22; 21:1-23.

Meise, M. 1936. Zur Systematik und Verbreitungsgeschichte der Haus- und Weidensperlinge, Passer domesticus (L.) und hispaniolensis (T.). Jour. ornithologie, 84:631-672.

Meister, N., and N. A. Tjumjakoff. 1928. Rye-wheat hybrids from reciprocal crosses. J.G., 20:233-245.

Metz, C. W. 1914. Chromosome studies in Diptera I. J.E.Z., 17:45-56.

Meurman, O. 1928. Cytological studies in the genus Ribes L. H., 11: 289-356.

——— 1929. *Prunus laurocerasus L.*, a species showing high polyploidy. J.G., 21:85-94.

Michaelis, P. 1933. Entwicklungsgeschichtlich-genetische Untersuchungen an Epilobium II. Z.i.A.V., 65:1-71, 353-411.

Missiroli, A., L. W. Hackett, and E. Martini. 1933. Le razze di *Anopheles maculipennis* e la loro importanza nella distribuzione della malaria in alcune regioni d'Europa. Riv. Malariologia, 12:1-56.

Moenkhaus, W. J. 1910. Cross fertilization among fishes. Proc. Indiana Acad. Sci.: 353-393.

Muller, H. J. 1925. Why polyploidy is rarer in animals than in plants. A.N., 59:346-353.

——— 1928a. The problem of genic modification. V.K.V., 1:234-260.

——— 1928b. The measurement of gene mutation rate in Drosophila, its high variability, and its dependence upon temperature. G., 13:279-357.

——— 1935. A viable two-gene deficiency. J.H., 26:469-478.

——— 1936. On the variability of mixed races. A.N., 70:409-442.

——— and F. Settles. 1927. The non-functioning of the genes in spermatozoa. Z.i.A.V., 43:285-312.

——— A. A. Prokofyeva-Belgovskaja, and K. V. Kossikov. 1936. Unequal crossing-over in the bar mutant as a result of duplication of a minute chromosome section. C.R. Acad. Sci. U.R.S.S., 1(10):87-88.

Müntzing, A. 1930. Outlines to a genetic monograph of the genus Galeopsis. H., 13:185-341.

——— 1932. Cyto-genetic investigations on synthetic Galeopsis Tetrahit. H., 16:105-154.

——— 1933. Quadrivalent formation and aneuploidy in *Dactylis glomerata*. Bot. Notiser: 198-205.

——— 1934. Chromosome fragmentation in a Crepsis hybrid. H., 19: 284-302.

——— 1935a. Cyto-genetic studies on hybrids between two Phleum species. H., 20:103-136.

——— 1935b. Chromosome behavior in some Nicotiana hybrids. H., 20:251-272.

Müntzing, A. 1936. The evolutionary significance of autopolypoidy. H., 21:263-378.

Murie, A. 1933. The ecological relationship of two subspecies of Peromyscus in the Glacier Park region, Montana. Occ. Papers Mus. Zool., Univ. Michigan, 270:1-17.

Navashin, M. S. 1929. Über die Veränderung von Zahl und Form der Chromosomen infolge der Hybridization. Z.Z.m.A., 6:195-233.

———— 1934. Chromosome alterations caused by hybridization and their bearing upon certain genetic problems. C., 5:169-203.

*Nawashin, S. 1912. Sur le dimorphisme nucléaire des cellules somatiques de Galtonia candicans. Bull. Acad. Imp. Sci. Petersbourg, VI série, 4: 373-385.

———— 1927. Zellkerndimorphismus bei Galtonia candicans Des. und einigen verwandten Monokotylen. Ber. Deut. Bot. Ges., 45:415-428.

Newman, H. H. 1914. Modes of inheritance in teleost hybrids. J.E.Z., 16: 447-499. 1915. Development and heredity in heterogenic teleost hybrids. J.E.Z., 18:511-576.

Newton, W. C. F., and C. Pellew. 1929. Primula kewensis and its derivatives. J.G., 20:405-467.

Nikoro, Z., S. Gussev, E. Pavlov, and I. Griasnov. 1935. The regularities of sex isolation in some stocks of Drosophila melanogaster. B.Zh., 4:569-585.

Nilsson, F. 1934. Studies in fertility and inbreeding in some herbage grasses. H., 19:1-162.

———— 1935. Amphidiploidy in the hybrid Festuca arundinacea × gigantea. H., 20:181-198.

Nishiyama, J. 1934. The genetics and cytology of certain cereals. VI. Chromosome behavior and its bearing on inheritance in triploid Avena hybrids. Mem. College Agr. Kyoto, 32:1-157.

Noble, G. K. 1934. Experiments with the courtship of lizards. Natural History, 34:1-15.

Oliver, C. P. 1932. An analysis of the effect of varying the duration of X-ray treatment upon the frequency of mutation. Z.i.A.V., 61:447-488.

———— and E. W. Van Atta. 1932. Genetic and cytological correlation of chromosome aberrations of Drosophila. P.VI.C.G., 2:145-147.

*Ormancey, 1849. Récherches sur l'étui penial considéré comme limite de l'espèce dans les Coleoptères. Ann. Sci. Nat., Zool., ser. 3, 12.

Osborn, H. F. 1927. The origin of species V. Speciation and mutation. A.N., 61:5-42.

Painter, T. S. 1928. A comparison of the chromosomes of the rat and mouse with reference to the question of chromosome homology in mammals. G., 13:180-189.

———— 1934. A new method for the study of chromosome aberrations and the plotting of chromosome maps in Drosophila melanogaster. G., 19: 175-188.

Pariser, K. 1927. Die Zytologie und Morphologie der triploiden Intersexe des rückgekreuzten Bastards von *Saturnia pavonia L.* und *Saturnia pyri Schiff.* Z.Z.m.A., 5:415-447.

*Pascher. A. 1916. Über die Kreuzung einzelliger Haploider-organismen Chlamydomonas. Ber. Deut. Bot. Ges., 34:228-242.

Pätau, K. 1935. Chromosomenmorphologie bei *Drosophila melanogaster* und *Drosophila simulans* und ihre genetische Bedeutung. Naturwiss., 23: 537-543.

Patterson, J. T., and H. J. Muller. 1930. Are "progressive" mutations produced by X-rays? G., 15:495-578.

Pellew, C., and E. R. Sansome. 1932. Genetical and cytological studies on the relations between Asiatic and European varieties of *Pisum sativum.* J.G., 25:25-54.

Peto, F. H. 1933. The cytology of certain intergeneric hybrids between Festuca and Lolium. J.G., 28:113-156.

Petrov, S. G. 1936. The population of fowl near Shabalino. B.Zh., 5:57-78.

Philiptchenko, Jur. 1934. Genetics of soft wheats. Ogiz, Moscow-Leningrad.

Phillips, J. C. 1915. Experimental studies of hybridization among ducks and pheasants. J.E.Z., 18:69-143.

—————— 1921. A further report on species crosses in birds. G., 6:366-383.

Philp, J., and C. L. Huskins. 1931. The cytology of *Matthiola incana R.Br.* especially in relation to the inheritance of double flowers. J.G., 24:359-404.

Piaget, J. 1929. Les races lacustres de la *Limnaea stagnalis L.* Bull. Biol. France Belgique, 63:424-455.

Pictet, A. 1926. Localisation dans une région du Parc national Suisse, d'une race constante de papillons exclusivement composée d'hybrides. Revue Suisse Zool., 33:399-406.

—————— 1928a. Le déterminisme des proportions numériques entre les divers composants d'une population mixte de Lépidoptères. Revue Suisse Zool., 35:214-246.

—————— 1928b. Les conditions du déterminisme des proportions numérique entre les composants d'une population polymorphe de Lépidoptères. Revue Suisse Zool., 35:473-505.

—————— 1936. La zoogéographie expérimental dans ses rapports avec la génétique. Mem. Musée Hist. Natur. Belgique, Série 2, 3:233-282.

Pinney, E. 1918. A study of the relation of the behavior of the chromatin to development and heredity in teleost hybrids. J. Morphology, 31:225-291.

—————— 1922. The initial block to normal development in cross-fertilized eggs. J. Morphology, 36:401-419.

Poll, H. 1910. Über Vogelmischlinge. Ber. V. Internat. Ornithol. Kongr. :399-468.

Poll, H. 1920. Pfaumischlinge (Mischlingsstudien VIII). Arch. mikr. Anat. Festschrift Hertwig :365-458.

Poole, C. 1931. The interspecific hybrid, Crepis rubra × C. foetida, and some of its derivatives. I. Univ. Calif. Publ. Agric., 6:169-200.

—— 1932. The interspecif hybrid, Crepis rubra × C. foetida, and some of its derivatives. II. Univ. Calif. Publ. Agric., 6:231-255.

Poulson, D. F. 1937. Chromosomal deficiencies and the embryonic development of Drosophila melanogaster. P.N.A.S., 23:133-137.

Poulton, E. B. 1908. Essays on evolution. Clarendon, Oxford.

Pratt, H. S. 1935. A manual of the common invertebrate animals exclusive of insects. Blakiston's, Philadelphia.

Quayle, H. J. (in press). The development of resistance in certain scale insects to hydrocyanic acid. Hilgardia.

*Rabinerson, A. I. 1925. Investigations upon the natural history of the Murman herring (Clupea harengus). Bull. Bureau. Appl. Ichtyology, Leningrad, 3.

Rancken, C. 1934. Zytologische Untersuchungen an einigen wirtschaftlich wertvollen Wiesengräsern. Acta Agralia Fennica, 29:1-92.

Renner, O. 1929. Artbastarde bei Pflanzen. Borntraeger, Berlin.

—— 1934. Die pflanzlichen Plastiden als selbständige Elemente der genetischen Konstitution. Ber. Mat. Phys. Klasse Sachsischen Akad. Wiss., 86:241-266.

—— 1936. Zur Kenntnis der nichtmendelnden Buntheit der Laubblätter. Flora, 30:218-290.

Rensch, B. 1929. Das Prinzip geographischer Rassenkreise und das Problem der Artbildung. Borntraeger, Berlin.

—— 1936. Studien über klimatische Parallelität der Merkmalsausprägung bei Vogeln und Saugern. Arch. Naturgesch., N.F., 5:317-363.

Richardson, M. M. 1935. Meiosis in Crepis II. Failure of pairing in Crepis capillaris (L) Wallr. J.G., 31:119-143.

—— 1936. Structural hybridity in Lilium martagon album × L. hansonii. J.G., 32:411-450.

Robertson, W. R. B. 1915. Chromosome studies. III. Inequalities and deficiencies in homologous chromosomes: their bearing upon synapsis and the loss of unit characters. J. Morphology, 26:109-141.

Robson, G. C., and O. W. Richards. 1936. The variations of animals in nature. Longmans Green, London.

Romaschoff, D. D. 1931. On the conditions of equilibrium in populations. B.Zh., 7:442-454.

*Rosenberg, O. 1909. Cytologische und morphologische Studien an Drosera longifolia × rotundifolia. K. Svensk. Vet. Handl., 43:1-64.

—— 1930. Apogamie und Parthenogenesis bei Pflanzen. Borntraeger, Berlin.

Roubaud, E. 1920. Les conditions de nutrition des Anophèles en France (*Anopheles maculipennis*) et la rôle du bétail dans la prophylaxie du paludisme. Ann. Inst. Pasteur, 34:181-228.

———— 1932. Les races trophiques de *l'Anopheles maculipennis* décelées par les élevages expérimentaux comparés. C.R. Acad. Sci. Paris, 194: 1694-1696.

Rybin, V. A. 1929. Über einen allotetraploiden Bastard von *Nicotiana tabacum* × *Nicotiana sylvestris*. Ber. Deut. Bot. Ges., 37:385-394.

———— 1927. Polyploid hybrids of *Nicotiana tabacum* L. × *Nicotiana rustica* L. Bull. Appl. Botany, 17:191-240.

Sacharow, W. W. 1935. Iod als chemischer Faktor, der auf den Mutationsprozess von *Drosophila melanogaster* wirkt. B.Zh., 4:107-112.

———— 1936. Iod als chemischer Faktor, der auf den Mutationsprozess von *Drosophila melanogaster* wirkt. Genetica, 18:193-216.

Saltykovsky, A. I., and V. S. Fedorov. 1936. Chlorophyll abnormalities in white mustard (*Synapis alba*). Bull. Appl. Bot., Ser. 2, 9:287-305.

Samjatina, N. D., and O. T. Popowa. 1934. Der Einfluss von Iod auf die Entstehung von Mutationen bei *Drosophila melanogaster*. B.Zh., 3: 679-693.

Sansome, E. R. 1931. Chromosome associations in Pisum. J.G., 25:35-54.

*Sansome, F. W., and S. S. Zilva. 1933. Polyploidy and vitamin C. Biochem. Journal, 27:1935-1941.

Sapehin, A. A. 1928. Hylogenetic investigations of the vulgare group of Triticum. Bull. Appl. Bot., 19:127-166.

Sapehin, L. A. 1928. Hylogenetics of durum wheat. Bull. Appl. Bot., 19: 167-224.

Satina, S., and A. F. Blakeslee. 1935. Fertilization in the incompatible cross *Datura stramonium* × *D. metel*. Bull. Torrey Bot. Club, 62:301-312.

Saveliev, V. 1928. On the manifold effect of the gene vestigial in *Drosophila melanogaster*. Travaux Soc. Natur. Leningrad, 63:65-88.

Sax, K. 1922. Sterility in wheat hybrids. II. Chromosome behavior in partially sterile hybrids. G., 7:513-552.

———— 1935. The cytological analysis of species hybrids. Bot. Review, 1:100-117.

———— and Edgar Anderson. 1933. Segmental interchange in chromosomes of Tradescantia. G., 18:53-94.

———— and H. J. Sax. 1924. Chromosomes in genus cross. G., 9:454-464.

Scheuring, L. 1929-30. Die Wanderungen der Fische. Ergebnisse Biol., 5: 405-691; 6:4-304.

Schmidt, J. 1917. Statistical investigations with *Zoarces viviparus*. J.G., 7:105- .

———— 1923. Racial investigations. V. Experimental investigations with *Zoarces viviparus* L. C.R. Travaux Lab. Carlsberg, 14, No. 9:1-14.

Schnakenbeck, W. 1931. Zum Rassenproblem bei den Fischen. Zeit. Morph. Ökologie Tiere, 21:409-566.

Schultz, J. 1926. Radiation and the study of mutations in animals. Biol. Effects of Radiation, 2:1209-1261.

——— and Th. Dobzhansky. 1933. Triploid hybrids between Drosophila melanogaster and Drosophila simulans. J.E.Z., 65:73-82.

Seiler, J. 1925. Zytologische Vererbungsstudien an Schmetterlingen I. J. Klaus Arch. Vererb. Sozialanthrop., Rassenhygiene, 1:63-117.

——— 1927. Ergebnisse aus der Kreuzung parthenogenetischer und zweigeschlechtlichen Schmetterlinge. B.Z., 47:426-446.

*Semenov-Tian-Shansky, A. 1910. Die taxonomische Grenzen der Art und ihrer Unterabteilungen. Berlin.

Serebrovsky, A. S. 1927. Genetic analysis of the population of domestic fowl of the Daghestan mountaineers. J. Exp. Biol. (Russian), 3:62-146.

——— 1929. A general scheme for the origin of mutations. A.N., 63:374-378.

——— 1935. Hybridization of animals. Biomedgiz, Moscow-Leningrad.

Shen, T. H. 1932. Cytologische Untersuchungen über Sterilität bei Mänchen von Drosophila melanogaster, etc., Z.Z.m.A., 15:547-580.

Singleton, W. R. 1932. Cytogenetic behavior of fertile tetraploid hybrids of Nicotiana rustica and Nicotiana paniculata. G., 17:510-544.

Sipkov, T. P. 1936. A contribution to the cytology of Agropyrum-Triticum hybrids. Bull. Appl. Bot., Ser. 2, 9:357-360.

Skovsted, A. 1929. Cytological investigations of the genus Aesculus L. H., 12:64-70.

——— 1935. Cytological studies on cotton. III. A hybrid between Gossypium davidsonii Kell. and G. sturtii F. Muell. J.G., 30:397-405.

Smith, L. 1936. Cytogenetic studies in Triticum monococcum L. and T. aegilopoides Bal. Univ. Missouri Agric. Exp. Sta., Res. Bull. 248:1-38.

Smith, S. G. 1935. Chromosome fragmentation produced by crossing over in Trillium erectum L. J.G., 30:227-233.

Sorokin, H. 1927. Cytological and morphological investigations on gynandromorphic and normal forms of Ranunculus acris L. G., 12:59-83.

Spett, G. 1931. Gibt es eine partielle sexuelle Isolation unter den Mutationen und der Grundform von Drosophila melanogaster Meig.? Z.i.A.V., 60:63-83.

*Spooner, G. M. 1932. An experiment on breeding wild pairs of Gammarus chevreuxi at a high temperature with an account of two new recessive types of red eye. J. Marine Biol. Ass., 18:337-354.

Stadler, L. J. 1929. Chromosome number and the mutation rate in Avena and Triticum. P.N.A.S., 15:876-881.

——— 1932. On the genetic nature of induced mutations in plants. P.VI.C.G., 1:274-294.

Stadler, L. J. 1933. On the genetic nature of induced mutation in plants. II. A haplo-viable deficiency in maize. Missouri Agr. Exp. Sta. Research Bull., 204:1-29.

Standfuss, M. 1896. Handbuch der paläarktischen Grossschmetterlinge für Forscher und Sammler. G. Fischer, Jena.

Steiner, H. 1935. Vererbungsstudien an Vogelbastarden I. Einfache, mono-hybride Mendelspaltung beim Artbastard von *Amadina erythrocephala* × *A. fasciata*. Verh. Schweiz. Naturf. Ges., 116:348-349.

Stern, C. 1929. Untersuchungen über Aberrationen des Y-Chromosoms von *Drosophila melanogaster*. Z.i.A.V., 51:253-353.

———— 1936. Interspecific sterility. A.N., 70:123-142.

Stone, L. H. A. 1933. The effect of X-radiation on the meiotic and mitotic divisions of certain plants. Ann. Bot., 47:815-825.

Storer, T. J., and P. W. Gregory. 1934. Color aberrations in the pocket gopher and their probable genetic explanation. J. Mammalogy, 15:300-312.

Strasburger, E. H. Über Störungen der Eientwicklung bei Kreuzungen von *Epilachna chrysomelina* F. mit *Epilachna capensis Thunb.* Z.i.A.V., 71:538-545.

Stubbe, H. 1930-32. Untersuchungen über experimentelle Auslösung von Mutationen bei *Antirrhinum majus*. Z.i.A.V., 56:1-38; 60:474-513.

———— 1933. Labile Gene. Bibliogr. Genetica, 10:299-356.

———— 1935. Das Merkmal acorrugata, eine willkürlich auslösbare, domi-nante und labile Genmutation von *Antirrhinum majus*. Nachr. Ges. Wiss. Göttingen, Biol., N.F., 2:57-88.

Sturtevant, A. H. 1915a. A sex-linked character in *Drosophila repleta*. A.N., 49:189-192.

———— 1915b. Experiments on sex recognition and the problem of sexual selection in Drosophila. J. Animal Behavior, 5:351-366.

———— 1917. Genetic factors affecting the strength of linkage in Droso-phila. P.N.A.S., 3:555-558.

———— 1918. An analysis of the effects of selection. Carnegie Inst. Wash-ington, Publ. 264:1-68.

———— 1920-21. Genetic studies on *Drosophila simulans*. G., 5:488-500; 6:179-207.

———— 1921. The North American species of Drosophila. Carnegie Inst. Washington, Publ. 301:1-150.

———— 1923. Inheritance of direction of coiling in Limnaea. S., 58:269-270.

———— 1925. The effect of unequal crossing over at the Bar locus of Drosophila. G., 10:117-147.

———— 1926. A crossover reducer in *Drosophila melanogaster* due to inver-sion of a section of the third chromosome. B.Z., 46:697-702.

———— 1929a. The genetics of *Drosophila simulans*. Carnegie Inst. Wash-ington, Publ. 399:1-62.

Sturtevant, A. H. 1929b. The claret mutant type of *Drosophila simulans;* a study of chromosome elimination and of cell lineage. Zeit. wiss. Zool., 135:323-355.

───── 1931. Known and probable inverted sections of the autosomes of *Drosophila melanogaster.* Carnegie Inst. Washington, Publ. 421:1-27.

───── and G. W. Beadle. 1936. The relations of inversions in the X-chromosome of *Drosophila melanogaster* to crossing over and disjunction. G., 21:554-604.

───── and Th. Dobzhansky. 1936a. Geographical distribution and cytology of "sex ratio" in *Drosophila pseudoobscura.* G., 21:473-490.

───── 1936b. Inversions in the third chromosome of wild races of *Drosophila pseudoobscura,* and their use in the study of the history of the species. P.N.A.S., 22:448-450.

───── and C. R. Plunkett. 1926. Sequence of corresponding third chromosome genes in *Drosophila melanogaster* and *Drosophila simulans.* B.B., 50:56-60.

───── and C. C. Tan. 1937. The comparative genetics of *Drosophila pseudoobscura and D. melanogaster.* J.G., Vol. 34:415-439.

Sukatschew, W. 1928. Einige experimentelle Untersuchungen über den Kampf ums Dasein zwischen Biotypen derselben Art. Z.i.A.V., 47:54-74.

Sumner, F. B. 1923. Results of experiments in hybridizing subspecies of Peromyscus. J.E.Z., 38:245-292.

───── 1924a. The stability of subspecific characters under changed conditions of environment. A.N., 58:481-505.

───── 1924b. Hairless mice. J.H., 15:475-481.

───── 1929. The analysis of a concrete case of intergradation between two subspecies. P.N.A.S., 15:110-120, 481-493.

───── 1930. Genetic and distributional studies of three subspecies of Peromyscus. J.G., 23:275-376.

───── 1932. Genetic, distributional, and evolutionary studies of the subspecies of deer mice (Peromyscus). Bibliogr. Genetica, 9:1-116.

───── and R. R. Huestis. 1925. Studies of coat-color and foot pigmentation in subspecific hybrids of *Peromyscus eremicus.* B.B., 48: 37-55.

Sutton, W. S. 1903. The chromosomes in heredity. B.B., 4:231-251.

Sveshnikova, I., and J. Belekhova. 1936. Translocations in an interspecific hybrid. Bull. Appl. Bot., Ser. 2, 9:63-70.

Sweschnikowa, I. 1928. Die Genese des Kerns in Genus Vicia. V.K.V., 2: 1415-1421.

Tan, C. C. 1935. Salivary gland chromosomes in the two races of *Drosophila pseudoobscura.* G., 20:392-402.

───── and J. C. Li. 1934. Inheritance of the elytral color patterns of the lady-bird beetle, *Harmonia axyridis Pallas.* A.N., 68:252-265.

Tedin, O. 1925. Vererbung, Variation und Systematik in der Gattung Camelina. H., 6:275-386.

Thompson, W. P. 1925. The correlation of characters in hybrids of *Triticum durum* and *Triticum vulgare*. G., 10:285-304.

———— 1931. Cytology and genetics of crosses between fourteen- and seven-chromosome species of wheat. G., 16:309-324.

Thorpe, W. H. 1930. Biological races in insects and allied groups. Biol. Reviews, 5:177-212.

Timofeeff-Ressovsky, H. 1935. Divergens, eine Mutation von *Epilachna chrysomelina F*. Z.i.A.V., 68:443-453.

———— and N. W. Timofeeff-Ressovsky. 1927. Genetische Analyse einer freilebenden *Drosophila melanogaster* Population. A.E., 109:70-109.

Timofeeff-Ressovsky, N. W. 1929. Rückgenovariationen und die Geno-variabilität in verschiedenen Richtungen. I. Somatische Genovariationen der Gene W, w^e, und w bei *Drosophila melanogaster* unter dem Einfluss der Röntgenbestrahlung. A.E., 115:620-634.

———— 1931. Reverse genovariations, and gene mutations in different directions. II. The production of reverse genovariations in *Drosophila melanogaster* by X-ray treatment. J.H., 22:67-70.

———— 1932. Verschiedenheit der "normalen" Allele der white-Series aus zwei geographisch getrennten Populationen vom *Drosophila melanogaster*. B.Z., 52:468-476.

———— 1933a. Rückgenmutationen und die Genmutabilität in verschiedenen Richtungen. III. Röntgenmutationen in entgegengesetzten Richtungen am forked-Locus von *Drosophila melanogaster*. Z.i.A.V., 64:173-175.

———— 1933b. Rückgenmutationen und die Genmutabilität in verschiedenen Richtungen. IV. Röntgenmutationen in verschiedenen Richtungen am white-Locus von *Drosophila melanogaster*. Z.i.A.V., 65:278-292.

———— 1933c. Rückgenmutationen und die Genmutabilität in verschiedenen Richtungen. V. Gibt es ein wiederholtes Auftreten identischer Allele innerhalb der white-Allelenreihe von *Drosophila melanogaster?* Z.i.A.V., 66:165-179.

———— 1933d. Über die relative Vitalität von *Drosophila melanogaster* Meigen und *Drosophila funebris Fabricius* unter verschiedenen Zucht-bedingungen, in Zusammenhang mit den Verbreitungsarealen dieser Arten. Arch Naturgesch., N.F., 2:285-290.

———— 1934a. The experimental production of mutations. Biol. Reviews, 9:411-457.

———— 1934b. Über den Einfluss des genotypischen Milieus und der Ausenbedingungen auf die Realisation des Genotyps. Nachr. Ges. Wiss. Göttingen, Biologie, N.F., 1:53-106.

———— 1934c. Über die Vitalität einiger Genmutationen und ihrer Kombinationen bei *Drosophila funebris* und ihre Abhängigkeit vom "genotypischen" und vom äusseren Mileau. Z.i.A.V., 66:319-344.

———— 1934d. Auslösung von Vitalitätsmutationen durch Röntgenbestrahlung bei *Drosophila melanogaster*. Strahlentherapie, 51:658-663.

Timofeeff-Ressovsky, N. W. 1935a. Über geographische Temperaturrassen bei *Drosophila funebris F.* Arch. Naturgesch., N.F., 4:245-257.

———— 1935b. Auslösung von Vitalitätsmutationen durch Röntgenbestrahlung bei *Drosophila melanogaster.* Nachr. Ges. Wiss. Göttingen, Biologie, N.F., 1:163-180.

———— 1937. Experimentelle Mutationsforschung in der Vererbungslehre. Theodor Steinkopff, Dresden und Leipzig.

————, K. G. Zimmer, and M. Delbrück. 1935. Über die Natur der Genmutation und der Genstruktur. Nachr. Ges. Wiss. Göttingen, Biologie, N.F., 1:189-245.

Tischler, G. 1906. Über die Entwicklung des Pollens und der Tapetenzellen bei Ribes-Hybriden. Jahrb. wiss. Bot., 42:545-578.

Toyama, K. 1912. On certain characteristics of the silk-worm which are apparently non-Mendelian. B.Z., 32:593-607.

Tschetwerikoff, S. S. 1926. On certain features of the evolutionary process from the viewpoint of modern genetics. J. Exp. Biol. (Russian), 2:3-54.

———— 1927. Über die genetische Beschaffenheit wilder Population. V.K.V., 2:1499-1500.

Turesson, G. 1922. The genotypical response of the plant species to the habitat. H., 3:211-350.

———— 1925. The plant species in relation to habitat and climate. H., 3: 147-236.

———— 1926. Studien über *Festuca ovina L.* H., 8:161-206.

———— 1929. Ecotypical selection in Siberian *Dactylis glomerata.* H., 12: 335-351.

———— 1930. The selective effect of climate upon plant species. H., 14: 99-152.

———— 1931. The geographical distribution of the alpine ecotypes of some Eurasiatic plants. H., 15:329-346.

Upcott, M. 1936. The parents and progeny of *Aesculus carnea.* J.G., 33: 135-149.

Vandel, A. 1928. La parthénogenèse géographique. Bull. Biol. France Belgique, 62:164-281.

———— 1934. La parthénogenèse géographique. Bull. Biol. France Belgique, 68:419-463.

Vavilov, N. I. 1928. Geographische Genzentren unserer Kulturpflanzen. V.K.V., 1:342-369.

Wachs, H. 1926. Die Wanderungen der Vögel. Ergebnisse Biol., 1:479-637.

Watkins, A. E. 1930. The wheat species: a critique. J.G., 23:173-263.

———— 1932. Hybrid sterility and incompatibility. J.G., 25:125-162.

Webber, J. M. 1930. Interspecific hybridization in Nicotiana XI. The cytology of a sesquidiploid hybrid between *tabacum* and *sylvestris.* U.C.P.B., 11:319-354.

Weier, F. 1935. Die Rassenfrage bei *Culex pipiens* in Deutschland. Zeit. Parasitenkunde, 8:104-115.

*Weldon, W. F. R. 1899. Presidential address. British Ass. Adv. Sci., Section D (1898):887-902.

Wettstein, F. von. 1924. Morphologie und Physiologie des Formwechsels der Moose auf genetischer Grundlage. Z.i.A.V., 33:1-236.

—— 1928. Morphologie und Physiologie des Formwechsels der Moose auf genetischer Grundlage II. Bibliot. Genetica, 10:1-216.

Wichler, G. 1913. Untersuchungen über den Bastard *Dianthus armeria* × *Dianthus deltoides* nebst Bemerkungen über einige andere Art Kreuzungen der Gattung Dianthus. Z.i.A.V., 10:177-232.

Wiegand, K. M. 1935. A taxonomist's experience with hybrids in the wild. S., 81:161-166.

Winge, O. 1917. The chromosomes, their number and general importance. C.R. Trav. Lab. Carlsberg, 13:131-275.

Winkler, H. 1916. Über die experimentelle Erzeugung von Pflanzen mit abweichenden Chromosomenzahlen. Zeit. Bot., 8:417-531.

Woolsey, C. I. 1915. Linkage of chromosomes correlated with reduction in numbers among the species of a genus, also within a species of the Locustidae. B.B., 28:163-187.

Wright, S. 1921. Systems of mating. G., 6:111-178.

—— 1925. The factors of the albino series of guinea-pigs and their effects on black and yellow pigmentation. G., 10:223-260.

—— 1929. Fisher's theory of dominance. A.M., 63:247-279.

—— 1930. The genetical theory of natural selection. J.H., 21:349-356.

—— 1931a. Evolution in Mendelian populations. G., 16:97-159.

—— 1931b. Statistical theory of evolution. Amer. Stat. Journ., Suppl.: 201-208.

—— 1932. The rôles of mutation, inbreeding, crossbreeding, and selection in evolution. P.VI.C.G., 1:356-366.

—— 1934. Physiological and evolutionary theories of dominance. A.N., 68:24-53.

—— 1935. Evolution in populations in approximate equilibrium. J.G., 30:257-266.

Wulff, E. V. 1932. Introduction to the historical geography of plants. Inst. of Plant Breeding, Leningrad.

Yarnell, S. H. 1931. Genetic and cytological studies on Fragaria. G., 16:422-454.

Zimmermann, K. 1931. Studien über individuelle und geographische Variabilität paläarktischer Polistes und verwandter Vespiden. Zeit. Morph. Ökol. Tiere, 22:173-230.

—— 1936. Die geographische Rassen von *Epilachna chrysomelina* F. und ihre Beziehungen zu *Epilachna capensis Thunb.* Z.i.A.V., 71:527-537.

INDEX

INDEX

Printed in the USA
CPSIA information can be obtained
at www.ICGtesting.com
JSHW011520221024
72172JS00014B/115

9 780231 054751